Immigration Reform and U.S. Agriculture

EDITED BY
PHILIP L. MARTIN
WALLACE HUFFMAN
ROBERT EMERSON
J. EDWARD TAYLOR
REFUGIO I. ROCHIN

Published Under the Auspices
of the Western Regional Coordinating Committee-76 (WRCC-76),
Immigration and U.S. Agriculture Project,
and with the Support of the Farm Foundation
and the Urban Institute's Program for Research on
Immigration Policy

University of California
Division of Agriculture and Natural Resources

Publication 3358

ORDERING INFORMATION:

To order this publication, contact

Publications
Division of Agriculture and Natural Resources
University of California
6701 San Pablo Avenue
Oakland, California 94608-1239

Telephone (510) 642-2431
within California call 1-800-994-8849
FAX (510) 643-5470

Publication 3358

ISBN 1-879906-20-1

© 1995 by the Regents of the University of California,
Division of Agriculture and Natural Resources

Printed in the United States of America.

The University of California, in accordance with applicable Federal and State law and
University policy, does not discriminate on the basis of race, color, national origin,
religion, sex, disability, age, medical condition (cancer-related), ancestry, marital status, citizenship, sexual orientation, or status as a Vietnam-era veteran or special dis-
abled veteran. The University also prohibits sexual harassment.

Inquiries regarding the University's nondiscrimination policies may be directed to the
Affirmative Action Director, University of California, Agriculture and Natural Resources, 300 Lakeside Drive, 6th Floor, Oakland, CA 94612-3560, (510) 987-0096.

1.5m-pr-3/95-WJC/LK

Foreword

The people of rural America have been the major concern of the Farm Foundation since its founding in 1933. Rural Americans include farmers and their families, as well as the nation's hired farm workers, the two-and- a-half million individuals who work on a farm for wages sometime during a typical year.

Most hired farm workers today are immigrants to the United States. After the third wave of immigration in the early 1900s, the percentage of immigrant workers in most U.S. industries fell from the 1930s through the 1970s. In those commodities and areas most dependent on hired workers, however, immigrants continued to be a significant portion of the work force.

During the 1970s and 1980s, the number of immigrants arriving illegally rose sharply. Most came for U.S. jobs and, although most were employed in nonfarm jobs, many were believed to have had at least a first job in agriculture, a traditional port of entry especially for immigrants from Mexico. After a protracted political struggle, the Immigration Reform and Control Act of 1986 (IRCA) recognized agriculture's historical dependence on immigrant workers. A separate legalization program has been included for farm workers and, if now-legal workers left agriculture, they could be replaced with additional immigrants.

IRCA's agricultural provisions were formulated with little input from researchers at land-grant universities. There were several reasons. Immigration reform legislation was considered by judiciary committees that were not inclined to turn to economists at land-grant institutions for advice. In many states there were few researchers knowledgeable about farm labor issues, a reflection of inadequate data and the political sensitivity of the issue. Finally, farm labor did not seem to be an important issue: It was often crowded out by marketing, environmental, and other concerns.

Immigration reform highlighted the importance of farm labor issues in many states. The Farm Foundation recognized that farm labor researchers needed to create a network to coordinate their efforts to monitor the impacts of immigration reform.

In 1987, the foundation supported what became the Western Regional Coordinating Committee-76 (WRCC-76). WRCC-76 has met annually since then, providing a forum for land-grant researchers and others to share data and analyses of the legislation's effects.

After five years of work, the committee held a conference in Washington D.C. to explore the effects of IRCA's agricultural provisions on farm workers, farmers, and the rural and national economies. More than 60 participants discussed the 30 papers presented at the March 1993 conference. Farm Foundation support enabled non-land-grant researchers to participate and for papers to be prepared for publication.

The papers in this volume paint a sobering picture. IRCA did not have its intended effect of stopping illegal immigration. Neither U.S. farmers nor Mexican workers are gradually lessening their historical dependence on one another. The papers in this volume review what has happened in the wake of IRCA, and they note that the United States will probably revisit the immigrant farm worker issue in the 1990s. This volume, and the continuing contributions of their authors, should prove useful in formulating sound immigration and farm labor policy.

— Walter Armbruster

Contents

Acknowledgments

This book is the collaborative effort of 30 researchers interested in immigration and farm labor. Their work was coordinated under the auspices of the Western Regional Coordinating Committee-76. The WRCC-76 researchers were first brought together with the support of the Farm Foundation in 1987, and from this initial meeting, researchers from a number of states and disciplines agreed to conduct studies concerning the effects of immigration reform on farmers and farm workers. The committee has met annually since 1988 and, under the guidance of James Zuiches, administrative advisor, is expected to continue its activities.

This book would not have been possible without the assistance of the WRCC-76 participants. Leslie Whitener, Shannon Hamm, and Jim Duffield made conference arrangements for the March 1993 meeting at which these papers were presented and discussed. The five editors went through each paper carefully, and the authors did an admirable job in responding to their suggestions for revision in a timely fashion. Robyn Brode did an admirable job copyediting the manuscript, and Lee Knous successfully coordinated the intricate and complex process of merging papers prepared with a variety of computer programs and typesetting the manuscript. Walt Armbruster played two supporting roles in this project: He chaired the most important session at the conference and the Farm Foundation provided support for researchers from non-land-grant universities to participate in the meeting.

The papers in this volume represent benchmark studies of the effects of immigration reforms on U.S. agriculture. No common methodology runs through them and a variety of data are analyzed, but they reach the common conclusion that IRCA did not have its intended effects on agriculture. While the authors do not agree on whether IRCA's failure to affect the farm labor market as expected is good or bad, they all expect the "farm labor issue" once again to dominate the headlines. These papers should provide valuable background for this coming debate.

The contributions of Philip Martin and J. Edward Taylor were developed in cooperation with the Urban Institute's Program for Research on Immigration Policy. We are indebted to Jeff Passel and Michael Fix for their support.

List of Contributors

Steven G. Allen is an economist in the Department of Agriculture and Resource Economics at North Carolina State University in Raleigh, North Carolina.

Andrew J. Alvarado is a sociologist at California State University's Center for Agricultural Business, Fresno, California.

Walter Armbruster is managing director of the Farm Foundation in Oak Brook, Illinois.

Luis Caraballo is director of the Office of Immigration Coordination, Executive Department, State of Oregon, in Salem, Oregon.

Richard W. Carkner is an Extension agricultural economist at the Washington State University station in East Puyallup, Washington.

Monica D. Castillo is a research specialist with the Bureau of Labor Statistics, U.S. Department of Labor, in Washington, D.C.

Tim Cross is an agricultural economist at Oregon State University in Corvallis, Oregon.

James Duffield is an agricultural economist at the U.S. Department of Agriculture, Office of Energy, in Washington, D.C.

Clyde Eastman is a rural sociologist at New Mexico State University in Las Cruces, New Mexico.

Robert D. Emerson is an agricultural economist in the Department of Food and Resource Economics, University of Florida, Gainesville, Florida.

Enrique E. Figueroa is an agricultural economist in Cornell University's Department of Agriculture, Resource and Managerial Economics in Ithaca, New York.

Susan M. Gabbard is a senior associate at Aguirre International in San Mateo, California.

David Griffith is an anthropologist at East Carolina University in Greenville, North Carolina.

Lewell Gunter is an agricultural economist in the Department of Agriculture and Applied Economics at the University of Georgia in Athens, Georgia.

Shannon Hamm is an agricultural economist at the U.S. Department of Agriculture, Economic Research Service, in Washington, D.C.

John J. Haydu is an agricultural economist at the University of Florida, Institute of Food and Agricultural Sciences, in Fort Lauderdale, Florida.

Monica L. Heppel is an anthropologist at the Inter-American Institute on Migration and Labor, Mount Vernon College, in Washington, D.C.

Alan W. Hodges is an agricultural economist at the University of Florida's Department of Food and Resource Economics in Gainesville, Florida.

James S. Holt is an agricultural economist at McGuiness & Williams in Washington, D.C.

Wallace E. Huffman is an agricultural economist in the Department of Economics, Iowa State University, Ames, Iowa.

Jeff Jacksich is an economic analyst at the Washington State Employment Department, Olympia, Washington.

Edward Kissam is an anthropologist in Sebastopol, California.

Philip L. Martin is an agricultural economist in the Department of Agricultural Economics, University of California at Davis, California.

Herbert O. Mason is an agricultural economist at the Center for Agricultural Business, California State University, Fresno, California.

Robert Mason is a sociologist at the Survey Research Center, Oregon State University, Corvallis, Oregon.

Richard Mines is an agricultural economist at the U.S. Department of Labor in Washington, D.C.

David S. North is a policy analyst at New TransCentury in Arlington, Virginia.

Victor J. Oliveira is an agricultural economist at the U.S. Department of Agriculture in Washington, D.C.

Leo C. Polopolus is an agricultural economist in the Department of Food and Resource Economics, University of Florida, Gainesville, Florida.

Gary L. Riley is the director of advanced learning technology and a research statistician at the School of Agricultural Sciences and Technology at California State University, Fresno, California.

Refugio I. Rochin is an agricultural economist at Michigan State University in East Lansing, Michigan.

Gil Rosenberg is a rural sociologist at the University of Kentucky in Lexington, Kentucky.

Howard R. Rosenberg is an Extension specialist in agricultural labor management at the University of California at Berkeley, California.

David Runsten is an agricultural economist at the California Institute of Rural Studies in Berkeley, California.

Daniel A. Sumner is an agricultural economist in the Department of Agricultural Economics at the University of California at Davis, California.

J. Edward Taylor is an agricultural economist in the Department of Agricultural Economics at the University of California at Davis, California.

Don Villarejo is director of the California Institute of Rural Studies in Davis, California.

Harry Vroomen is an agricultural economist at the U.S. Department of Agriculture in Washington, D.C.

Leslie A. Whitener is a rural sociologist at the U.S. Department of Agriculture, Economic Research Service, in Washington, D.C.

Glenn Zepp is an agricultural economist at the U.S. Department of Agriculture, Economic Research Service, in Washington, D.C.

Introduction

PHILIP L. MARTIN AND J. EDWARD TAYLOR

Agriculture is sometimes considered a crown jewel of the American economy. A half-century of rapid gains in farm productivity permit Americans to devote a smaller percentage of their incomes to food than the citizens of any other nation. Most of these U.S. farm productivity gains occurred in the grain and livestock agriculture exemplified by Iowa corn and hog farms.

Agriculture is not a homogeneous industry. Small farms that sell produce to New England tourists have little in common with Iowa farmers, and congressional debates on the future of the wheat program are of vital interest to Kansas wheat farmers but largely irrelevant to California peach growers. Interest in congressional immigration debates is reversed; they are a top priority of growers but of little interest in midwestern family farming areas. The American agriculture success story is often painted in such broad strokes that these differences between agriculture's various subsectors are overlooked.

The chapters in this volume examine the impact of the Immigration Reform and Control Act of 1986 (IRCA) on American agriculture. Although they cover all commodities grown with hired farm workers in the major farm labor states, the focus is on farms that employ hired workers to produce fruits, vegetables, and horticultural specialties (FVH) such as flowers and nursery products. Such farms comprise less than 10 percent of all U.S. farm employers, but they pay 40 percent of total farm wages, hire 60 percent of the nation's migrant and seasonal farm workers, and account for almost all of agriculture's interest in immigration reform.

FVH farms share in the glow of the agricultural success story, but they have had a persistent problem in securing seasonal farm workers. Over a century ago, California farmers lamented the lack of labor willing to be seasonal workers, and turned to immigration as a solution to the labor problem. The *California Farmer* in 1854 asked, "Where shall the laborers be found?" to be seasonal workers, and offered an answer: "The Chinese . . . [will] be to California what the African has been to the South" (Fuller 1940, p. 19,802). Immigrants are today the majority of all farm workers in each of the areas and commodities examined in this volume.

Immigration policy is linked to agriculture today because immigrants constitute nearly all of the entrants to the seasonal fruit and vegetable work force.[1] Among the 140 million Americans who have paid employment sometime during a typical year, only about 3 percent do farm work, and many of these farm workers are teenage hired hands on midwestern family farms. But of the almost 6 million immigrant workers who were in the United States legally by the end of the 1980s, up to 1.5 million, or one in four, did at least some farm work.[2] If the immigration patterns of the 1980s continue in the 1990s, up to one-quarter of working-age U.S. immigrants who arrive during the decade may be Mexicans whose initial U.S. employment is in fruit and vegetable agriculture.

A Century of Controversy

Foreign workers and labor-intensive agriculture have been linked uneasily for over a century, especially in California, the state that produces about 40 percent of the nation's fruits and vegetables. Farmers worried about both the availability and the cost of seasonal farm workers. Immigration solved both problems: A California farm spokesperson in 1872 observed that hiring seasonal Chinese workers who housed themselves and then "melted away" when they were not needed made them "more efficient . . . than Negro labor in the South [because] it [Chinese labor] is only employed when actually needed, and is, therefore, less expensive" than slaves, since no capital outlay was required to purchase them; they boarded themselves while employed and reproduced abroad; they were available when needed but they were paid only for the time they were actually employed; and, at the end of the season they "moved on, relieving [the] employer of any burden or responsibility for his [worker's] welfare during the slack season" (Fuller 1940, pp. 19,809, 19,824).

Fruit and vegetable agriculture, especially in the West, soon rested on immigrant foundations. When labor unions and urban interests sought to stop Chinese immigrants in the 1880s, farmers responded that, without the Chinese, "there would not now be one fruit tree or grape vine in the state where there are now ten" (Fuller 1940, p. 19,812).

When Chinese immigration was nonetheless stopped in 1882, the *California Farmer* noted that "the [fruit] crop of the present year [1883], although deemed a short one, taxed the labor capacity of the State to the utmost. . . If such was the situation this year, what will it be when the numerous trees now just coming into bearing will be producing full crops?" (Fuller 1940, p. 19,813).

The availability of Chinese immigrant farm workers was soon capitalized into land values, giving landowners an incentive to preserve an influx of immigrants for agriculture. As early as 1888, for example, California land used for fruit production, where the wages paid to Chinese workers were $1.00 to $1.25 per day, was worth $200 to $300 per acre. Land used to produce grain and hay, by contrast, was worth only $25 to $50 per acre, and the wages paid to white workers on these farms were $2 to $3 daily (Fuller 1940, p. 19,816). Without Chinese or similar workers, farmers noted, wages might have to be raised or farmers would have to switch to less labor-intensive crops. In either case, land values would fall, explaining why farmers and their allies had an economic incentive to lobby government to preserve access to immigrant farm workers.

There is a relationship between immigration, farm structure, and food prices, and the position of farm workers near the bottom of the American job ladder. There have been many discussions of alternatives to immigrant farm workers. Diversifying the crops grown so that there would be less of a seasonal peak need for labor; mechanizing labor-intensive tasks; and offering the wages, housing, and benefits needed to attract American workers have all been discussed, but there has never been any agreement on exactly how such adjustments could be made in a manner that protected both American farmers and farm workers. Instead, a succession of immigrant workers have been available: the Japanese, Mexicans, Dust Bowl refugees to California, and in recent decades Mexicans again.

The availability of immigrant farm workers who accommodated themselves to seasonal farm work has had consequences that persist today. First, farmers persuaded themselves and many policy makers that fruit and vegetable agriculture by its very nature required an immigrant work force. Varden Fuller concluded that farmers soon became accustomed to the fact "that with no particular effort on the part of the employer, a farm labor force would emerge when needed, do its work, and then disappear — accepting the terms and conditions offered, without question" (Fuller 1991, p. vii).

Agriculture's need for immigrant workers became as much an article of faith as fair prices, and this "truth," as well as higher land values, are not the only effects of agriculture's historical dependence on migrants from abroad. A third persisting effect is the prominent role of intermediaries such as farm labor contractors (FLCs) in agricultural labor markets. Seasonal workers are often hired in gangs or crews, and bilingual intermediaries are needed to link English-speaking employers and non-English-speaking workers. These FLCs have evolved from agents of workers to independent businesses interested in maximizing the difference between what a farmer pays to have a job done and what a worker receives. FLCs seem to have found it easier to extract profits from vulnerable newcomers than to bargain for higher commissions from farmers.

IRCA and Agriculture

Farmers were the only major group of U.S. employers who acknowledged their dependence on illegal alien workers and asserted that alien farm workers were necessary to their survival. They threatened to oppose immigration reform unless they were assured that a replacement alien work force would be available.

During congressional debate on farmer-proposed temporary worker programs, it was asserted that illegal aliens were 50 to 70 percent of the seasonal farm work force, and the USDA estimated that 300,000 to 500,000 illegal aliens were employed sometime during the year in agriculture. The Current Population Survey data analyzed by the USDA did not support these assertions: The CPS report on the characteristics of hired farm workers estimated that only 326,000 Hispanics did farm work for wages sometime during 1985 and only 137,000 Hispanic farm workers did 75 to 249 days of farm work (Oliveira and Cox 1988). Not all of these 137,000 SAW-eligible Hispanic farm workers in the CPS were illegal aliens; many were U.S. citizens and legal immigrants.

Despite these data ambiguities and the reluctance to initiate another discredited *bracero* program to obtain alien farm workers, farmers succeeded in getting the U.S. House of Representatives in 1984 to include a guest worker program for agriculture in pending immigration reform legislation, and the Senate in 1985 included a similar program. As a result, immigration reform was expected to be stymied by the apparently intractable opposition of unions and farm worker advocates to guest workers, fighting against the farmers who had succeeded in getting each house of Congress to approve a guest worker program. In a last-minute compromise, farmers and farm worker advocates agreed to legalize currently illegal farm workers under a Special Agricultural Worker (SAW) legalization program and to monitor the farm labor market to determine whether farm labor shortages developed.

The manner in which the agricultural worker issue was settled, in what became IRCA, meant that no careful studies were done beforehand of how its agricultural provisions might affect U.S. agriculture. Indeed, predictions of IRCA's effects had to take into account IRCA's three major agricultural provisions: deferred sanctions enforcement and search warrants, the SAW legalization program, and the H-2A and Replenishment Agricultural Worker (RAW) foreign worker programs, and their interaction. These agricultural provisions are summarized briefly below.

Enforcement

Until IRCA, Immigration and Naturalization Service (INS) enforcement in agriculture usually involved the border patrol driving into fields and apprehending aliens who tried to run away. Farmers pointed out that the INS was required to obtain search warrants before inspecting factories for illegal aliens, and they argued that the INS should similarly be obliged to show evidence that illegal aliens were employed on a farm before raiding it. IRCA extended the requirement that the INS have a search warrant before raiding a work place for illegal aliens from nonfarm to agricultural work places.

This search warrant provision was won by the farmers' argument that farms should be treated like factory work places. Farmers also argued that farms, unlike factories, were extraordinarily dependent on unauthorized aliens and that employer sanctions should not be enforced while the legalization program for farm workers was underway. Sanctions enforcement was thus deferred in most of crop agriculture until December 1, 1988.

SAWs

IRCA created two legalization programs[3]: a general program that granted legal status on the basis of U.S. residence since 1982 and the farm worker SAW program that granted legal status to illegal aliens who had done at least 90 days of qualifying farm work in the 12 months ending May 1, 1986. It was easier for illegal alien farm workers to achieve legal status under the SAW program than it was for nonfarm illegal aliens to satisfy the residence requirement. SAW applicants, for example, could have entered the United States illegally in early 1986, left after doing 90 days of farm work, and then applied for SAW status from abroad. The SAW application process was also easier: Most SAW applicants submitted only an affidavit from an employer asserting that the worker named in the letter had, for example, picked tomatoes for 92 days between June and September 1985. The burden of proof then shifted to the INS to disprove the alien's claimed employment.

In addition to easier qualification rules, the definition of qualifying work was stretched repeatedly. Aliens had to be employed at least 90 days in Seasonal Agricultural Services (SAS). SAS was defined by commodity (perishable) and activity (field work); SAW applicants were illegal aliens who performed or supervised field work during the 12 months ending May 1, 1986 that was related to planting, cultural practices, cultivating, growing, and harvesting "perishable" fruits and vegetables of every kind and "other perishable commodities." The definition of perishable commodity was stretched first by the USDA and then by the courts to include virtually all plants grown for human food (except sugarcane) and many nonedible plants, such as cotton, Christmas trees, cut flowers, and Spanish reeds. Field workers included all of the paid hand- or machine-operator workers involved with these SAS commodities, the supervisors of field workers and equipment operators, mechanics who repair machinery, and pilots who spray crops. These elastic definitions of perishable and field work meant that a variety of aliens could qualify for legal U.S. residence, including an illegal alien investor-manager of a farm and the illegal aliens employed there.

The major surprise of the SAW program was that 1.3 million aliens applied for SAW status, almost three-fourths as many as applied for the general legalization program, even though it was widely asserted that only 15 to 20 percent of the undocumented workers in the United States in the mid-1980s were employed in agriculture. The number of SAW applicants was almost four times the estimated 350,000 illegal aliens employed in U.S. agriculture in the mid-1980s.

SAW applicants turned out to be mostly young Mexican men. Their median age was 29 and half were between 20 and 29. A few limited surveys found that SAWs who had an average of five years of education earned between $30 and $35 daily for 100 days of farm work in 1985–86.

RAWs and H-2A

Farmers expected IRCA to stop the influx of illegal aliens, and they feared that newly legalized SAWs would leave seasonal farm jobs for year-round urban jobs. For this reason, farmers argued that they would need access to legal alien farm workers. Americans, they said, could not be persuaded to do seasonal farm work, and farmers should be spared the choice of breaking the law by hiring illegal aliens (if they were available) or seeing their crops rot in the fields.

There are two major types of alien worker programs: contractual programs that tie a foreign worker to a particular job vacancy and noncontractual programs that admit foreign workers and give them work permits to hunt for jobs. Western growers argued that contractual programs require an impossible amount of labor planning and coordination: Under the H-2 (now H-2A) contractual foreign worker program, farm employers have to develop job descriptions, determine the number of people to be hired, guarantee a minimum wage, arrange for free housing, and then attempt to recruit Americans before they receive permission to bring in foreign workers. Western growers argued that ever-changing weather and crop conditions require them to have a noncontractual or free-agent program.

Farm employers had argued consistently that newly legalized SAWs would quit doing farm work, so they obtained a Replenishment Agricultural Worker (RAW) program to admit immigrant workers after October 1, 1989, if the exit of SAWs was likely to lead to farm labor shortages. The number of RAWs was to be determined by a complex two-part formula. First, RAWs would enter the United States and roam the countryside looking for farm jobs. Then, RAWs who did at least 90 days of qualifying farm work for three years would be eligible to become legal immigrants, and five years of such work would enable RAWs to become naturalized United States citizens.

No RAW workers were admitted during the four-year life of the program, and the number of H-2A farm workers has shrunk since 1986.

Mixed Signals

The combination of employer sanctions and the SAW and RAW programs sent mixed and, to a certain degree, opposite signals to farmers and workers. Employer sanctions were meant to encourage farmers to lessen their dependence on unauthorized immigrant workers by introducing a new cost and risk in the form of penalties and, for repeated violations of the law, imprisonment of employers who knowingly hired such workers.

Employer sanctions should have dramatically changed the way labor-intensive farms do business. Restricted to hiring from a pool of legal workers, farmers were expected to have to take steps to attract and retain legal workers through some combination of wage and benefit incentives and improved working conditions. Denied access to a U.S. job by such changed employer behavior, and told about these changes by the efficient information networks that connect villages and towns in Mexico and Central America with relatives and friends in the United States, prospective unauthorized immigrants should have opted to stay at home. With most employers complying, enforcement resources should then have been targeted to the handful of employers and workers not in compliance, further increasing their risks and costs.

In terms of the familiar supply-and-demand curves used by economists, employer sanctions should have decreased both the supply and demand for unauthorized immigrant labor. If legal and illegal labor could be represented by different supply-and-demand schedules, the demand for legal workers should have shifted to the right on the familiar supply-demand cross. This, in turn, should have exerted upward pressure on the wages, benefits, and work-place amenities for legal workers, possibly inducing an increase in the number of legal workers willing to work in farm jobs. In this manner, immigration reform should have resulted in higher earnings and better working conditions. Labor costs should have risen and, to the extent that these costs are passed on in the form of higher prices, food prices should have been higher.

IRCA's SAW and RAW provisions sent farmers an entirely different message. The SAW program was designed essentially to legalize the status quo in the farm labor market. Farmers and labor contractors were anxious to secure a work force and, in order to have a large base of SAWs so as to maximize the number of RAWs[4] who could be imported, farmers encouraged their workers to obtain legal status — indeed, there were stories of farmers requiring U.S. citizen workers to buy false SAW papers in order to be hired.

The final number of SAWs granted legal status almost certainly exceeded by a wide margin the number of workers who actually performed 90 days of farm work between May 1985 and May 1986. The roughly 1 million SAWs that were approved, plus an additional 80,000 general legalization applicants with farm worker occupations, implies that the U.S. government granted legal status to an equivalent of 55 percent of the estimated 2 million persons employed sometime during the year on crop farms — almost 1 percent of the entire U.S. labor force. This suggests a great deal of undetected fraud among SAW applicants: A California study concluded that the number of SAW applications in that state was more than three times greater than the number of workers, legal or illegal, who worked 90 days or more on California farms during 1985–86 (Martin, Taylor, and Hardiman 1988).

The SAW program dramatically increased the size of the legal farm work force. It also strengthened the networks of family ties that link villagers in Mexico and Central American with U.S. jobs by legalizing individuals who anchored these networks in the United States. Village studies in Mexico uniformly find that, except possibly for a temporary interruption caused by uncertainty among workers around the time IRCA was enacted, migration to the United States has continued and even accelerated since 1986 (Taylor 1992). At the same time, the RAW program assured farmers that if labor shortages did develop, new legal foreign workers would be made available to do farm work.

These three considerations — the large number of SAW legalizations, the continuation and perhaps acceleration of illegal immigration, and the assurance of RAW workers if needed — created incentives for farmers to continue and even to expand their labor-intensive operations without making efforts to attract and retain legal workers. Meanwhile, the SAW and RAW programs sent a message to foreigners that work as an illegal immigrant in agriculture is a means to qualify for future worker legalization programs. Anecdotes abound about villagers in Mexico who migrated to U.S. farm jobs as a way to "queue up" for the hoped-for next legalization, and about urban workers who switched to seasonal farm work to accumulate SAW or RAW "credit." Over 90 percent of the illegal aliens on the RAW registry compiled by the INS in case RAWs became necessary were presumably unauthorized workers who gave U.S. addresses to be notified in the event of labor shortages.

IRCA's Effects on Agriculture

The fact that the SAW–RAW compromise was a last-minute agreement meant that its effects on U.S. farmers and farm workers were not debated extensively. However, IRCA should have had at least three effects in agriculture: Farm workers today should be legal U.S. workers; the efforts of farmers to retain newly legalized SAW workers should have led to noticeable increases in labor costs; and the presumed replacement of aging SAWs with H-2A or RAW workers should have led to a great deal more data and understanding of the "real" supply and demand for farm workers at national and local levels.

IRCA has not had these effects in agriculture. Most pre-IRCA studies concluded that 20 to 30 percent of all seasonal farm workers were illegal aliens (Mines and Martin 1986). Employer and worker testimony and studies today suggest that, despite the legalization of illegal alien farm workers, 20 to 40 percent of the seasonal work force is once again illegal. This percentage is even higher in the activities that are traditional ports of entry, such as the California raisin harvest.

Before IRCA, illegal alien farm workers were often called undocumented workers because they did not have documents that authorized them to work in the United States. IRCA changed unauthorized farm workers into "documented illegal aliens." Today's "documented illegals" have work authorization documents (most carry and give to employers a copy of an I-551 permanent U.S. immigrant visa or green card and a driver's license), although these documents are generally purchased for $30 to $50 and are not issued by the U.S. government. IRCA may well be remembered as a stimulus to illegal immigration because IRCA has spread work authorization documents and knowledge about them to very poor and unsophisticated rural Mexicans and Central Americans, encouraging first-time entrants from these areas.

The spread of fraudulent documents among even poor and rural workers abroad, making it easier for them to enter the United States illegally, is a major feature of the post-IRCA farm labor market. Despite apparent pre-IRCA levels of illegal alien employment, there were far fewer farm worker apprehensions in the early 1990s than there were in the early 1980s. The conversion of enforcement from the pursuit of aliens in the field to a paper chase through farm employer offices seems to have made it easier rather than harder to employ unauthorized workers. In the early 1980s, studies called attention to the demise of unions in California, the scaling back of fringe benefits for farm workers, and the apparent stability of farm worker earnings — they seemed to be stuck at about $5 hourly after 1981 (Martin, Mines, and Diaz 1985; Smith and Coltrane 1981). In the early 1990s, pre-IRCA legal farm workers, newly legalized SAWs, and a continuing influx of illegal aliens have produced an ample supply of seasonal workers, giving farmers little incentive to improve wages and working conditions. Worker surveys and testimony confirm that 90 percent of the new entrants to the farm work force are immigrants. In most instances, farmers have not raised wages, improved working conditions or housing, or otherwise made adjustments to retain these workers.

Recent research, including the studies in this volume, overwhelmingly points to a surplus of farm workers since IRCA's complete implementation in agriculture after December 1, 1988, and little effort by farm employers to lessen their dependence on immigrant workers; to improve wages, job stability, and working conditions; or to switch to less labor-intensive crops or techniques.

These findings do not imply that employer sanctions have not had an impact on farm workers or employers. A number of studies reveal changes in the structure of farm labor markets that may be related to efforts by farmers to shift the new costs and risks imposed by employer sanctions to labor-market intermediaries. As the studies in this volume confirm, workers hired through intermediaries such as farm labor contractors (FLCs) are among the most recent entrants and are thus more likely to be unauthorized workers.

If they were faced with a labor shortage, most employers report that they would hire even more workers through FLCs. FLCs thus seem to be perceived by farm operators to have special abilities to recruit immigrant workers and to evade immigration and labor law enforcement. In addition, competition between FLCs to do harvesting and other seasonal tasks on farms has reduced the commissions or fees that FLCs can charge farmers. Since farmers hiring workers directly would incur the same payroll tax and hiring costs that FLCs must pay, farmers rationally turn hiring, and the associated risks of enforcement, over to FLCs.

The ample supply of farm workers has discouraged adjustments on the part of farmers that might have led to fewer and better farm jobs. Instead, the IRCA-wrought influx of workers has encouraged reductions in (real) hourly earnings and hours of work and has often increased the costs associated with seasonal employment in agriculture, such as establishing a temporary place to live, arranging for rides to work, and handling items that range from check cashing to entertainment. Seasonal farm workers have always had to deal with these issues, but the changes in the farm labor market since IRCA seem to have made them pay more of these costs, especially in the western states, where workers typically live off the farm in small towns.

This third effect of IRCA — the shifting of more of the risk of employment in a seasonal industry backward in the production process to its most vulnerable element, recently arrived immigrant workers — is perhaps the most important, but it has not been easy to document. In the wake of IRCA, there was a great deal of interest in how immigration reforms would affect the one U.S. industry that got special treatment, and there was a rash of worker and employer surveys. But these ad hoc surveys did not lead to a continuing federal and state commitment to collecting reliable farm labor data. In what is perhaps a symbolic indicator of the exclusion of farm workers from data collection efforts, the so-called Westat survey of 5,000 newly legalized aliens 3 and 6 years after the change in their legal status did not include farm workers.

The Contractor Phenomenon

Illegal immigration continues to bring unauthorized workers into the farm work force, wages and earnings are falling in real terms, and there is no firm data basis to determine just how immigration reform is changing the farm labor market in the 1990s. But one theme, echoed throughout the chapters that follow, is that intermediaries seem to be playing an ever larger role in the farm labor market. These intermediaries have many names — crew boss, foreman, and farm labor contractor — but, from a recently arrived immigrant's point of view, the intermediary is the boss. Many of these intermediaries are not registered as FLCs, and some would not legally be considered FLCs, but all play the essential FLC role of recruiting and supervising farm workers.

Since IRCA, there appears to have been a trend among farmers away from hiring seasonal workers directly and toward hiring them through intermediaries such as farm labor contractors. This trend was already underway in the mid-1980s, but appears to have accelerated since IRCA's passage. As farmers hired more seasonal workers through FLCs, their presence was felt in what would otherwise be incongruous labor market events. In Monterey County, California, for example, vegetable production from the nation's salad bowl rose, the number of workers hired directly by vegetable farms fell, and employment reported by FLCs soared. Such local labor market trans-formations, multiplied many times, help to explain why it is estimated that in California perhaps one-half of all seasonal farm work was per-formed by FLC workers in 1992.

Hiring workers through FLCs could in theory be a solution to the puzzle of how to provide workers with stable employment in spite of the seasonality of labor demands on individual farms (Fisher 1952). FLCs could operate in a way similar to state employment services or union hiring halls, moving workers from farm to farm by arranging a sequence of contracts with different farmers. Economies of scale in recruitment and hiring would reinforce the FLCs' potential role in increasing labor market efficiency. Growers in this scenario would concentrate on cultivating crops, while FLCs would hone their com-parative advantage in labor recruitment and management in this "Kelly Girl" model of the farm labor market. Indeed, if employer sanc-tions could succeed in closing the border to new flows of immigrants into agriculture, the expansion of FLC employment could be compat-ible with an improvement in employment stability, wages, and work-ing conditions for farm workers.

Labor market studies, however, contradict this model. FLC em-ployment is expanding, but this expansion appears to be accompa-nied by a deterioration, or at best no change, in farm worker real wages, employment stability, and working conditions. In California, for example, farm worker real wages fell between 1986 and 1992, and the average earnings and retention among FLC workers were the low-est of all agricultural employees. Moreover, the instability of farm employment, as reflected in the probability of workers changing their principal employer from one quarter to the next, increased signifi-cantly between 1986 and 1991 (Taylor and Thilmany 1992).

The switch to FLCs reflects a rational response by farmers to the contradictory incentives effectuated by immigration reforms and FLCs' comparative advantage. Farmers could have responded to employer sanctions in either of two ways. First, they could have re-duced their reliance on unauthorized workers by switching to less labor-intensive crops or technologies or by providing wage, employ-ment, and other incentives to attract and retain legal workers. This strategy would have required an investment in recruiting and screen-ing workers. These represent fixed costs per worker, which may not be worth undertaking if workers are needed only for a short time. It also requires an increase in variable costs, in the form of incentive packages to retain these workers.

Alternatively, farmers could shift many of the fixed costs of hiring legal workers and the risks of hiring unauthorized workers onto labor intermediaries. Labor contractors normally are considered to be employers in their own right, responsible for complying with employer sanctions under IRCA. Although in theory a farmer could be held responsible for hiring an unauthorized worker through a labor contractor, in practice it is extremely difficult to demonstrate that a farmer has done so knowingly. To date, no farmer has been fined or imprisoned for knowingly hiring an unauthorized worker through a labor contractor. FLCs are a risk buffer between their farmer-clients and immigration authorities, and most farmers recognize this.

Labor contractors, in turn, have a demonstrated comparative advantage in recruiting immigrant workers at low wages and with few benefits, sometimes offering these workers services they need as new immigrants, at a cost, and then recruiting new immigrants to take these workers' place as soon as they have other job options. FLCs are often more than employers. They may house and feed workers, provide them with transportation to the fields, and even offer credit in the form of advance wages. The fees they charge for these services, which often are required informally whether workers desire them or not, quickly erode FLC workers' take-home pay, making hourly wages or earnings a poor indicator of the true remuneration for working for an FLC.

Interviews with FLCs reveal that these worker charges often are critical to the FLC's success. Competition between FLCs is often described as cut-throat, the result of too many FLCs in pursuit of a limited number of farmer contracts in some areas. Some studies reveal that the margins FLCs charge farmer-clients are insufficient to generate a profit above and beyond the minimum wage, employment taxes, and other fees mandated by state and federal governments (Martin and Vaupel 1987; Taylor and Thilmany 1992). Some of the most forceful complaints about unscrupulous FLCs come from other labor contractors who claim to be making a good-faith effort to comply with labor and immigration laws.

In California, most FLCs are ex–farm workers themselves, and many acquired legal status under the SAW program. IRCA may have increased both the demand for FLC services by farmers and, through the SAW legalization program, the supply of FLCs to provide these services. The shift to FLCs represents a fundamental and far-reaching change in the structure of farm labor markets in the United States. In one sense, it indicates that the farm labor market is becoming more like the nonfarm labor market, since the trend there is also to hire more temporary and permanent workers through specialized intermediaries such as Manpower and so-called employee leasing firms. But agriculture is becoming more dependent on FLC intermediaries without first upgrading their decades-long shady reputation.

Summary and Themes

Documented illegals, FLCs, and reduced earnings have not given farmers incentives to plan for a future of fewer and better-paid legal workers, and most have not made these plans. Unlike U.S. manufacturing industries that raised labor productivity sharply by reducing employment in the face of lower-wage competition during the 1980s, labor-intensive U.S. agriculture expanded and retained its dependence on hand-harvest workers, suggesting that there would be farm labor shortages if immigration laws were strictly enforced. In state after state, the story is similar: Washington apple acreage up 25 percent since IRCA; 3,000 acres of citrus planted in remote areas of California since the mid-1980s; 1,000-acre blocks of oranges planted in areas of southern Florida that have neither workers nor the infrastructure for seasonal workers. When asked about the labor assumptions that went into these plantings, the responses were that seasonal labor would no doubt continue to be available "as it always has been."

Agriculture has traditionally opened side or back doors to generally unskilled immigrants and nonimmigrants. In the debate over the agricultural provisions of IRCA, the major issue was whether the foreign workers that farm employers wanted Congress to make available to them should be immigrants or nonimmigrants. IRCA provided agriculture with both types of workers, as well as unintended mechanisms for obtaining additional "documented illegal" workers.

The papers in this volume suggest that IRCA is but the latest example of a federal policy meant to alter the farm labor market, whose unintended consequences were far more significant than its stated goals. Four major themes run throughout the book. First, instead of eliminating illegal aliens from the areas where they were most prevalent, IRCA seems to have accelerated the spread of such workers throughout the United States. In many cases, falling wages and deteriorating conditions have set in motion forces that are likely to turn the traditional revolving-door farm labor market faster when nonfarm jobs become available: Farm workers are likely to leave the farm work force as soon as they can, creating a vacuum that, without effective enforcement, will attract additional immigrants.

Second, IRCA apparently relieved farmers of their concern that seasonal workers would not be available. Especially in California, which had enough union activities during the 1970s to make labor a prime concern, IRCA seems to have pushed labor down on the list of employer concerns, below water availability, taxes, and marketing. Indeed, by making labor so readily available, IRCA increased the competitive advantage of U.S.-grown fruits and vegetables in international markets, explaining the apparent paradox that Mexico has emerged as a major new market for U.S.-grown fruits and vegetables that are harvested by Mexican immigrants. New plantings of labor-intensive crops in areas that lack the amenities necessary to obtain temporary foreign workers testify to the widespread "field-of-dreams" conviction among farmers — that if there is a crop to be harvested, workers will come.

The third theme that runs through these papers is that the farm labor market is changing in ways that further remove "farmers" from their seasonal workers. The reason is that FLCs, crew bosses, and similar intermediaries have become, in the eyes of many workers, "the" employer. In many cases, such intermediaries are a necessity for language reasons; however, intermediaries have the advantage for farmers of absorbing any risks associated with the labor and immigration law violations that are acknowledged to be widespread.

The fourth theme hinted at in these papers is that the traditional farm labor problem may turn into a rural poverty problem. Immigrants who were seasonal farm workers traditionally left rural areas after 10 to 15 years, either to return to their country of citizenship or to go to U.S. cities. However, the SAW legalization program, and the family unification that seems to be taking place in its wake, may re-create the rural poverty that many thought had been reduced when small black and white farmers and farm workers left the farm in the 1950s and 1960s. The successes with those rural poor may be harder to replicate in the future, since today's immigrant workers have less education, relative to the average American, than did these earlier rural poor, and they often do not speak English.

The papers in this volume paint a sobering picture. Immigration reforms did not set in motion the gradual adjustments to a legal farm work force that, at a minimum, IRCA was expected to produce. Instead, the Grand Bargain of legalization and sanctions seems to have succeeded so well in one dimension — legalization — that it has laid the groundwork for the continued immigration that has swamped sanctions enforcement. When the country once again considers what to do about the farm labor problem, these papers should provide a useful reminder that goals and outcomes are often very different in agriculture.

NOTES

1. Seasonal fruit and vegetable farms employ between 1 and 2 million workers sometime during a typical year. Almost all of the 200,000 to 300,000 entrants to the work force are immigrants; a few are the U.S.-born children of immigrants.

2. About 600,000 legal immigrants entered the United States annually during the 1980s, and 50 percent or 300,000 joined the labor force, including about 5 percent (150,000 of 3 million) with farming occupations. About 80,000 farm workers were legalized under the general legalization program, and all of the 1.3 million SAW applicants should have done farm work in the mid-1980s, suggesting that about 1.5 million immigrants who arrived during the 1980s had at least one farm job.

3. These programs are often distinguished by the type of INS application form completed by aliens seeking legal status. General program applicants file an I-687 form and SAWs file an I-700 form.

4. The number of SAW approvals established a ceiling on the number of RAWs that could be approved under the law.

Part I. IRCA and Agriculture: National Issues and Effects

This part summarizes several of the general outcomes of the Immigration Reform and Control Act of 1986 (IRCA). First, rather than eliminating the illegal worker problem, the administration of public policies during the post-IRCA period resulted in illegal immigration continuing at a similar rate to the year immediately preceding IRCA.

Second, the net effects of IRCA on farm wage rates and employment are still to be determined, but current evidence suggests that the trend in nominal wage rates for farm labor did not change significantly.

Third, IRCA seems to have accelerated the switch of employers obtaining farm labor from farm labor contractors rather than hiring workers directly themselves. There are forces coming from both the supply and demand sides of the market for this change. New survey results also document a relatively large and growing share of farm labor for crops that is supplied by Hispanics, and the settling of these people has increased the demand for social services in rural and urban areas.

Fourth, the production of labor-intensive crops in the United States generally increased during the post-IRCA period, providing further evidence that significant shortages of crop labor did not occur in the six years since passage of IRCA. Thus, IRCA has had more unintended than intended effects to date.

1

IRCA and Agriculture: Hopes, Fears, and Realities

PHILIP L. MARTIN

Introduction

Agriculture is the oldest and most widely dispersed industry in the United States. However, the farmers and farm workers who produce food and fiber are largely invisible to most Americans: About 2 percent of the nation's population live on farms and 2 percent of the nation's wage and salary workers are employed on farms.[1] Most Americans do not know any farmers or farm workers; they see agriculture only through the lens of a grocery store.

Farmers and farm workers are considered the keystone, but farming is actually the smallest part of the three-part food-and-fiber sector. The first part are the industries that feed and clothe Americans and foreigners; it includes banks, chemical companies, and equipment manufacturers that supply agriculture with credit, fertilizers, and machinery. The second part is the production of crops and livestock on farms, and the third — the largest — is the trucking, retailing, and restaurant activities that process and distribute food and fiber to consumers. There were 21 million Americans employed in this food-and-fiber sector of the American economy in 1990, or one in six American workers.[2]

The farmers who are at the center of this sector are generally held in high esteem. Thomas Jefferson, the third president, had an agrarian ideology that made farmers the backbone of democracy. In his *Notes on Virginia*, Jefferson wrote that "Those who labor in the earth are the chosen people of God . . . corruption of morals in the mass of cultivators is a phenomenon of which no age or nation has furnished an example" (Koch and Peden 1944, p. 280). According to this agrarian ideology, agriculture is the most basic and important industry in the nation, rural life is superior to urban life, and self-sufficient family farmers are the guarantors of democracy. This ideology continues to influence American attitudes and policies today. It helps to explain, for example, why farming is described as both a business and an exemplary way of life.

[21]

Farmers are also praised because agriculture is an industry in which the United States is a world leader. The most important expression of this U.S. leadership is at the grocery store: Only about 10 percent of what Americans spend is devoted to food, versus 20 percent in Germany, 36 percent in Japan, and 51 percent in India (USDC, *Statistical Abstract,* 1993, p. 833). The United States is also a major food exporter: Exports account for about one-fourth of annual farm production and farming is one of the few sectors of the American economy that has run a trade surplus for decades.

This general glow of success usually shines on the entire heterogeneous agricultural industry. Most American farms are family farms — defined under one USDA program as those that can operate with less than the equivalent of one and one-half year-round hired hands.[3] There are relatively few family farms — as defined by their dependence on hired labor — in the subsector of U.S. agriculture that is most closely associated with immigrant workers, the 75,000 U.S. farms that hire workers to produce fruits, vegetables, and horticultural specialties (FVH) such as flowers and nursery products. The 75,000 number is deceptive, because the largest 10 percent of these FVH farms account for 80 percent of U.S. fruit and vegetable production and employment. It is true that most U.S. farms, as well as most fruit and vegetable operations, are small family-run operations, but seasonal factories in the fields account for most of agriculture's perennial quest for immigrant workers.

Immigration policy and immigrants are linked to fruit and vegetable agriculture because immigrants constitute almost two-thirds of the industry's current work force and nearly all the entrants to the seasonal fruit and vegetable work force.[4] Among the 140 million Americans who have paid employment sometime during a typical year, only about 3 percent do farm work, and many of these farm workers are teenage hired hands on midwestern family farms. But of the almost 6 million immigrant workers who were in the United States legally by the end of the 1980s, up to 1.5 million or 25 percent did at least some farm work, reflecting the enormous appetite of seasonal farm factories for new workers.[5] If 1980s immigration patterns continue in the 1990s, up to one-quarter of the working-age U.S. immigrants who arrive during the decade may be Mexicans whose initial U.S. employment is in fruit and vegetable agriculture.

Immigration Reform: Exceptions for Agriculture?

When the congressional hearings that eventually culminated in IRCA began in 1981, the positions of farm worker and farmer advocates were not well developed. United Farm Workers (UFW) representative Stephanie Bower, for example, testified on September 30, 1981 that the UFW supported "imposing sanctions on employers who hire illegal aliens . . . [but since] laws covering farm workers have been rarely enforced . . . we strongly urge that a large budget for staff and operations be allocated to enforce sanctions (U.S. Senate 1981, p. 78). The UFW also supported issuing counterfeit-proof social security cards to all workers, including farm workers, to verify their legal right to work in the United States.

The National Council of Agricultural Employers (NCAE) testified that it had not yet developed a position on employer sanctions, but that if there was to be a sanctions law, "we [agriculture] must have some means to offset a worker shortfall if there are not enough U.S. workers to fill the needs of agricultural employers" (U.S. Senate 1981, p. 125). The NCAE offered two reasons why sanctions might lead to farm labor shortages: "They [illegal alien farm workers] will move to other jobs where they may get 12 months out of the year employment" . . . and many of "those people" (illegal aliens) "do not want amnesty," so if they must "choose amnesty" in order to work in the United States, "they may just opt [to go] back to Mexico" (U.S. Senate 1981, p. 125).

The so-called Simpson–Mazzoli Immigration Reform Bill, introduced in 1982, included sanctions, amnesty, and a streamlined H-2 program that essentially wrote then-current U.S. Department of Labor (DOL) regulations governing the admission of temporary alien farm workers for temporary U.S. jobs into law. Users of H-2 alien farm workers seemed generally satisfied, but western farm employers were not. They argued that they could not plan their need for seasonal labor in the manner prescribed by the H-2 program because they produced perishable commodities whose harvesting window was very sensitive to unpredictable weather conditions. Western farmers also acknowledged that they might have difficulty getting the DOL to certify their need for H-2 alien farm workers because many could not offer the free housing required for such migrant workers. Farmers also feared that the UFW and other worker advocates would urge the DOL not to approve their requests for alien workers on the grounds that U.S. workers were available.

In January 1983, representatives of most U.S. farm employers met in Dallas, Texas, to decide whether to press for further changes in the H-2 program to accommodate western growers or to seek a new foreign worker program. The decision was made to seek a new foreign worker program. The Farm Labor Alliance (FLA), a coalition of about 22 farm organizations, was created to press for such a program in Congress.[6] The FLA decided to work through Tony Coehlo (D-CA), then the "number-three" person in the Democratic hierarchy of the House of Representatives. Thomas Hale of the California Grape and Tree Fruit League in Coehlo's district became president of the FLA (Farm Labor Alliance 1986).

The FLA wanted a flexible guest worker program under which legal nonimmigrant workers would be confined while in the United States to farm work, and U.S. farmers would not be required to go through a certification process to employ them. The justifications offered for such a free-agent program included the unpredictable need for labor and the argument that a free-agent program would have the flexibility necessary to accommodate the migrant farm workers who followed the harvest from south to north. Both Senator Simpson and Representative Mazzoli opposed such a free-agent guest worker program, arguing that such a program was hard to justify in legislation designed to reassert control over immigration. It might be hard for them to find their own housing, they argued, and, as free agents, it would be difficult to regulate them in a manner that would not undermine the wages and working conditions of U.S. farm workers.

Representatives Panetta (D-CA) and Morrison (R-WA) introduced the FLA's P-visa guest worker program as an amendment to the Simpson–Mazzoli bill in the House in 1984. To the surprise of many observers, the House approved the Panetta–Morrison guest worker program in June 1984, in what the *New York Times* described as one of the year's top ten political stories. In the Senate, Simpson tried to appease the FLA with a streamlined H-2 program for agriculture, a three-year transition program under which farmers (and only farmers) could use 100, 66, and 33 percent of their base year's employment of illegal aliens, and a Commission on Agricultural Workers (CAW) to study farm labor issues further.

The Panetta–Morrison program was dropped by the Senate-House Conference Committee that met during fall 1984 to resolve differences between the House- and Senate-passed versions of immigration reform legislation, but the conference broke up over the issue of reimbursing state and local governments for the cost of providing services to newly legalized aliens. In 1985, Simpson reintroduced immigration reform legislation in the Senate, and once again offered farmers a streamlined H-2 program, a transition program, and a CAW. The FLA got Senator Wilson (R-CA) to offer another version of the Panetta–Morrison program as an amendment to immigration reform, and the amendment initially failed. However, after Wilson agreed to cap the number of guest workers at 350,000, the amendment passed.

When the House considered immigration reform in 1986, Representative Rodino (D-NJ) asserted that he would try to block legislation that included a Wilson-type guest worker program for agriculture. Rodino's position appeared to doom immigration reform legislation. However, during the summer of 1986, Representative Schumer (D-NY) negotiated a compromise farm worker legalization program with Representative Panetta (D-CA), representing employer interests, and Representative Berman (D-CA), representing worker interests.

The Schumer compromise created the Special Agricultural Worker (SAW) and Replenishment Agricultural Worker (RAW) programs — SAW to legalize and stabilize the current illegal alien work force and RAW to admit probationary immigrant farm workers if newly legalized SAWs quit doing farm work and their exit led to farm labor shortages. Worker advocates agreed to the SAW-RAW system because the SAWs were free to leave agriculture; these advocates were concerned primarily about H-2-type workers who were dependent on their U.S. farm employers to stay employed in the United States. SAWs, their theory went, would have the power to say no to unfair wages and working conditions and, after illegal immigration stopped, the power to negotiate higher farm wages.

Farmers won assurances that the relatively easy requirement for SAW legalization would permit them to continue to employ their current workers, and then RAWs could enter the United States if labor shortages developed. The SAW-RAW compromise was accepted by congressional leaders, who then instructed members to leave it alone if they wanted to see IRCA enacted.

This brief history illustrates how the farm labor concerns and proposals discussed in pre-IRCA hearings played only a small role in the SAW-RAW compromise that resulted. The Reagan administration and especially the Immigration and Naturalization Service (INS) were opposed to a SAW-RAW type of immigrant farm worker program; they preferred a nonimmigrant program. Since IRCA included both the probationary immigrant and the nonimmigrant H-2A program for agriculture, and because Congress could not agree on which type of program agriculture should have, CAW was created so that there would be a forum to discuss how many and what type of long-run foreign worker programs (if any) should be developed for agriculture.

Table 1.1. Legalization Applicants

Characteristic	LAW*	SAW**
Median age at entry	23	24
1. Age 15 to 44 (%)	80	93
2. Male (%)	57	82
3. Married (%)	41	42
4. From Mexico (%)	70	82
5. Applied in California (%)	54	52
Total applicants	1,759,705	1,272,143

Source: USINS 1992, pp. 70-74.
*Persons filing I-687 legalization applications.
**Persons filing I-700 legalization applications. An additional 80,000 farm workers received legal status under the general or pre-1982 legalization program.

IRCA's Effects in Agriculture

The effects of the SAW-RAW compromise on U.S. farmers and farm workers were not debated extensively in Congress, although the possible effects of other immigration reforms on agriculture had been discussed throughout the early 1980s. It is possible to discern a range of expected IRCA effects on farmers and farm workers, but most suggest that IRCA should have had at least two effects in agriculture: Farm workers should be legal U.S. workers and farmers should have raised wages somewhat and improved employment practices in order to retain newly legalized SAW workers.

IRCA has not had these hoped-for effects in agriculture. There are several reasons, including a continued influx of unauthorized workers and less rather than more effective enforcement of immigration and labor laws in agriculture.

Continuing Illegal Immigration

One of the most dramatic changes in the farm labor market due to IRCA is the switch in the fields from undocumented workers to falsely documented workers. Illegal immigrants who do farm work are usually among the poorest and least sophisticated such workers in the United States. Before IRCA, the INS typically enforced immigration laws by driving through fields and apprehending those workers who ran, disrupting harvesting for the employer and causing the worker to lose wages until he could re-enter the United States and find a job.

IRCA converted immigration enforcement from the pursuit of aliens in the field to a paper chase. There are far fewer farm worker apprehensions today despite apparent pre-IRCA levels of illegal alien employment. Employers and aliens evade paper-chase enforcement in many ways, such as having valid work authorization papers attached to each I-9 employment verification form, but different workers actually working in the fields.[7] IRCA may well be remembered as a stimulus to illegal immigration for spreading work authorization documents and knowledge about them to very poor and unsophisticated rural Mexicans and Central Americans, encouraging first-time entrants from these areas.[8]

In the early 1990s, pre-IRCA legal farm workers, newly legalized SAWs, and a continuing influx of illegal aliens have produced an ample supply of seasonal workers, giving farmers little incentive to improve wages and working conditions. Study after study confirms the fact that most farmers have not raised wages, improved working conditions or housing, or otherwise made adjustments to recruit and retain legal workers (Martin and Taylor 1990). Instead, the IRCA-wrought influx of workers has in some cases encouraged reductions in wages and farm worker earnings.[9]

Most farm workers have been paid the same rate per hour worked or unit of work done since the late 1980s, but farm worker earnings have nonetheless declined in many instances because, with more workers, each does fewer hours of work. Farm worker profiles indicate that a typical seasonal worker is available for farm work about 40 weeks per year, and finds work during 20 to 25 weeks for 700 to 1,000 hours of work per season or year. At $5 hourly, seasonal farm work generates individual earnings of $3,500 to $5,000 annually.[10]

FLCs

Seasonal farm workers have also been affected by the post-IRCA tendency of farm operators to hire more seasonal workers through FLCs. In California, the "market share" of FLCs appears to have risen from about one-third of all job matches in the early 1980s to over half in the early 1990s (Martin and Taylor 1990). Employees of FLCs are worse off in several ways, including the tendency of FLCs to pay lower wages to recently arrived immigrants. Many FLCs condition employment on worker acceptance of housing away from the work site, and then they charge workers for both housing and rides to work. Housing away from the farm, usually in barracks-style accommodations, costs each worker $25 to $35 weekly, and then the private rural taxis (*raiteros*) that provide rides to work sites typically charge each worker $3 to $5 daily. A worker getting $200 weekly (40 hours at $5), often has $50 or 25 percent less take-home pay if he or she is employed by a FLC because of these housing and ride-to-work charges.

Hiring documented illegal workers through FLCs means that few farmers have planned for a future of fewer and better-paid legal workers. Unlike the U.S. manufacturing industries that shrank in the face of lower-wage competition during the 1980s, labor-intensive U.S. agriculture expanded, usually in ways that guarantee farm labor shortages in the 1990s. In state after state, the story is similar: Washington apple acreage up 25 percent since IRCA; 3,000 acres of citrus planted in remote areas of California since the mid-1980s; 1,000-acre blocks of oranges planted in southern Florida. When asked about the labor assumptions that went into these plantings, many of which will not need harvest workers until the mid-1990s, the answers are a sheepish "we didn't think about labor" or "we assumed seasonal labor would be available at the minimum wage, as it always has been."

The Commission on Agricultural Workers

The SAW program and related IRCA provisions affecting agriculture were not subject to extensive public hearings and debate. There was a great deal of uncertainty about their probable effects on farmers and farm workers, and this uncertainty, as well as the four-year life of the RAW program, gave the Commission on Agricultural Workers (CAW) its mandate to review the effects of IRCA and especially its SAW provisions on the farm labor market.

The commission was charged with "reviewing" nine questions and conducting "an overall evaluation of the special agricultural provisions" of the Immigration Reform and Control Act of 1986 (IRCA) (Figure 1.1). These questions ask about the impact of the SAW program on, among other things, U.S. farm workers and their wages as

Figure 1.1. CAW's Mandate

1. Conduct an Overall Evaluation of IRCA's Special Agricultural Worker (SAW) Provisions and Recommend Appropriate Changes.

2. Review the Impact of the SAW Program:
   - on the wages and working conditions of U.S. farm workers
   - on the adequacy of the supply of labor
   - on the ability of agricultural workers to organize
   - on the international competitiveness of U.S. crops
   - on whether SAWs stay in the farm work force
   - on whether employer sanctions affected the supply of farm labor.

3. Determine the Extent:
   - to which agriculture relies on a temporary work force
   - of unemployment and underemployment among U.S. farm workers
   - to which farmers are unable to find workers because they do not use modern labor-management techniques
   - to which the supply of farm labor is adequate
   - to which certain geographic regions need special programs to secure labor.

well as the extent of unemployment among farm workers in light of IRCA and other factors affecting the farm labor market.

On the basis of case-study research and hearings, the commission made findings that can be grouped into three categories (Figure 1.2). First, in its overall evaluation of the SAW program, the commission concluded that the majority of SAW-eligible undocumented workers gained legal status, but through such a flawed worker- and industry-specific legalization program that Robert Suro in the *New York Times* described the SAW program as "one of the most extensive immigration frauds ever perpetrated against the U.S. government."[11] Second, the commission found that, although the SAW program legalized many undocumented farm workers, the continued influx of illegal workers prevented newly legalized SAWs from obtaining im-

Figure 1.2. CAW Findings

1. Overall Evaluation of the Special Agricultural Worker (SAW)
   Provisions

   • The majority of SAW-eligible workers gained legal status;
   there was also significant fraud in the sense that nonqualified
   aliens achieved legal status through the SAW program.
   • The SAS work force in 1992–93 continues to include a significant share of undocumented workers.
   • The combination of U.S. citizen and immigrant farm workers,
   SAWs, H-2A workers, and undocumented workers produced
   labor surpluses that make it unnecessary for farmers to improve wages or working conditions in order to obtain a seasonal work force.
   • The SAW legalization program was flawed — limiting
   eligibility to workers ignored the reality that workers have
   families and that U.S. immigration policy favors family unification. Limiting eligibility to one subsector of one industry led to
   pressures for a far more expansive definition of SAS than was
   intended. Inadequate documentation requirements encouraged
   fraudulent applications.
   • The probationary immigrant RAW program that was not
   implemented was also flawed. The formula for determining
   whether a national labor shortage was expected included
   answers to hypothetical questions and was thus open to
   political manipulation. Most RAW applicants appear to be
   undocumented workers already working in the United States,
   so their "admission" would not add to the supply of farm
   workers or ensure that these now legal workers go to areas
   experiencing labor shortages.

2. IRCA, SAW, and Farm Labor

   • Employer sanctions have not reduced significantly the
   employment of undocumented farm workers.
   • The SAW program granted legal status to over 1 million
   mostly young Mexican men. Almost all of those who have
   been observed working in agriculture since IRCA continue to
   be farm workers.

- The SAW program did not reverse the decline in real farm wages that began in the early 1980s. It may have contributed to a more uniform average farm wage by adding to the labor supply and reducing the premium over the minimum wage that many western farmers once paid, and by spreading Mexican SAW farm workers throughout the United States.
- The SAW program contributed to the oversupply of farm workers, dampening pressures on farmers to improve housing and working conditions in order to attract and retain seasonal workers. Legal status encouraged some SAW workers to make demands, but the oversupply of workers limited their ability to achieve wage or working condition improvements.
- The SAW program had little direct impact on the international competitiveness of U.S. crops. By stabilizing wage and benefit costs, it may have contributed to making the United States a net exporter of labor-intensive crops.

3. Farm Workers and Farmers

- Seasonal farm workers are concentrated on a few large farms in a few states that grow fruits and nuts, vegetables and melons, and horticultural specialty (FVH) commodities.
- These farms can obtain seasonal workers for temporary tasks, and so they offer only temporary jobs and depend on temporary or seasonal workers to be available to fill them.
- The supply of farm workers is adequate and does not need to be supplemented further with foreign workers except for the current H-2A program.
- Almost all less-than-year-round farm workers experience unemployment during a typical year. Those farm workers who obtain most of their income from farm work average about 25 weeks of such work and earn $5000. They are unemployed about 20 weeks. Unemployment in farm worker communities ranges from 10 to 30 percent.
- Many farm employers do not have modern personnel systems to recruit and train workers, to identify and encourage the return of the best workers, and to make hiring and layoff decisions on the basis of objective criteria such as seniority.
- Farm labor markets vary by geography, but there is no need for special programs for particular geographic areas.

provements in wages and benefits from farmers. Third, the commission reported that the farm labor market continues to leave the average farm worker with below-poverty-level earnings.

The commission recommended that federal and state governments could take steps to develop a legal farm work force, to improve social services for farm workers and their families, and to improve the enforcement of labor laws (Figure 1.3). In response to IRCA's failure to reduce illegal immigration into the United States, the commission recommended more border and interior enforcement and a fraud-proof work authorization card. To combat declining real wages, the absence of benefits like health insurance, and the exclusion of some farm workers from federal and state programs that would make them eligible for unemployment benefits and workers compensation insurance, the commission recommended that the federal government provide more services to farm workers and their children and that farm workers be covered under protective labor laws. Finally, the commission recognized that federal and state agencies today have only a limited ability to enforce farm labor laws, and recommended that enforcement efforts should be better coordinated and targeted.

Figure 1.3. CAW Recommendations

1. To Have a Legal Farm Work Force

   • Make a renewed effort to reduce illegal immigration with better border and interior enforcement; develop a fraud-proof work authorization document.
   • Do not extend the RAW program or begin any new supplementary foreign worker program.
   • Re-examine the H-2A temporary farm worker for temporary U.S. job program.
   • Do not have another worker- and industry-specific legalization program, despite significant levels of undocumented worker employment today.

2. To Improve Services for Farm Workers

   • Improve Employment Service worker-job matching services by having the ES plan migrant itineraries and distribute information to workers about available farm jobs.
   • Educate farmers about the need for and benefits of modern labor-management practices for farm workers, such as systems to employ fewer workers for longer periods.

- End the exclusion or differential treatment of farm workers under unemployment insurance, workers compensation, and labor relations laws.
- Provide more funding and flexibility to promote the availability of public and private farm worker housing.
- Develop programs for farm worker children so that they are not disadvantaged because of their parents' occupation.
- Establish a federal council to coordinate the services available to farm workers and their children.

3. To Improve the Enforcement of Labor Laws in Agriculture

- Improve coordination between federal and state agencies so that employers who violate labor laws do not gain a competitive advantage.
- Create a task force to target enforcement efforts on the most serious violations and violators.
- Increase the bonding and testing requirements for farm labor contractors (FLCs).
- Make farm operators who use unlicensed FLCs solely liable for any federal labor law violations the FLCs commit while employing farm workers in their operations.
- Make federal and state agencies that enforce labor laws in agriculture liable for fines levied on employers who were told by the agencies that they were in compliance.

The Commission also made several recommendations that may undermine some of these efforts to reduce farm worker poverty. For example, it recommended that the same regulations govern the importation of nonimmigrant farm and nonfarm workers. This recommendation was made in the spirit of ending agricultural exceptionalism, the practice of exempting farmers from labor laws under the theory that family farmers should not be burdened by excessive formal regulation. However, the employer certification and worker protections in the agricultural H-2A program are there because of past problems with employer violations of the agreements they signed to employ nonimmigrant farm workers. While the commission developed no evidence that the nonfarm H-2B program needs these extra procedures and protections, it heard considerable testimony that current protections in the H-2A program are inadequate. Thus, the intent of this recommendation can only be to substitute weaker H-2B procedures for stronger H-2A rules, and this flies in the face of the evidence that was developed. The H-2A program may need to be reviewed, but probably not with the goal of substituting H-2B criteria for H-2A criteria.

The Unfinished Agenda

Most of the commission's recommendations are a useful step in the right direction, but they fail to deal with the root causes of the farm labor problem and the legacies of IRCA's agricultural provisions. The federal government has permitted immigration to be a subsidy for the labor-intensive fruit and vegetable subsector of U.S. agriculture. This immigrant labor subsidy encourages the expansion of an industry in which the majority of workers earn below-poverty-level incomes. Farm worker poverty is recognized widely, but there is usually an avoidance of the question of who is responsible for the farm worker's plight.

Many people believe that poor farm workers are the price that must be paid for cheap food. But the relationship between cheap farm workers and cheap food is not so clear. The numbers seem to belie this conventional wisdom. Two-thirds of the nation's farm work is done by farmers and their families. Hired workers do only one-third of the nation's farm work. Immigrant farm workers, the poorest hired workers, do about two-thirds of the work done by hired workers. This means that, if there were no immigrant farm workers, almost 80 percent of the nation's farm work would be done without them. In other words, holding down the wages of seasonal farm workers, while it impoverishes more than 1 million American workers, holds down the average family's food bill only a little. Even in the case of the fruits and vegetables that immigrant workers harvest, farm wages typically account for less than 10 percent of the retail price of a head of lettuce or a pound of apples. Doubling farm wages, and thus practically eliminating farm worker poverty, would raise retail food prices by less than 10 percent.

Retail food prices might not even increase if the U.S. government aimed to increase rather than depress farm wages. The farmers who relied on Mexican *bracero* workers in the early 1960s argued that "the use of *braceros* is absolutely essential to the survival of the tomato industry" (Calif. Senate 1961, p. 105). What happened when they none-theless disappeared? The termination of the *bracero* program in 1964 accelerated the mechanization of the harvest in a manner that qua-drupled production to 10 million tons between 1960 and 1990. Cheaper tomatoes permitted the price of ketchup and similar products to drop, helping to fuel the expansion of the fast-food industry.

Not only is it inefficient to hold down food prices by holding down farm worker wages; it is also morally wrong.[12] Why should immigration exceptions hold down the wages of farm workers, who average $5000 annually, in order to lower food prices for nonfarm workers, whose average earnings are $25,000 per year?

The commission identified but did not deal effectively with the dynamic element that is gradually reducing the government's ability to regulate the farm labor market and reduce farm worker poverty: farm labor contractors (FLCs). Farm labor contractors are the inter-mediaries who, for a fee, recruit, transport, and supervise farm work-ers. Since IRCA was enacted in 1986, the share of all seasonal job matches made by FLCs has increased. Today it exceeds 50 percent in many harvest labor markets. Worker, farmer, and agency testimony, as well as research, suggest that FLCs are practically a proxy for the employment of undocumented workers and egregious or subtle vio-lations of labor laws.

The expansion of FLC activities in the wake of IRCA has helped to lower wages and incomes in rural America. FLCs are perhaps the most important dynamic actors in bringing the new immigrants to the United States. As they play the role of nineteenthh-century ship captains in recruiting, transporting, and employing new arrivals, their activities promise to bring into rural communities some of the needi-est immigrants — relative to the average American — that have ever arrived in the United States.

The FLC may be fundamentally flawed in a labor market awash with immigrant workers. A contractor operates between a farmer and a farm worker, but the power of the two over the contractor is very different. Farmers typically know what the going overhead or com-mission is, and thus FLCs are unlikely to extract an extra-high fee from them. Newly arrived immigrants, on the other hand, may not know the minimum or the prevailing piece-rate wage, so that FLCs can turn what appears to be a money-losing deal with farmers into a profit-making deal by extracting money from workers. As the U.S. Industrial Commission observed in 1901, "the position of the contractor . . . is peculiarly that of an organizer and employer of immigrants. . . . He holds his own mainly because of his ability to get cheap labor . . . [and he] succeeds because he lives among the poorest class of people, knows them personally, knows their circumstances, and can drive the hardest kind of bargain with them" (U.S. Industrial Commission 1901, pp. 320-321).

IRCA's most important legacy may be the foundation it laid for a new wave of rural poverty in the United States. The rural poverty that many thought was reduced when millions of small white and black farmers left agriculture in the 1950s and 1960s may be recreated through immigration in the 1990s.

This rural poverty may prove even more difficult to reduce. Most of the immigrants arriving in rural America are young men with a primary school education or less. They are seasonal farm workers for only 10 or 15 years so that, at age 35 to 40, many can no longer find jobs harvesting fruits and vegetables. In the past, most returned to Mexico and sent their sons to replace them in U.S. fields and orchards. Today, more appear to be settling in the United States.

The SAW program was worker and industry specific. Most of those legalized under the SAW program were young men from Mexico. Many of these men have families, and some of them have brought their families to the United States without authorization. Immigration policies developed in deference to family ties mean that most of these large Mexican families will be tolerated in the United States. The SAW program does not necessarily exclude family members; it merely puts them in an illegal status until the reality that they are here and will not be deported is eventually accepted and they, too, are legalized.

However, while these mixed families — mixed in the sense that the members have a variety of immigration statuses — wait for this reality to be accepted, their presence raises important social policy questions. In some SAW families living in the United States, the father is a permanent resident alien, the mother and older children are unauthorized, and there are one or more U.S.-citizen babies. The SAW father is barred from some federal assistance programs for five years, and other members of the family may be ineligible for most services, but the U.S. citizen family member is not barred from any service on grounds of immigration status. The SAW program produced far more mixed-family situations than the general legalization program because 82 percent of the SAW applicants were men.

As SAWs bring their families to the United States, they raise the issue of how social policies should deal with families, some of whose members are legal and others of whose members are unauthorized.

Nowhere is this mixed-family issue more prominent than in agriculture, primarily because only agriculture had a worker- and industry-specific legalization program.

The most common summary of the effects of IRCA on U.S. agriculture is: good intentions gone awry. Worker and employer representatives agreed on a Grand Bargain that legalized illegal workers and established two programs through which legal foreign workers could be admitted in the event of labor shortages. This compromise helped farmers far more than workers. There is thus one less option available to Congress to deal with the century-old issue of the proper role of alien farm workers in U.S. agriculture.

NOTES

1. In 1990 there were 4.6 million farm residents, 1.4 million farm families, and 2.1 million farms. Farm worker data are even more confusing: There were an average 900,000 farm workers, according to the Current Population Survey, the source of most labor force data. But farm work is seasonal, and the number of indiviuals who are employed on farms sometime during a typical year is about three times larger, or 2.5 to 3 million (USDC, Statistical Abstract, 1993, pp. 642-644).

2. According to the USDA, farming employs about 2.5 million people or 3 percent of the 125 million U.S. labor force. These food-and-fiber industries generated 15 percent of the nation's $5.5 trillion GNP in 1990. The USDA subtracts from annual farm sales of $170 billion the value of the inputs farmers purchase from the nonfarm sector, and for this reason, the USDA estimates that agriculture accounted for $71 billion in value added — 1.4 percent of the $5 trillion GNP — in 1989.

The USDA estimates that 2.5 million persons were employed in agriculture in 1989, or 3 percent of U.S. employment. The Economic Report of the President, 1992 reports that 3.2 million persons were employed in agriculture, or that 2.8 percent of civilian employment (117 million) was in agriculture.

The USDA estimates that 5.4 million workers are employed in industries that provide inputs to farmers. Almost three-quarters provide farmers services such as banking, research, and insurance. Another 13.2 million are employed to manufacture and distribute food products, including 10 million who are employed in food retailing and restaurants.

3. There is no official definition of a family farm.

4. Farm worker data are notoriously unreliable. Seasonal fruit and vegetable agriculture appears to employ between 1 and 2 million workers annually, and almost all of the 200,000 to 300,000 new work force entrants each year are immigrants. There are some U.S. citizen children of immigrant farm workers who also enter the farm work force, but most are farm workers only as teenagers.

5. About 600,000 legal immigrants entered the United States annually during the 1980s, and 50 percent or 300,000 joined the labor force, including about 5 percent (150,000 of 3 million) with farming occupations. About 80 percent of the 1.8 million general legalization applicants were in the labor force (1.44 million), including 80,000 farm workers (5 percent). All of the 1.3 million SAW applicants should have done farm work in the mid-1980s. The INS indicated in March 1991 that it would approve only 910,000 or 71 percent of these SAW applications.

6. The FLA included 22 farm organizations, from the California Grape and Tree Fruit League to the Washington Asparagus Growers. Many farm employers on whose behalf the FLA lobbied belonged to two or more of the organizations that in turn comprised the FLA.

7. Interviews during summer 1990 indicated that many employers of seasonal farm workers, especially FLCs, leave hiring to a field supervisor or foreman, so that, at 6 or 7 a.m., when work begins, I-9 forms are completed, each worker provides a copy of his or her work authorization, and then the I-9 and the copy of the green card and driver's license are returned to the office to be available for INS inspection.

8. Anecdotal evidence suggests that the aliens know their green cards are fraudulent, since many continue to cross the U.S. border without inspection and use their green cards only to get a U.S. job. U.S. identification papers, including a driver's license and green card, can be purchased at weekend flea markets in California for $25 to $50.

9. Survey data suggest that most seasonal workers are paid either the minimum wage for work paid on an hourly basis (when the quality of work done is more important than the speed with which the work is done) or piece rate wages, when the quantity of work done is the major criterion. Hourly wages rose with the minimum (from $3.35 to $4.25 in California on July 1, 1988; from $3.35 to $3.80 in the United States on April 1, 1990 and to $4.25 on April 1, 1991), but piece rate wages have generally remained unchanged since IRCA, and in some instances have declined.

10. The DOL National Agricultural Worker Survey (NAWS), which probably includes a disproportionate share of long-season and year-round workers who do most of the days of farm work done by hired workers, reports that median earnings in 1989 were $7,500 for 37 weeks of work. These SAS workers were young (median age 30), male ($\frac{3}{4}$), foreign born ($\frac{3}{4}$), and poorly educated (median six years of education; $\frac{3}{4}$ cannot read English well). SAWs are about 40 of the NAWS sample, and post-IRCA "documented illegals" are about 10 percent.

11. Reprinted in the *Sacramento Bee*, November 12, 1989, p. A1.

12. Using immigrant workers to lower fruit and vegetable prices, if successful, increasingly subsidizes foreign rather than U.S. consumers as the United States becomes a major exporter of these commodities. By some measures, about one-fifth of the $25 to $30 billion of U.S. FVH production is exported. Importing workers to hold down labor costs and spur exports seems to work against the long-run comparative advantage of the United States.

2

Trends in Labor-Intensive Agriculture

SHANNON HAMM, VICTOR J. OLIVEIRA, GLENN ZEPP, AND JAMES DUFFIELD

Introduction

The supply of labor-intensive crops, mainly fresh fruits and vegetables, increased during the 1970s, 1980s, and 1990s. Labor accounted for a third of total fruit and vegetable expenses. Growers were able to increase supplies of fruits and vegetables mainly through yield gains and continued availability of labor. Fruits and vegetables are still primarily hand harvested and packed, although the technology to support increasing mechanization has advanced. Fruit and vegetable output will probably continue to rise faster than population growth due to the growing per capita demand.

Fruit and vegetable crops are coined "labor-intensive crops" because they have the highest labor expenses of any broad class of agricultural crops. These crops are important to study because growers have feared their incomes would be reduced by the passage of the Immigration Reform and Control Act of 1986 (IRCA). IRCA's main intent was to stop the flow of illegal immigrants to the United States, which could have either increased wage rates or limited the use of labor. But the economic data for the fruit and vegetable industry appear to suggest another picture, since both supply and demand have continued to increase since 1986. The report to Congress from the Commission on Agricultural Workers (CAW) states that many of the intended impacts were never realized, while some unintended impacts materialized (USCAW 1993).

Specialized fruit and vegetable farms reported in 1990 that their labor-cost share of total production expenses was roughly three times as large as for all farms. Because these commodities are so highly perishable, they have mainly been harvested and packed by hand. This chapter describes a 23-year trend in fruit and vegetable acreage, production, trade, and use from 1970 to 1992. The economic data are analyzed during the 1970–86 and 1987–92 periods. Because impacts on labor from IRCA would be likely to affect acreage, yields, and production post-1986, a new trend line is used to analyze any possible changes. A description and a categorization of the use of hand/mechanical harvesting are also provided to better identify which crops depend on labor the most.

Expansion

Fruit and vegetable output has gained over the past 23 years, despite relatively constant acreage in both fruits and vegetables. In 1970, fruit and vegetable harvested acreage totaled 3.9 million, while by 1991 the acreage totaled 3.26 million (Table 2.1). However, aggregation masks very different sectorial trends. Total vegetable acreage has trended upward, while total fruit acreage has gone down.

Fruit and vegetable production occurs throughout the United States, with the largest acreages being in California and Florida (Table 2.2). As with the rest of U.S. agriculture, the number of fruit and vegetable farms has declined, while their size has increased.

Harvested vegetable acreage for fresh and processing vegetables remained fairly constant during the 1970s and the first half of the 1980s; then acreage expanded from 1986 to 1992 at about half a percent per year (Figure 2.1). The increase in acreage is primarily a result of increased use of processing tomatoes and frozen potatoes. Total harvested vegetable area increased post-IRCA, so IRCA's intent of reducing the pool of labor could not be observed (Figure 2.1).

However, fresh vegetable harvested area increased 15 percent over the 23-year period, again with nearly all the increase transpiring in the 1980s (Table 2.1). The added acreage reflects an increase in demand for fresh vegetables, which was realized through technological adoption. Technology adoptions have improved yields. The major changes introduced by growers were new varieties, the increased use of plastic mulch, and conversion to field packing. These changes have allowed growers to maximize technological adoption and utilize workers primarily for harvest activities.

Processing vegetable acreage remained fairly constant during the 1970s and 1980s. However, acreage increased 26 percent in 1989, the single largest year increase. Processing vegetables, unlike fresh, are virtually 100 percent mechanically harvested (Table 2.3). Thus, any impacts on the labor supply would not be immediately reflected in area, since the demand for labor is mainly at the processing plant.

Table 2.1.--Total Area Harvested of Labor Intensive Crops, 1970-92

Year	Fruit				Vegetables				Grand total
	Citrus	Noncitrus	Tree nuts	Total	Fresh	Processing	Other 1/	Total	
				1,000 acres					
1970	1,144	1,812	344	3,299	974	1,407	3,337	5,719	9,018
1971	1,194	1,781	366	3,340	937	1,437	3,132	5,506	8,846
1972	1,186	1,756	384	3,326	960	1,460	2,984	5,404	8,730
1973	1,202	1,758	399	3,359	981	1,593	3,016	5,590	8,949
1974	1,212	1,788	422	3,422	957	1,645	3,368	5,970	9,392
1975	1,216	1,828	443	3,487	934	1,744	3,137	5,816	9,302
1976	1,198	1,894	456	3,548	946	1,529	3,183	5,658	9,206
1977	1,180	1,908	483	3,571	956	1,533	3,031	5,519	9,091
1978	1,161	1,922	519	3,603	998	1,496	3,294	5,788	9,390
1979	1,150	1,901	556	3,606	1,012	1,539	3,075	5,626	9,232
1980	1,162	1,914	566	3,642	1,002	1,332	3,501	5,835	9,477
1981	1,148	1,904	561	3,614	994	1,264	3,931	6,189	9,802
1982	1,124	1,881	579	3,585	939	1,250	3,579	5,769	9,353
1983	1,092	1,925	599	3,615	935	1,197	2,798	4,931	8,546
1984	1,008	1,957	624	3,589	1,072	1,370	3,166	5,608	9,196
1985	899	1,981	657	3,538	1,071	1,392	3,238	5,701	9,238
1986	819	1,988	670	3,476	1,070	1,239	3,175	5,484	8,960
1987	826	1,999	675	3,501	1,122	1,312	3,402	5,837	9,337
1988	833	2,013	686	3,532	1,130	1,342	2,955	5,426	8,959
1989	848	2,009	687	3,543	1,146	1,475	3,290	5,910	9,454
1990	852	2,011	689	3,552	1,120	1,548	3,826	6,494	10,046
1991	850	1,982	660	3,492	1,066	1,570	3,696	6,333	9,824
1992	884	1,961	659	3,504	1,113	1,446	3,167	5,725	9,230

1/ Includes dry edible beans and peas, potatoes, and sweet potatoes.

Source: USDA, ERS, July 1992; USDA, ERS, September 1992.

Table 2.2. Ranking of Area for Labor-Intensive Crops, 1987

	Land in Orchards			Land in Vegetables	
States	Area Acres	Share Percent	States	Area Acres	Share Percent
California	2,152,664	47	California	882,741	25
Florida	762,068	17	Wisconsin	328,902	9
Texas	208,568	5	Florida	311,659	9
Washington	241,423	5	Minnesota	207,342	6
Michigan	161,567	4	Texas	201,702	6
Georgia	149,014	3	New York	150,054	4
New York	124,432	3	Washington	144,097	4
			Oregon	142,236	4
			Michigan	139,145	4
			Arizona	98,138	3

Source: USDC 1989.

Figure 2.1. Harvested Vegetable Area Rises Post-IRCA, 1970–92

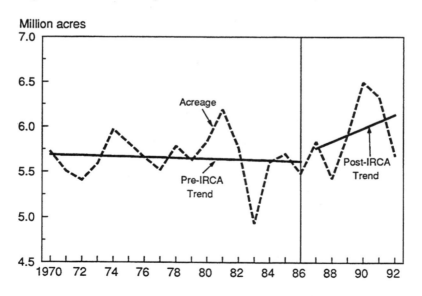

Table 2.3. Level of Hand Harvesting for Major U.S. Horticultural Crops

Percent of acreage hand harvested	Vegetable Crops		
A. Vegetable Crops			
76-100	Artichoke	Asparagus	Broccoli *	Cabbage
	Cantaloupe	Cauliflower	Celery	Cucumber *
	Lettuce	Green onions	Collards	Escarole
	Okra	Eggplant	Endive	Mushrooms
	Romaine	Kale	Squash	Rhubarb
	Parsnip	Peppers	Ginger	Watercress
	Rutabaga	Jerusalem artichoke	Turnip	
51-75	Sweet potatoes	Mustard greens	Parsley	Swiss chard
	Turnip greens			
26-50	Dry onions	Pumpkins *	Tomato*	
0-25	Carrots	White potatoes	Lima beans*	Snap beans*
	Sweet corn*	Spinach*	Horseradish	Red beet*
	Peas*	Garlic	Brussels sprouts	
	Malanga	Boniato	Radish	
B. Fruit and Berry Crops			
76-100	Apple	Apricot	Avocado	Sweet cherry*
	Grape*	Kiwi	Kumquat*	Peach*
	Lychee	Mango	Nectarine	Pomegranate
	Pear*	Persimmon	Pineapple*	Gooseberry*
	Strawberry	Wild blueberry*	Current*	Lime
	Orange*	Grapefruit*	Lemon*	
		Olive*	Tangerine	
		Tangelo*		
51-75	Red raspberry*			
26-50	Prune*	Blackberry*	Highbush blueberrry	Black raspberry
0-25	Tart cherry*	Date	Fig	Cranberry

*More than 50 percent of crop is processed.

Source: Thompson, 1992.

Tomatoes, the most important processing crop and the first processing vegetable to convert to mechanical harvesting, increased in
area from 1989 to 1991. The increase was due largely to a temporary
rapid expansion in tomato-processing capacity, which has been reversed due to oversupply, low product prices, and firms exiting the
industry. The 11 percent decline in 1992 reflects oversupply conditions and not a tight labor supply (Figure 2.1).

Bearing fruit area grew only slightly — 1.4 percent per year —
over the 23-year period (Table 2.1). A precipitous decline in area during the second half of the 1980s resulted from less citrus area (Figure
2.2). This is a direct result of freezes. Freezes have hit both California
and Florida during the past two decades, with most of them occurring in the second half of the 1980s. California and Florida are the
two most important citrus-producing states. Thus, the significant
decline in area of citrus trees is a direct result of freezes (Table 2.1).
Since IRCA passed in 1986, nursery stock area has increased, which
implies that growers are not expecting any future labor shortages.
The Commission on Agricultural Workers concluded that, since IRCA,
growers generally have been able to make planting decisions and
assume that workers would be available (USCAW 1993).

Figure 2.2. Bearing Fruit Area Trends Down Post-IRCA, 1970–92

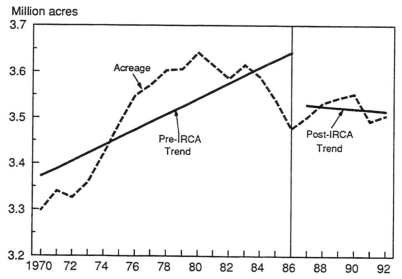

Driven by domestic demand that is rising faster than population growth and strong export demand, domestic fruit and vegetable production has trended upward during the 1970s, 1980s, and 1990s (Table 2.4). Production gains have come largely through increased yields and the ample availability of labor, among other resources.

Growth in fresh vegetable use is primarily attributed to the expansion of retail produce square footage, increased food service use, and increased advertisements.

The major source of higher yields was the introduction of more prolific hybrid varieties, many of which exhibit improved disease resistance as well as increased fruit set. Other factors important in boosting yields (or reducing losses) include such technological innovations as

- drip irrigation

- plastic mulches

- row covers (helps limit losses during hard freezes)

- more effective pesticide sprays for tree fruit

- increased use of irrigation in most states, including for freeze protection

- better harvesting and handling equipment (reduces losses in the field)

- improved cooling and packing equipment (reduces losses during packing and shipping).

A shift of production from lower-yielding areas to higher-yielding areas also has contributed to higher U.S. average yields. Such shifts result from competitive pressures, which favor production in desirable areas with lower per unit costs. For example, much of the fall potato production moved from lower-yielding eastern states to higher-yielding western states because of lower water, electric, and labor costs.

Imports and Exports

Despite the overall annual growth in labor-intensive output, fruit and vegetable production is largely seasonal and must be supple-

Table 2.4--Fruit and Vegetable Use by category, 1970-91

Year	Fruit						Vegetables						Grand Total
	Fresh 1/	Canning 2/	Freezing 3/	Dehyd 4/	Juice 5/	Total	Fresh 6/	Canning 7/	Freezing 8/	Dehyd/chips 9/	Pulses 10/	Total	
						Pounds, farm weight							
1970	79.4	17.8	3.8	10.4	84.8	196.1	177.0	94.6	41.6	29.4	6.9	349.5	545.6
1971	80.5	18.1	3.9	9.9	86.9	199.3	170.8	101.2	43.4	29.5	7.1	352.0	551.3
1972	75.0	16.3	4.0	7.8	95.7	198.7	172.7	98.5	43.6	29.1	6.7	350.6	549.3
1973	77.8	17.7	3.9	10.2	93.0	202.6	169.7	91.9	48.4	29.4	7.8	347.2	549.8
1974	79.0	17.6	2.9	9.2	103.1	211.9	163.7	93.1	49.5	30.2	5.7	342.2	554.1
1975	85.0	16.0	3.3	10.5	98.7	213.5	166.6	91.9	50.9	30.2	7.1	346.7	560.2
1976	84.0	16.0	3.2	9.9	102.2	215.4	166.1	97.4	55.7	32.1	6.9	358.2	573.6
1977	81.3	17.1	3.6	9.5	104.2	215.8	167.6	95.9	57.6	27.6	6.9	355.6	571.4
1978	83.2	17.3	3.6	8.5	106.4	219.1	163.0	91.2	57.0	28.3	6.0	345.5	564.6
1979	82.2	17.8	2.9	9.6	117.0	229.5	167.3	95.4	53.6	27.5	6.9	350.7	580.2
1980	87.2	17.1	3.3	9.6	110.6	227.7	167.7	94.2	49.8	26.0	5.8	343.5	571.2
1981	84.3	14.6	3.1	10.2	105.3	217.5	162.8	88.3	56.2	27.2	6.1	340.6	558.1
1982	85.4	15.9	3.3	10.5	123.6	238.7	172.8	88.0	52.2	27.3	7.0	347.3	586.0
1983	90.5	14.8	3.2	10.9	100.5	220.0	168.7	89.0	53.7	27.6	7.0	346.0	566.0
1984	88.9	14.3	3.3	12.2	118.5	237.2	179.6	94.6	61.2	28.1	5.6	369.1	606.3
1985	86.8	14.8	3.6	11.9	113.6	230.7	178.2	91.7	62.5	28.8	7.7	368.9	599.6
1986	93.1	14.8	4.0	11.4	123.4	246.7	179.5	91.9	62.2	28.9	7.4	369.9	616.6
1987	97.5	15.7	4.4	12.8	113.1	243.4	184.7	91.3	64.6	28.2	5.8	374.6	618.0
1988	97.4	15.5	4.3	13.8	118.8	249.8	189.3	87.2	61.5	27.4	7.6	373.0	622.8
1989	98.8	15.4	5.1	13.2	98.8	231.2	196.2	94.5	64.8	28.6	6.3	390.4	621.6
1990	92.6	15.7	4.7	13.8	110.5	237.3	190.8	97.1	68.4	30.2	6.5	393.0	630.3
1991	80.6	14.4	3.2	12.7	110.4	221.4	182.5	98.1	72.0	29.8	8.1	390.5	611.9

1/ Includes oranges, tangerines, tangelos, lemons, limes, grapefruit, apples, apricots, avocados, bananas, cherries, cranberries, figs, grapes, kiwifruit, mangos, nectarines, peaches, pears, pineapples, papayas, plums and prunes, strawberries, olives, persimmons, pomegranates, and other fruit. 2/ Includes apples, apricots, sweet and tart cherries, olives, peaches, pears, and plums and prunes. 3/ Includes blackberries, blueberries, raspberries, strawberries, other berries, apples, apricots, cherries, grapes and pulp, peaches, prunes, plums and other miscellaneous fruit. 4/ Includes apples, apricots, dates, figs, peaches, pears, prunes, and raisins. 5/ Includes orange, grapefruit, lemon, lime, apple, grape, and prune. 6/ Includes asparagus, broccoli, carrots, cauliflower, celery, sweet corn, honeydew, lettuce, onions, tomatoes, artichokes, garlic, eggplant, potatoes, cucumbers, bell peppers, cabbage, snap beans, cantaloupe, and watermelon. 7/ Includes asparagus, snap beans, carrots, sweet corn, pickles, green peas, tomatoes, potatoes, and mushrooms. 8/ Includes asparagus, snap beans, broccoli, carrots, cauliflower, sweet corn, green peas, tomatoes, and potatoes. 9/ Includes potatoes. 10/ Includes dry peas and lentils and dry edible beans.

Source: USDA, ERS, July 1992; USDA,ERS, September 1992.

mented by imports during the winter months (Table 2.5). However, an increasing share of total supply is coming from imports. In 1986, 19 percent of the total fruit and vegetable shipments were from imports, while by 1991 imports accounted for 23 percent of the total.

The bulk of fresh produce imports comes during the winter and early spring months when domestic output is restricted to the bellwether states (Table 2.5). Climate is the most important factor determining this seasonality. As such, most of the fruits and vegetables are harvested in the southern-tier areas of Florida, Texas, Arizona, and southern California.

Total value of imports in constant dollars (1987=100) for fruit and vegetables also increased between 1970 and 1991 (Table 2.6). Total U.S. 1991 imports of fruits and vegetables held a real value of $4.2 billion, up from $1.7 million in 1970. The average annual real growth in vegetable imports was strong, rising about 4 percent per year to $1.6 billion in 1991. The largest component in vegetable import value was tomatoes from Mexico. Fruit imports grew at approximately the same rate, with a real value of $4.1 billion in 1991 (1987=100). The most important U.S. fruit import is bananas, which arrive mainly from Costa Rica. Most U.S. fresh fruit and vegetable imports originate from countries south of the U.S. border, such as Mexico, Ecuador, and Chile, and enter the country during the winter.

In terms of real value, U.S. fruit and vegetable exports also grew over the past 22 years (Table 2.6). The real value of U.S. vegetable exports grew 5 percent per year to reach $1.8 billion in 1991. The real value of fruit and tree nut exports grew 4 percent per year over the 22-year period and reached $2.1 billion in 1991. Total fruit and vegetable exports exceeded imports in only 9 of the past 22 years.

The value of vegetable exports, in real terms, displayed a stronger rate of growth during the 1970s than in the 1980s, at around 7 percent per year versus 1.2 percent per year. Growth slowed during the 1980s due primarily to the over-valued dollar. The most important exports are potatoes (mostly frozen), fresh tomatoes, canned sweet corn, and fresh head lettuce. Canada buys the largest amount of U.S. vegetable exports, especially of fresh vegetables during the winter and spring. In 1991, Canada accounted for 45 percent of the total vegetable export value. Japan, the second leading destination, accounted for 16 percent.

The real value of fruit exports grew at an annual average growth rate of 7 percent per year during the 1970s. During the 1980s, real export value growth declined 1 percent per year. The most important reason for the decline was the less favorable exchange rate. Leading

Table 2.5--Seasonal Shipments of Produce, 1986-92

A. Domestic

Years	Jan	Feb	Mar	Apr	May	Jun	Jul	Aug	Sept	Oct	Nov	Dec	Annual
							1,000 tons						
1986	1,818	1,622	1,981	2,103	2,677	2,311	2,106	1,913	1,678	1,807	1,892	1,922	23,829
1987	1,912	1,717	1,868	2,102	2,355	2,486	2,264	1,955	1,852	1,960	1,865	2,003	24,338
1988	1,962	1,844	2,106	2,383	2,491	2,606	2,142	2,070	1,815	2,021	2,001	2,210	25,649
1989	1,925	1,780	2,080	2,266	2,687	2,695	2,199	1,999	1,836	1,958	2,032	2,090	25,548
1990	1,699	1,561	2,044	2,157	2,673	2,740	2,178	2,004	1,907	2,005	2,093	2,102	25,163
1991	1,990	1,773	2,046	2,142	2,414	2,301	2,316	2,126	1,851	2,091	2,205	2,026	25,278
1992	2,047	2,105	2,262	2,557	2,827	2,610	2,176	1,953	1,879	2,111	1,984	2,065	26,576

B. Imported

Years	Jan	Feb	Mar	Apr	May	Jun	Jul	Aug	Sept	Oct	Nov	Dec	Annual
							1,000 tons						
1986	654	618	727	640	444	413	383	320	326	351	455	438	5,768
1987	609	676	814	665	532	515	354	333	347	357	478	498	6,178
1988	680	774	822	700	555	401	377	395	406	402	452	495	6,459
1989	676	766	808	763	641	444	444	405	366	445	471	557	6,787
1990	820	808	941	767	635	485	432	401	393	456	471	578	7,186
1991	728	840	932	912	656	557	451	447	429	457	506	606	7,522
1992	757	717	837	797	650	555	427	427	418	455	543	125	6,708

Source: USDA, Ag. Mkt. Serv., *Fruit and Vegetable Shipments*

Table 2.6. Value of U.S. Fruit and Vegetable Trade, 1970–91

Year	Vegetables and preparations 2/			Fruit and preparations 3/			Total vegetables and fruits		
	Exports	Imports	Net trade	Exports	Imports	Net trade	Exports	Imports	Net trade
				Millon dollars (deflated, 1987=100)					
1970	587.7	695.4	-107.7	939.9	1,025.9	-86.0	1,527.6	1,721.4	-193.7
1971	540.0	685.1	-145.1	931.6	1,022.2	-90.5	1,471.6	1,707.3	-235.7
1972	600.8	774.8	-174.0	1,084.8	1,040.4	44.5	1,685.6	1,815.2	-129.6
1973	857.6	877.7	-20.1	1,276.8	1,096.4	180.4	2,134.4	1,974.1	160.3
1974	1,014.0	780.0	234.1	1,309.1	1,068.8	240.3	2,323.2	1,848.8	474.4
1975	987.4	655.3	332.1	1,403.7	1,035.2	368.5	2,391.1	1,690.4	700.6
'1976	1,251.6	745.3	506.3	1,456.6	1,161.8	294.8	2,708.2	1,907.1	801.1
1977	1,100.5	1,053.3	47.2	1,475.8	1,325.6	150.3	2,576.4	2,378.9	197.5
1978	1,138.5	1,230.3	-91.9	1,661.2	1,516.3	144.9	2,799.7	2,746.6	53.1
1979	1,125.3	1,117.4	7.9	1,702.3	1,609.0	93.3	2,827.6	2,726.4	101.2
1980	1,578.0	1,096.2	481.7	1,841.6	1,420.2	421.3	3,419.5	2,516.5	903.1
1981	1,861.2	1,244.0	617.2	1,879.8	1,643.3	236.5	3,741.1	2,887.3	853.7
1982	1,342.8	1,247.0	95.8	1,617.2	1,891.8	-274.6	2,960.0	3,138.8	-178.8
1983	1,073.9	1,158.5	-84.6	1,521.2	1,849.0	-327.8	2,595.1	3,007.5	-412.4
1984	1,064.8	1,393.1	-328.2	1,346.2	2,342.6	-996.5	2,411.0	3,735.7	-1,205.5
1985	947.9	1,385.6	-437.7	1,238.8	2,637.5	-1,320.4	2,186.7	4,023.1	-1,733.6
1986	1,085.2	1,573.5	-488.2	1,328.7	2,431.8	-1,068.9	2,413.9	4,005.3	-1,542.0
1987	1,116.8	1,479.6	-362.8	1,464.6	2,511.3	-1,046.7	2,581.4	3,990.9	-1,409.5
1988	1,271.4	1,483.6	-212.2	1,672.4	2,588.6	-916.3	2,943.8	4,072.3	-1,172.5
1989	1,416.0	1,848.2	-432.2	1,672.0	2,444.3	-772.4	3,088.0	4,292.5	-1,306.9
1990	1,736.5	1,622.5	114.0	2,026.7	2,743.1	-716.4	3,763.2	4,365.6	-602.5
1991	1,798.0	1,579.7	218.3	2,061.1	2,585.1	-524.0	3,859.1	4,164.9	-305.8

1/ Exports were not adjusted for a known undercount of exports to Canada prior to 1990. 2/ Includes fresh and processed vegetables, melons, mushrooms, and pulses. Excludes olives and hops. 3/ Includes fresh and processed fruit (including juices) and olives. Excludes melons, tree nuts and wine.

Source: USDA, ERS, *Foreign Agricultural Trade*.

fruit exports are almonds, fresh grapefruit, and apples. The most important trading partners for fruit are Canada, Japan, and Hong Kong. In 1991, the value of exports of fresh citrus, apples, and almonds accounted for 32 percent of the total value of fruit exports.

Consumption

Population gains (up 22 percent) and an increasing preference for fiber, low fat, and natural sources of vitamins and minerals in the diet have spurred demand for fruits and vegetables during the 1970s, 1980s and 1990s (Table 2.4 and Figure 2.3). Increased export demand caused by consumption trends in Canada that are similar to those in the United States and increased overseas promotion of U.S fruits and vegetables also spurred production. Americans, on average, ate about 612 pounds of fruits and vegetables in 1991, up from 546 pounds in 1970. The most popular items in 1991 were fresh and frozen potatoes, processing and fresh tomatoes, orange juice, and apples (Figure 2.4). Growth in produce use is primarily attributed to expansion of retail produce square footage in the late 1980s, increased food service use, and increased advertisements (Love 1991).

Figure 2.3. Index of Per Capita Fruit and Vegetable Use and Population

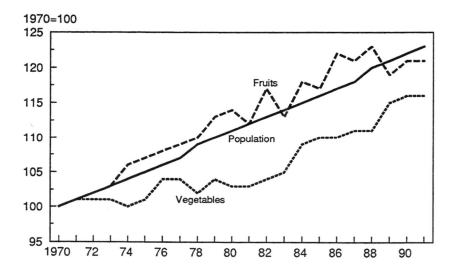

Figure 2.4. 1991 Fruit and Vegetable Use

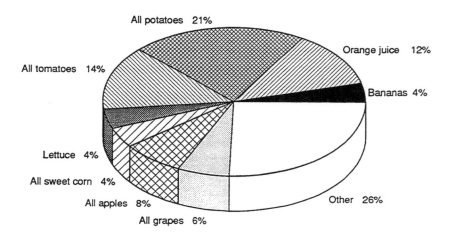

Total 612 pounds per person, farm weight

Retailers offer most major fruits and vegetables the year round by augmenting domestic supplies with imports. The result has been increased consumption. For example, domestic supplies of grapes end in the fall and the source of winter supply is from countries such as Chile. This has resulted in fresh grape consumption increasing from 2.5 pounds in 1970 to 7.3 pounds in 1991. Southern-hemisphere countries also supply other soft fruits, including peaches, plums, and nectarines, during the winter season.

Mechanization

Even with the move to more automated field-packing systems, labor remains the key component behind field technology. Stumbling blocks to complete automation in the field exist; the key factors are minimizing the damage and loss to fresh fruits and vegetables and maximizing the gathering of properly matured produce (Thompson 1992).

Mechanization has been widely adopted for processing vegetables and some processing fruits (Table 2.3). Of the major fruits and vegetables grown in the United States, the preponderance of commodities for fresh are primarily harvested by hand. The exception to the fresh rule is potatoes and snap beans, which can be harvested mechanically to meet the fresh market-quality criteria.

The primary advantages of hand over mechanical harvesting are selective and multiple harvesting, minimized damage, flexibility in harvest scheduling and rate, and minimum outlay of capital. All of these factors have combined to slow the adoption of mechanizing fruit and vegetable harvest beyond that of processing vegetables and tree nuts (Thompson 1992).

The development of mechanized cultural and harvesting systems progressed steadily after 1940 and probably reached a peak of intensity during the 1960s and early 1970s. An estimated 63 percent of vegetable production and 11 percent of fruit production was harvested mechanically in the early 1980s (Brown et al. 1983). Those percentages have not changed very much ten years later. Early development was spurred by labor shortages caused by a sudden outflow of workers to support World War II efforts. Wages for hired farm workers increased faster than the cost of machinery during the war years (Table 2.7).

Table 2.7. Ratio of Wages for Hired Farm Workers Index to Prices Paid for Farm Machinery Index, 1934–90 1/

Year	Ratio	Year	Ratio
1934	69	1963	167
1935	72	1964	167
1936	76	1965	171
1937	84	1966	178
1938	82	1967	183
1939	82	1968	189
1940	84	1969	202
1941	97	1970	206
1942	120	1971	205
1943	154	1972	204
1944	183	1973	208
1945	204	1974	208
1946	213	1975	182
1947	203	1976	174
1948	184	1977	171
1949	159	1978	170
1950	153	1979	170
1951	158	1980	163
1952	163	1981	160
1953	165	1982	154
1954	163	1983	148
1955	165	1984	143
1956	164	1985	143
1957	163	1986	150
1958	161	1987	153
1959	165	1988	147
1960	165	1989	152
1961	164	1990	151
1962	166		

1/ Index of hired farm wages (1910-14) divided by the index of prices paid by farmers for machinery.

Source: USDA, NASS, *Farm Labor* 1975 ff.

Labor remains a major component of the costs for producing many fruits and vegetables, especially those intended for the fresh market, which prefers produce with an unblemished physical appearance. Mechanized handling usually causes a higher incidence of physical damage than hand harvesting. Mechanization has consequently shifted to fresh-use fruits and vegetables more slowly than for processed.

A decline in the cost of labor relative to the cost of capital during the late 1970s and the 1980s undoubtedly contributed to slowing the trend to greater mechanization. The fact that the remaining hand-labor jobs are the most difficult to mechanize also accounts for the slowdown in mechanization.

The remainder of this chapter describes the status of harvest mechanization for the major fruits and vegetables and the outlook for additional mechanization. The information utilizes materials from G. K. Brown and Donald L. Peterson.

Fruits

Prototype harvester systems using limb shakers, trunk shakers, air shakers, and either catching frames or pickup machines technically exist for use on early and midseason (single-crop) oranges for processing. A successful method of selectively harvesting mature Valencia oranges, lemons, and grapefruit that may simultaneously have mature and immature (next-crop) fruit on the tree has not been perfected. Shake harvesting causes substantially more physical damage to the fruit than hand picking. The difficulty in sorting damaged fruit from unblemished reduces the feasibility of using mechanical harvesting on oranges for fresh use. Adoption of the present systems will not occur unless labor costs rise substantially relative to the costs of machinery.

Brown and others estimated that U.S. grape growers harvested 25 percent of their acreage with mechanical shake harvesting in 1983. That percentage has changed little since then. Growers of raisin grapes and grapes for the fresh market harvest all their acreage by hand picking. Juice grape growers and some of the wine grape growers machine harvest. Further adoption of mechanical harvesting for wine grapes will occur only if labor costs rise sharply.

Apple growers mechanically harvest less than 5 percent of the apple crop, all for processing. Excessive damage caused by the shake-harvest systems used on apples prevents wider acceptance for the processing industry and prevents its use for fresh-market apples. Additional mechanization will occur only if severe shortages of harvest workers or other factors drive labor costs higher.

Vegetables

Growers mechanically harvest all potatoes and sweet corn, tomatoes, green peas, cucumbers, and snap beans for processing. Some fresh-market sweet corn and snap beans are also mechanically harvested.

Although mechanical systems have been available since the early 1970s, growers still cut all head lettuce by hand. Worker productivity in lettuce harvesting has been raised by using labor-saving aids; paying on a crew-incentive, piece-rate basis; and using improved culture practices to obtain a higher yield with fewer cuttings.

Public Support

Changing attitudes toward government support for fruit and vegetable mechanization research, which may have peaked in the 1960s and 1970s, have led to diminished public support during the 1980s. Declining real wages after the mid-1970s reduced the economic incentive to adopt additional mechanization, thereby eroding industry support for public funding.

Heightened social concern for farm workers also contributed to the loss of political support for public funding. Many people believed that mechanization research primarily benefited large corporate farms and contributed to displacing farm workers from their jobs, causing hardships among workers' families. Therefore, they believed that such research should not be supported with public funds. Also, the opinion that the private sector could more efficiently support agricultural research, such as fruit and vegetable mechanization, gained prominence during this period and may have contributed to the decline in public support.

A lengthy lawsuit against the University of California for using public funds to support mechanization research probably marked a turning point in support for this research. Following the lawsuit, the U.S. Department of Agriculture cut most of its funding for mechanization research, and state experiment stations never again allocated much research money to mechanization research.

Outlook

Fruit and vegetable output will probably continue to rise faster than population growth due to the growing per capita demand. During the 1970s and 1980s, fruit and vegetable output expanded faster than the population, primarily due to the adoption of higher-yielding varieties and the growth in exports. The key factors for expanded output through the year 2000 are likely to be continued growth in exports, government-sponsored advertising and promotion, and continued movement toward healthier lifestyles.

During the 1990s, the fruit and vegetable industry (in cooperation with the National Cancer Institute and the federal government) is attempting to double fruit and vegetable per capita use by the year 2000 (Love 1991). Although the goal may be hard to fully achieve, increased emphasis and media attention for fruits and vegetables will undoubtedly result in gains in per capita use (Love 1991). Several issues face the industry, one of which is the impact of the North American Free Trade Agreement (NAFTA). Mexico, as shown, has been an important supplier of many fruits and vegetables. Because of its geographical proximity to the United States and favorable weather, it has a unique comparative advantage. The research to date, while noting the problems of rapid production expansion in Mexico, cannot predict what will happen to the costs of production relative to prices received (Cook et al. 1992).

Labor availability is also related to Mexico, since it is the major supplying country. If Mexican fruit and vegetable output expands rapidly, labor demand may increase, causing wage rates to rise relative to U.S. wage rates. While this scenario may occur, it is unlikely to cause similar increases in U.S. wage rates, due to the ample labor pools created out of the Immigration Reform and Control Act of 1986.

However, labor shortages still remain a concern for growers. Growers were concerned that IRCA was going to shrink the labor pool. The report to Congress from the Commission on Agricultural Workers stated that IRCA did not cause many of the expected results, such as controlling the flow of illegal aliens. Instead, it actually created a fraudulent document industry, whereby illegal immigrants could buy some type of documentation that would receive less scrutiny from employers. The result has been an explosion of fraudulently documented workers in the U.S. fruit and vegetable industry. CAW paved the way for continued availability of fruit and vegetable workers by recommending modification to the H-2A program and expanding the definition of seasonal agricultural workers to include sugarcane and sod workers.

3

IRCA's Effect on Farm Wages and Employment

JAMES DUFFIELD, LEWELL GUNTER, AND HARRY VROOMEN

Some had hoped that IRCA, the Immigration Reform and Control Act of 1986, would result in a smaller but more stable farm work force. By keeping undocumented workers out of the work force, it had been envisioned that wages and working conditions would improve. However, as time has passed, it appears that IRCA is not having these positive effects. Immigrant laborers continue to come into the country illegally and work in agriculture. Indeed, it has been reported that illegal migration from Mexico has increased, and that more seasonal workers than usual are remaining in the United States in the hope of another amnesty program or an extension of the Special Agricultural Worker program. Consequently, some experts surmise that IRCA has stimulated unauthorized immigration and caused a labor surplus in the United States. A labor surplus tends to drive wages down.

The possible effects of IRCA on farm labor supply and wages can be summarized in two alternative scenarios: (1) the law has worked as intended and reduced the farm labor supply, causing wages to rise, or (2) the law has failed to control the influx of undocumented labor and farm labor supply has increased, causing wages to fall. The objective of this chapter is to identify farm labor trends and examine the data supporting each of these scenarios.

NASS Farm Labor Data

This study uses employment and wage data from the National Agricultural Statistics Service (NASS) of the U.S. Department of Agriculture (USDA, NASS 1991). NASS has been reporting farm labor information since 1866. Before 1975, a nonprobability mail survey was the primary method of collecting the data. Volunteers reported the number of workers on their farms and gave information about the prevailing wage rate. Since 1975, the wage and employment estimates reported in the *Farm Labor* publication have been based on regional probability surveys (USDA, NASS 1975ff).

During 1981–89, NASS cut back the farm labor surveys because of budget reductions. Thus, the data on both employment and wages contain some missing observations. In order to provide a continuous time series, we estimated these missing data, as described in the *SAS/ETS Users Guide*. There is no way to know how close these estimates are to the actual values if the data had been collected. However, they are consistent with observed data.

Farm Employment Trends

Labor use on U.S. farms started dropping dramatically in the 1940s, reducing the number of farm operators and family workers. NASS reported almost 7.9 million annual average family workers in 1945, versus only about 4 million in 1965. The number of hired workers also declined, but not as much as the number of family workers, because farm enlargement increased the need for hired workers on bigger farms. From 1945 to 1965, the number of annual average hired farm workers declined from about 2 million to about 1.5 million.

According to NASS's quarterly farm employment estimates from 1950 to 1992 in the United States, with the exception of July employment, the downward trend in hired workers appears to be leveling off in recent years. There is a variation in employment associated with seasonality. January has the lowest farm employment, there is a slight rise in April, employment peaks in July, and employment falls in October.

The downward trend in July-to-July employment does not hold true for all production regions. Employment of seasonal workers in July has been stable in regions that specialize in labor-intensive crop production (Figure 3.1).

Looking at Figure 3.1, it does not appear that employment trends have deviated much after IRCA's passage in 1986. Moreover, statistical tests have given no indication that hired labor employment trends in any of the regions or at the U.S. aggregate level have been significantly altered since IRCA's passage. These findings suggest that the employer sanctions issued under IRCA have resulted in no reduction of agricultural employment. These data indicate only employment, or the matching of supply and demand for labor, but they do

Figure 3.1. Regional Hired Farmworkers, July 1975–92

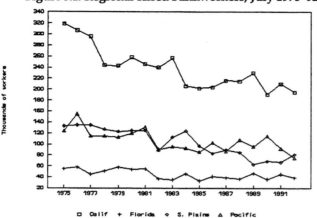

Note: Southern Plains includes Oklahoma and Texas. The Pacific region includes Oregon and Washington.

not support the hypothesis that IRCA has increased the farm labor supply by stimulating unauthorized immigration. The statistical analysis used to test the impact of IRCA on the farm labor market is in Vroomen and Duffield (1992).

Farm Wage Trends

A plot of U.S. aggregate farm labor data shows nominal farm wages increasing, with a seasonal pattern throughout the study period (Figures 3.2 and 3.3). Notice that the seasonal pattern in the early years, shown in Figure 3.2, is less variable than the post-1975 data shown in Figure 3.3. Also, the October estimate in the earlier years is consistently the lowest quarterly wage reported for the year, but the July estimate is consistently the lowest quarterly wage after 1975. One possible reason for these inconsistencies is the difference in the sample basis for the wage estimates in the two time periods. NASS switched to a probability survey in 1975, and apparently this corrected a bias in the wage data collected in the old survey. As shown in Figure 3.3, the U.S. farm wage estimate is expected to fall to its lowest level in April and July of each year because a large number of low-paying seasonal workers enter the work force during these months, which lowers the average hourly wage. Conversely, the U.S. average wage is highest in October when the farm work force includes a greater percentage of higher-paid full-time employees.

Figure 3.2. Hourly Wage for U.S. Hired Farmworkers, 1950–75

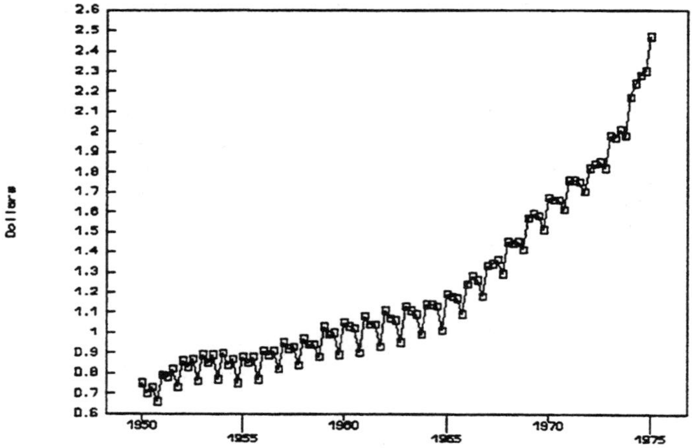

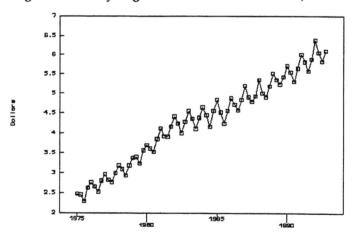

January, April, July, and October of each year

Figure 3.3. Hourly Wage for U.S. Hired Farmworkers, 1975–92

January, April, July, and October of each year

With the exception of seasonal dips, the nominal farm wage rose consistently throughout the data period. The trend in U.S. wage rates did not appear to change after the passage of IRCA. Although not shown here, plots of the regional data also failed to detect any major change in wage rates as a result of IRCA (see Vroomen and Duffield 1992 for regional graphics). Statistical tests performed on the wage data provided no evidence that IRCA had a significant effect on nominal farm wage trends.

Summary and Conclusions

It appears that the established patterns in U.S. farm employment and wage data have not been influenced significantly by the IRCA program. Our results provide no evidence that IRCA has caused farmers to adjust to a smaller work force. At the same time, our findings do not suggest that farm employment has increased significantly due to a labor surplus. While our findings show that IRCA has had no effect on employment or wages, it could have impacted farm employment or wages in local areas within certain regions. The NASS data are aggregated to describe the U.S. and regional labor markets, and any local impacts may be overwhelmed by a lack of significant effects in the rest of the region. Thus, our results may not apply to some local labor markets.

In addition, it has been hypothesized that a large part of IRCA's effect would be on the employment of contract workers. Our analysis excludes contract labor because the NASS farm labor survey was designed primarily to focus on labor hired directly by farm operators.

4

Farm Worker Demographics: Pre-IRCA and Post-IRCA Field Workers

SUSAN M. GABBARD AND RICHARD MINES

Introduction

The Immigration Reform and Control Act of 1986 (IRCA) generated an increase in the amount and completeness of data on farm worker demographics. To analyze the impact of IRCA on farm workers, several major information-gathering efforts were launched by the federal government. In addition to the data collection mandated by IRCA, universities, states, and private foundations also sponsored research initiatives that collected data on farm worker demographics.

The aim of this chapter is to summarize information on characteristics of farm workers from the National Agricultural Workers Survey (NAWS). While some points presented here are unique to the NAWS, many findings can be confirmed by other IRCA-sponsored surveys and case studies.

Before IRCA, the image of farm workers was that most were casual and temporary workers, including Anglo housewives, students, and other short-term workers (Fuller and Mason 1977). Additional seasonal labor came from two sources: (1) Mexican farmers, who supplemented their income with short trips to the United States, and (2) U.S.-based migrant families, many of them southern blacks or Texas-based Mexican-Americans, who followed the crops along three geographic streams similar to the migrants depicted in John Steinbeck's *The Grapes of Wrath* (1939).

A different image of field workers has emerged from analysis of the NAWS. The NAWS findings include:

1. Latino immigrants and their descendants do most of the field work in the United States. Many of these farm workers have low educational levels and do not speak English. Some continue to be undocumented. Increasingly, the national farm labor force is part of a process of continued Latino migration and settlement.

[63]

2. Although field work continues to be casual, temporary, and short term, farm workers work most of the year in agriculture and depend on farm work for most of their income.

3. Migrants are still a crucial seasonal labor supply for U.S. agriculture. Domestic follow-the-crop migration is not the dominant migration pattern. Instead, most migrants are now international migrants who move back and forth between jobs in the United States and their homes in Mexico.

4. Abundant labor is available at the relatively low farm wage of $5 to $6 per hour. Since farm workers depend on farm work for most of their income, the result is under employment, unemployment, and poverty.

Methodological Issues

During the mid-1980s, researchers and policy makers began to acknowledge that the farm labor force was changing and that these changes were not reflected in national farm labor data. They noted that the prevailing image and the national data on farm workers often did not match the results of case studies and local reports.

Most of the national data on farm workers came from administrative data on production, general national population surveys, or program data. Prior to IRCA, findings from these sources provided a confusing picture. According to the production data, the number of field workers should have been increasing. By contrast, the number of farm workers reported in the USDA's Quarterly Agricultural Labor Survey (QALS) was decreasing (USDA, NASS 1991). Meanwhile, data from the Hired Farm Work Force (HFWF), a biennial supplement to the Current Population Survey (CPS), was at variance with regional surveys. For example, the UC-EDD survey (Mines and Martin 1986) reported that California workers were 87 percent Latino at a time that the HFWF reported Pacific Region farm workers as 44 percent Latino (Oliveira and Cox 1989). These figures were clearly incompatible. Further doubt was cast on the HFWF because it reported that only 14 percent of the crop labor force worked in the Pacific Region (Oliveira and Cox 1989), while the Census of Agriculture (COA) reported that 40 percent of all crop labor hours were worked in the Pacific Region (Martin and Martin 1992).

Two major factors seem to account for the decline in data quality. First, the farm work force was increasingly drawn from hard-to-survey populations. Two overlapping populations were of particular concern: recent and undocumented immigrants and employees of farm labor contractors. Second, certain tax laws tended to artificially increase the number of farm family dependents counted as hired farm workers, particularly in areas where cash grains predominate.

Before IRCA, the two most important national sources for data on farm workers characteristics were the Hired Farm Work Force and the Decennial Census. By the mid-1980s, research had been conducted showing that the census and the CPS had difficulties counting hard-to-survey populations (Fein 1989). The list of populations considered hard to survey begins to sounds like a description of the farm labor force: low income, low literate, limited English proficient, minority, immigrant, undocumented, and migrant.

Reports on the HFWF acknowledge the undercount problem. They note that many farm workers are not in the United States in December and that undocumented aliens are less likely to be counted (Oliveira and Cox 1989). One indicator of the difficulty encountered by the CPS in sampling Latinos is that the standard errors for Latinos in the 1987 HFWF are 40 percent higher than those of other groups (Oliveira and Cox 1989).

With these sampling problems in mind, the U.S. Department of Labor (DOL) designed the NAWS to find farm workers proactively, using culturally sensitive survey methods and drawing on state-of-the-art research on hard-to-count populations. In designing the NAWS, DOL combined the most promising elements from several methodological approaches.

- DOL anchored both the number and the geographic distribution of farm crop workers to the USDA QALS. This made a national survey financially feasible by removing the issue of determining a population estimate.

- DOL used more efficient sampling methods, such as sampling with probabilities proportional to size, to expand the sample size. The NAWS surveyed 2,500 crop workers annually, whereas the 1987 HFWF generally had less than 1,500 observations covering both crop and livestock workers.

- The NAWS used a rolling sample that included data collected at different times of the year to overcome the data problems of transitory workers and fixed-point surveys.

- Sophisticated weighting schemes were implemented to prevent the underrepresentation of part-time workers and workers who worked for only part of the year.

- DOL opted to find farm workers proactively via their place of work (similar to employment estimates in the QALS). This is less expensive than a household survey and overcomes the undercount problems inherent in traditional household surveys for immigrant populations. It was thought that employed farm workers would most easily be found at work.

- The NAWS paid farm workers for their participation. This increased worker participation rates for both the initial and the longitudinal data.

The NAWS contains four years of national longitudinal data on farm workers. To date, the NAWS has interviewed more than 10,000 workers and collected data on more than 100,000 jobs and out-of-work periods. In addition, follow-up data were collected on over 4,500 NAWS participants who were initially interviewed between October 1988 and April 1990. Overall, the NAWS successfully obtained follow-up data from 75 percent of the people before the follow-up component was eliminated.

The NAWS found a higher proportion of Latinos in the farm labor force than earlier surveys. In 1990, the survey found that 71 percent of Seasonal Agricultural Service (SAS) farm workers were Latinos (Mines, Gabbard, and Boccalandro 1991). This is a dramatic increase over the previous 33 percent Latino figure for fruit, vegetable, and horticultural workers (Oliveira and Cox 1989).[1] Although not all of the difference between the HFWF and the NAWS can be accounted for, much of the difference can be explained by differences in survey methods (Gabbard, Kissam, and Martin 1993).

Farm Worker Characteristics

The NAWS found that most farm workers are foreign-born Latinos; most crop workers depend on agriculture for their livelihood; and most migrants are Mexican-based back-and-forth migrants. The farm labor force can best be understood as part of the process of a continual Latino migration and settlement. Farm workers are no longer primarily Anglo laborers but rather Latino migrants. Most of the crop labor force is young, foreign born, Latino, and male. Most have families to support. Crop workers typically have low levels of education, lack English skills, and have few other job skills (Table 4.1).

Age cohort analysis of the NAWS data supports a view that the number of Latino farm workers has been increasing over time. There have been increased numbers of Latinos and decreased numbers of Asians and blacks in more recent farm worker cohorts. Farm workers over 50 years old are mostly U.S.-born (55%) and three-fourths are non-Latino (Mines, Gabbard, and Samardick 1993). Immigrant Latinos make up only about 40 percent of farm workers over 50, but about 60 percent of farm workers between 18 and 35.

The settled population shows the impact of historic migration patterns. Settled farm worker populations in the western United States and in Florida are predominantly Latinos. Those in the Midwest and the rest of the East, where Latino immigration is more recent, are predominantly non-Latino whites. These settled populations are seasonally increased by migrants, who make up 42 percent of the farm labor force and are almost all Latinos (Gabbard, Mines, and Boccalandro forthcoming).

Table 4.1. Demographic Characteristics of SAS Workers, 1990

Characteristics	% of SAS Workers
Latino	71%
Foreign born	62%
Legally authorized workers	88%
SAWS	29%
Migrants	42%
Male	71%
Under 35 years old	65%
Married	64%
Parents	54%
Family incomes below the poverty level	50%
Use need-based social services	18%
Less than eight years of education	53%
Spanish speaking	65%
Speak English well	40%

Source: Mines, Gabbard, and Boccalandro 1991.

Farm Workers Are Not Casual Workers

Casual workers, available for only one or two seasonal jobs, have declined precipitously over the last few decades. Whereas these workers once may have comprised most of the labor market, now only 12 percent of the labor force are occasional workers who work in agriculture for a few weeks per year and whose incomes are not dependent on agriculture. Furthermore, these occasional workers do only 5 percent of agricultural work in perishable crops (Mines, Gabbard, and Samardick 1993).

Local occasional workers have practically disappeared and currently comprise only 6 percent of the crop labor force (Mines, Gabbard, and Samardick 1993). These are mainly nonfarm workers who infrequently do farm work. Rural development accompanied by increased female labor force participation has reduced the supply of local students, housewives, and others willing to do short-term farm work. In addition, family patterns have changed. Even in rural areas, women who formerly might have been housewives now often work outside the home. Non-Latino, rural teenagers are also more likely to be employed outside agriculture (for example, by local fast-food restaurants) than to be employed on farms. Short-term international migrants are also scarce.[2] Foreign-based occasional workers make up only 2.5 percent of the total farm labor force. Furthermore, these workers do only 1 percent of all farm work (Mines, Gabbard, and Samardick 1993).

Most farm workers employed in short-term tasks are not occasional workers. Rather, they are "periodic" workers who either piece together multiple (more than two) short work periods or combine short-term tasks with one or more long-term tasks to extend their work year. Periodic workers account for 39 percent of the farm labor force and do 90 percent of the short-term work.

Periodic workers are not casually attached to the farm labor force, but depend on farm work for most of their income. In fact, U.S. farm work is often their only source of earned income. Farm wages may be supplemented by unemployment payments and an occasional stint of nonfarm work. However, for most periodic workers, the bulk of their work year is spent in agriculture.

Regardless of whether they do short-term tasks, most field workers spend many years in farm work. On average current field workers have been doing farm work in the United States for ten years. A large proportion of the SAWS have remained in agriculture seven or more years after their qualifying employment. Three-quarters of crop workers expect to remain in farm work for more than five years and for as long as possible.

Most Migrants Are Immigrants

Transnational migration is now more important than U.S.-based migration (Gabbard, Mines, and Boccalandro, forthcoming). Without migrants, many farmers would have to change farming practices. Migrants are crucial to seasonal peak labor demand, since they make up 66 percent of the harvest labor force. Local labor, which accounts for 58 percent of the farm labor force, works mainly on long-term tasks (62%).

Almost all migrant farm workers are Latino immigrants or members of their families (86% are immigrants, 11% are U.S.-born Hispanics whose parents were farm workers, and 1.5% are married to foreign-born Latinos). U.S.-based Latino migrants comprise only about one-quarter of the migrant population. Most U.S.-based migrant farm workers begin as transnational migrants and later move their home base to the United States. If this current trend continues, many U.S.-based migrant Latino families will eventually become settled immigrants.

While most farm workers are not initially accompanied by their families when doing U.S. farm work, many eventually move with or are joined by their families in the United States. Among those who have not immigrated or do not intend to, most transnational migrants spend the majority of the year in the United States and support themselves and their families on their U.S. farm wages.[3]

Farm Worker Poverty

Minimal estimates suggest a crop worker population of 2.25 million, but at peak employment, only 1.5 million jobs (see Figure 4.1). After subtracting workers employed elsewhere and workers not in the country, approximately 300,000 out-of-work farm workers reside in the United States at peak season. Although many farmers and workers try to extend the season by coordinating jobs, on average farm workers only work for half the year. Three-quarters of farm workers say they would willingly do more farm work if it were available. Most of the remaining quarter are employed the year round (Gabbard, Mines, and Boccalandro, forthcoming).

Figure 4.1. Employment Status of Farm Workers, 1989–91

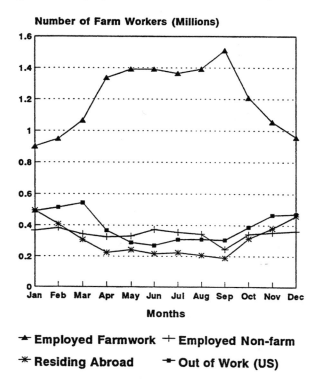

Number of Farm Workers (Millions)

Months

➤ **Employed Farmwork**   ＋ **Employed Non-farm**

✴ **Residing Abroad**   ➤ **Out of Work (US)**

IRCA may have inadvertently increased the U.S. farm labor surplus. In 1990, one-third of the SAS work force reported that they had been legalized under IRCA. In addition, a growing minority of workers continued to be undocumented (Mines, Gabbard, and Samardick 1993). The SAW program, by helping legalize many farm workers, may have strengthened international migrant networks, given that employer sanctions failed to stop undocumented hiring.

The surplus would not be so serious if farm workers were a casual labor force whose main source of support lay elsewhere. Over half of all crop workers live in poverty and 70 percent of the children of migrant farm workers live in poverty.

Summary

As a result of using improved methodology, the prevailing image of farm worker demographics has changed. First, the NAWS has shown that the crop labor force is mostly Latino and contains almost equal numbers of settled and migrant workers. Unless immigration trends change radically, the future crop labor force will become increasingly Latino. Second, the farm labor force is no longer composed of mostly casual or temporary workers. Only 12 percent of the farm labor force fits the description of a casual work force. The remaining workers (88%) depend on U.S. farm work to support themselves and their families. Third, transnational migrants account for three-quarters of all migrant farm workers and spend most of the year in the United States. U.S.-based migrant families are best characterized as recently arrived immigrant Latino families, many of whom will eventually become settled workers. Finally, the current labor surplus has serious consequences, in part because farm workers are not casually attached to farm work. When workers try to support themselves and their families in the face of chronic excess supply, the outcome is poverty, unemployment, and underemployment.

NOTES

1. This figure was calculated from the table containing ethnic breakdown by crop.

2. The definition of a worker who spends at least one month abroad as foreign based allows for an upper-bound estimate of the proportion of workers who are indeed foreign based. Conversely, the proportion of U.S.-based workers expressed herein is a lower bound. It is not unusual for Mexican workers employed in a seasonal industry such as agriculture to go to their native country during the off-season.

3. See Gabbard, Mines, and Boccalandro (forthcoming) for details.

Part II, Section A.
IRCA's Effects in Eastern States

Part II, Section A considers the effects of IRCA in various states, focusing on how immigration reform has affected farmers and farm workers. A common issue throughout this discussion is that farm employment in Florida, New York, Michigan, Washington, Kentucky, and North Carolina reflect not only the effects of IRCA, but also the environmental and marketing opportunities of each state.

Within these states most hired farm workers are employed in the fruit, vegetable, and horticultural (FVH) specialty sector. Also, an increasing number are being employed in nursery/flower production. The sector's products and the seasonal nature of FVH production and nursery business define the agricultural labor market. Where the patterns of production are dominant in flowers and FVH specialties, as in Florida, the patterns of farm worker employment appear to have been more affected by IRCA.

In other states, such as Kentucky and North Carolina, for example, IRCA's influence is evident in a couple of ways. Where more mechanized techniques can be installed, growers are still following labor-intensive practices, especially on larger farms. And in some areas, there is an apparent reversal in the trend toward more mechanical

[73]

harvesting. IRCA may have provided the supply of labor to produce these findings.

Not all states have the environmental and marketing conditions to shift rapidly into the more valued FVH. Consequently, the rates of change in labor utilization have varied according to the ability of farmers to round up required inputs, especially of labor. For farm producers with limited historical experience with immigrant workers (and immigration laws and regulations), the adjustment toward more labor-intensive farm production appears to be progressing more slowly in the aftermath of IRCA.

Farm workers have been affected by IRCA also. More workers are moving to more nontraditional areas in the East as farmers are adjusting to the enlarged pool of workers. For instance, more workers are now found in Michigan, Kentucky, and New York. In New York there has been a noticeable influx of Mexican, Mexican-Americans, and Central American farm workers, particularly into the central and western areas. Concomitantly, there appears to be a decline in the number of workers brought under the H-2/H-2A program. In Chapter 6, on New York's apple pickers, Figueroa questions the need for and intent of the H-2A program. For farm producers (especially in the southern states) who have more traditional experiences with immigrant workers, we have not found the effects of IRCA to be clearly pronounced. Wages are not depressed in any uniform pattern from state to state.

The chapters in this section show that the effects of IRCA have not been easy to study from state to state. In fact, there are several contrasts to be noted. One difference is in the diversity of environmental and marketing opportunities within and between states. The states discussed here produce over 50 different types of FVH specialties. Labor utilization and payments to labor are not uniform among the states or even from crop to crop. It should also be noted that factors such as the Canadian Free Trade Agreement, the Caribbean Basin Initiative, and NAFTA have raised and lowered expectations for different farmers in different states. Apple producers in New York and Washington envision expanded market opportunities for exports. California and Florida producers of tropical fruits (namely of avocados and tomatoes) are planning for the adverse effects of NAFTA.

The chapters that follow also show that farmers are adjusting differently to their state labor laws and policies, which are more pronounced in some states than others. Some states like New York, for example, have issued recommendations concerning farm labor policy that farmers have begun to consider. There, legislation has been proposed to regulate farm labor contractors and farm worker housing, and to protect farm workers from wage and payment abuses. As a consequence, not all states advance similar and far-reaching legislation that would alter the demand for and supply of farm labor. But where states have imposed new farm labor measures, the effects of IRCA are more difficult to discern as an isolated event.

What these chapters have in common are three basic conclusions:

- Each study reports a general failure of IRCA's measures to be enforced. As a result, IRCA has neither stemmed the flow and employment of undocumented aliens in agriculture nor caused employers to alter their traditional patterns of employment. In effect, in each of the chapters there is a reported increase in the supply of labor and a growing employment of Mexican workers, of which it cannot be proven with certainty that most are undocumented.

- Each study reports an increasing "Mexicanization" of agriculture, a pattern that is closely connected to the increased consumer demand for FVH specialties. As a result, some studies conclude that there is not likely to be a significant shift to mechanized agriculture as long as the cost of employing workers is relatively low. Moreover, related to this conclusion is the observation that SAWs (mostly from Mexico) are now being diffused from such traditional settlement areas as California, Texas, and Florida — which are experiencing pronounced labor surpluses — to the midwestern and eastern states — which are more likely to provide jobs for seasonal farm workers. As noted in most state studies, this diffusion is facilitated by farm labor contractors who provide the jobs and links with farmers.

- Each study reports a difficulty in gauging the direct effects of IRCA on farmers and farm workers. Aside from the limitations of data, each state has a variety of new laws and measures that have bearing on the supply of and demand for agricultural labor.

Historically, the southern states have developed a reliance on Mexican and other immigrant workers. This type of dependency on labor is also spreading north and across the United States. Any future study of IRCA and its effects should take into consideration the plight of workers in Mexico and other sending nations. As long as the agricultural economies of Mexico and the United States are closely tied together by labor, we do not expect that studies of independent states themselves will provide long-term answers to the effects of immigration laws.

Little research is underway that will determine the extent to which SAWs will continue to work in seasonal agriculture. Nor is any post-IRCA study in progress that will provide insight into worker decisions to stay or leave the agricultural labor market.

5

IRCA and Agriculture in Florida

LEO C. POLOPOLUS AND ROBERT D. EMERSON

Introduction

Because of the dominance of horticultural crops in Florida's diverse agricultural output mix, labor resources are particularly crucial to the level and growth of agricultural output over time. Thus, federal policies, such as the Immigration Reform and Control Act of 1986 (IRCA), which can influence farm labor supplies, have a potentially tremendous impact on Florida's competitive position in agricultural commodity markets.

The significance of labor resources in Florida agriculture is revealed by the fact that, while Florida ranks as the ninth largest state in the union in terms of cash farm receipts, it ranks third among 50 states in farm labor expenditures. Farm labor expenditures of farm employers totaled $722 million in 1987, up 51.5 percent from the level of farm labor expenditures in 1982 (U. of Florida 1992, p. 307). Also, farm labor contractors account for additional payroll for farm workers in Florida.

The purpose of this chapter is to provide an overview of agricultural labor markets in Florida, along with a preliminary and cursory assessment of the impact of IRCA on these markets.

Farm Structure

Crop agriculture has dominated livestock agriculture in terms of cash farm receipts and agricultural payroll for decades. Since the 1950s, crop agriculture has accounted for at least 70 percent of total cash farm receipts in Florida, while livestock and livestock product agriculture have accounted for less than 30 percent of cash farm receipts over the same period. Citrus and vegetables have consistently been the major commodity groups over the past several decades. However, foliage and other ornamental crops have become increasingly important, even though accurate data were not collected in 1971 and earlier years (Table 5.1).

Only one-third of Florida's 36,556 farms had farm sales of at least $10,000. Of the 14,667 farms with annual sales of over $10,000 in 1987, almost 10,000 farms were classified as crop farms. However, most seasonal farm employment was accounted for by Florida farms with annual sales of over $100,000. Thus, hired labor expenditures are concentrated with the 5,000 or so Florida farms with annual cash receipts of over $100,000.

Farm Employment

In Florida, agricultural output has increased rather dramatically over the past four decades. The associated increased labor demand has been dampened somewhat by technological developments in production systems, such as weed control and harvest mechanization for green beans, celery, sweet corn, potatoes, and sugarcane, to name a few. The net effect of these conflicting forces of increased agricultural output and labor-saving technology and adoption has been a slight overall increase in total agricultural labor demand over time in Florida.

Much of the buoyancy in farm labor demand in Florida is due to the particular commodity mix, with crops accounting for almost 80 percent of total cash farm receipts. Citrus, vegetables, ornamentals, and sugarcane are important crops produced in Florida, each with significant agricultural labor needs.

The diversity of agricultural labor requirements in terms of skills, seasonality, and location is due to the diversity of commercial agricultural production in Florida. More than 100 major agricultural commodities are produced and marketed by Florida farmers, with a farm value exceeding $6 billion annually in 1992.

Number of Farm Workers

Three problems result in a woefully inadequate statistical data base for the agricultural labor market in Florida. First is the casual nature of the farm labor market, in which entry and exit by individuals are relatively free, even in a given crop season. Second, there is the seasonal character of most farm labor markets in Florida. Third, a sizable contingent of seasonal farm workers are employed by inde-

Table 5.1. Percentage of Cash Farm Receipts in Florida by Commodity
Groups, 1950–90

Commodity Group	1950-54	1960-64	1971	1990
Value of Cash Receipts ($ Billion)	0.53	0.88	1.42	5.71
Total Crops & Livestock %	100.0	100.0	100.0	100.0
Crops	76.3	75.5	71.1	77.9
Oranges	26.8	31.2	23.5	19.1
Grapefruit	7.0	5.6	7.5	4.3
Veg. & Melons	24.0	19.4	19.3	21.9
Sugarcane	2.1	2.9	4.8	7.4
Ornamentals	n.a	n.a	n.a	9.3
Subtotal	59.9	59.1	55.1	62.0
Other Crops	16.4	16.4	16.0	15.9
Total Livestock & Products	23.7	24.5	28.9	22.1
Cattle/calves	6.7	7.7	11.5	6.8
Milk	8.0	9.9	9.9	7.4
Broilers	2.2	0.8	1.7	2.7
Eggs	3.0	4.3	4.0	2.3
Subtotal	19.9	22.7	27.1	19.2
Other Livestock	3.8	1.8	1.8	2.9

Source: Polopolus and Emerson 1975; U. of Florida 1992.

pendent farm labor contractors, and thus are not tallied according to estimates of employment by "farmers."

A resulting rough number of farm workers in Florida is about 100,000 in a peak month, when the employment of farmers and agricultural service firms (labor contractors) are combined. Actually, because of the casual and seasonal nature of Florida's hired farm work force, it has been estimated that as many as 300,000 people may be involved in some type of agricultural employment over a given calendar year (Polopolus 1991).

While the percentage varies seasonally, roughly two-thirds to three-fourths of all farm workers in Florida in the fall, winter, and spring months are hired, as opposed to unpaid family farm workers. During the slack summer months, however, the percentage of hired workers to the total number of workers on farms can drop to almost 50 percent (Table 5.2).

Table 5.2. Workers on Farms in Florida by Season, 1980, 1985, and 1990

	Jan	April	July	October
	\- 1,000 \-			
	1980			
All farm workers	93	117	82	84
Hired Workers	70	90	54	57
% Hired to total	75.3	77.0	65.9	67.9
	1985			
All farm workers	83	104	49	54
Hired Workers	63	82	33	41
% Hired to total	76.0	78.9	67.4	75.9
	1990			
All farm workers	90	76	66	76
Hired Workers	65	57	36	48
% Hired to total	72.2	75.0	54.6	63.2

Source: USDA, NASS 1991.

Additional farm workers are hired by independent farm labor contractors. According to one study, the number of these agricultural service workers in Florida varied from 35,000 during January of 1980 to 3,800 in July of 1985 (USDA, NASS 1991). These estimates, however, tend to be unreliable.

One must also factor in the number of temporary, seasonal foreign workers brought into Florida each year for the sugarcane harvest to arrive at the overall total of farm workers. While the number of these H-2A workers has declined recently because of the increased mechanization of Florida's sugarcane harvest, an estimated 4,000 foreign workers were involved with the sugarcane harvest in the 1992–93 season. Table 5.3 provides biweekly and monthly estimates of foreign and domestic seasonal farm workers in Florida for 1991. Peak employment (foreign and domestic) was 80,627 for January 15, 1991. The level of employment dropped to a low of 26,828 for the August 15, 1991 tally.

Table 5.3. Seasonal Hired Agricultural Workers in Florida, Biweekly or Monthly, Foreign and Domestic, 1991

Date	Seasonal Hired Workers		
	Foreign	Domestic	Total
January 15	8,355	72,272	80,627.00
January 31	8,231	59,023	67,254.00
February 15	7,417	54,734	62,151.00
February 28	8,086	59,505	67,591.00
March 15	7,277	62,639	69,916.00
March 29	7,144	60,373	67,517.00
April 15	-	54,502	54,502.00
May 15	-	44,570	44,570.00
June 15	-	30,607	30,607.00
July 15	-	28,481	28,481.00
August 15	-	26,828	26,828.00
September 15	-	30,056	30,056.00
October 15	3,226	40,481	43,707.00
October 31	6,569	49,465	56,034.00
November 15	,6,797	50,997	57,794.00
November 30	6,922	59,354	66,276.00
December 15	6,865	62,958	69,823.00
December 31	6,851	67,452	74,303.00

Source: U. of Florida 1992.

While January 15 is the peak period of employment for seasonal farm employment in Florida overall, individual commodities have peak employment demand at various times throughout the calendar year. For example, the peak employment period for Florida tomatoes is October 15, while the peak employment period for Florida strawberries is March 15 (Table 5.4).

Table 5.4. Peak Employment of Seasonal Hired Workers in Florida by Selected Crops, 1991

Crop	Peak Employment Seasonal Workers	Peak Date
Citrus	34,461	Jan 15
Flowers/Nursery	14,958	Jan 15
Sugarcane	10,764	Nov 30
Tomatoes	6,965	Oct 31
Strawberries	6,275	Mar 15
Leafy/Mixed/Tropical Vegetables	5,130	Dec 31
Peppers	2,700	Jan 15
Watermelons	1,672	Jun 15
Celery	1,650	May 15
Lemons & Limes	1,155	Aug 15
Tobacco	1,120	Aug 15
Sweet Corn	1,038	May 15
Cucumbers	801	Dec 15
Green Beans	715	Dec 15
Squash	715	Mar 15
Sod	700	May 15

Source: U. of Florida 1992.

In contrast with fruits and vegetables, employment on dairy, beef, and poultry farms is quite uniform throughout the year. Also, employment in Florida's ornamentals industries is only slightly seasonal.

Both farmers and workers strive to elongate the harvest/production season for increased efficiency of resources or annual earnings. Unfortunately for many horticultural crops, product prices are so volatile that economic losses can be sustained if products are marketed out of phase with viable market windows. Thus, the combination of climate and economics dictate the harvest season for these crops, tending to perpetuate the seasonality of hired farm labor demand.

Farm workers in Florida are geographically concentrated in the peninsular part of the state, as opposed to the panhandle. The top 11 counties of agricultural employment are also among the most populous: Palm Beach, Orange, Lake, Dade, Hillsborough, Polk, Lee, Collier, Volusia, Hendry, and Manatee. This juxtaposition of farms so close to urban centers creates additional competition for farm workers.

Potential Changes in Florida Farm Employment

A number of major factors, excluding IRCA, could greatly influence farm employment conditions in Florida over the next decade. They include:

- general level of economic activity

- international trade policies, particularly NAFTA

- environmental constraints and restrictions

- mechanization or technological developments in production and harvesting.

The level of economic activity in Florida has historically affected farm labor markets. When construction and tourist activity, particularly, are strong, farm labor markets are negatively affected in terms of availability and rate. Conversely, when the state's economy is stagnating or undergoing recession, there are ample farm labor supplies and wage rates remain relatively flat.

While Florida agriculture has experienced international competition within domestic markets for decades, the expected ratification of the North American Free Trade Agreement (NAFTA) with Mexico is likely to worsen Florida's competitive position in fresh citrus and vegetable markets in the United States. NAFTA is also expected to have an adverse economic impact on Florida's sugarcane and peanut industries (Polopolus 1992). The Dunkel version of the General Agreement on Tariffs and Trade (GATT) could also have an adverse economic impact on Florida's fruit, vegetable, sugar, and peanut industries.

Overall, trade liberalization policies such as NAFTA and GATT will have the effect of reducing tariffs, eliminating Section 22 import controls, weakening price support levels, and otherwise providing incentives for increased foreign imports. If these policies reduce production levels of Florida agricultural commodities, the level of farm employment will obviously be adversely affected.

Environmental controls and restrictions can also have the effect of reducing or limiting agricultural production in Florida by making production more costly. These controls and restrictions include programs to reduce phosphorus in water runoff, programs to reduce point and nonpoint sources of the pollution of lakes and rivers, and the elimination of pesticides and fungicides previously used for commercial agricultural production. An example of a potential threat to Florida agriculture is the expected removal of methyl bromide for use in Florida agriculture. While alternative chemicals are being tested and may prove to be almost equally effective, the elimination of methyl bromide will have dramatic impacts on Florida's vegetable production systems (Polopolus 1992). Reduced agricultural production from environmental controls and regulations will have an obvious and negative impact on the farm labor market in Florida.

The mechanization issue is much less certain. It is a fact that the Florida sugarcane industry is undergoing a dramatic transformation from mostly manual harvesting methods a few years ago to roughly a 50 percent level of mechanical harvesting in the 1992–93 season. Some growers have achieved 100 percent mechanical harvesting, while other growers have a mixture of hand harvesting and mechanical harvesting. The impetus toward mechanical harvesting in Florida has been affected by the uncertain regulatory environment of the H-2A program and the huge litigation costs, real and potential, associated with the H-2A program.

Mechanization of Florida citrus and tomato crops is even less certain than the case for Florida sugarcane. With decreased citrus prices and higher harvesting and assembly costs, there is renewed interest in adopting efficiencies in citrus harvesting and assembly. Tomato and strawberry harvesting continue to be very labor intensive. Earlier research in developing fresh tomato varieties that are amenable to mechanical harvesting have not received much attention lately. And the emphasis of the strawberry breeding program is toward disease resistance, not development of varieties for once-over mechanical harvesting. This research strategy for both tomatoes and strawberries would be more appropriate for industries oriented toward processed product markets.

Labor Force

Earnings/Wage Rates

Annual earnings data for seasonal farm workers are not readily available. The available data on annual earnings of agricultural workers in Florida are from employers covered by state and federal unemployment insurance programs. These data include seasonal farm workers, as well as year-round workers of agricultural employers, plus workers covered by forestry and fisheries employers. Agricultural workers in Florida earned $13,040 in 1990, up 5.9 percent from the $12,309 level in 1989 (Table 5.5). Earnings in agriculture are below those of workers in construction, services, and manufacturing, but slightly above those in retail trade.

Wage rate data are more readily available regarding seasonal farm workers in Florida. There is some seasonal variation in wage rates, with rates in the winter or spring months tending to be slightly higher than rates paid to hired farm workers in the summer months. Also, wage rates have risen consistently over time. For example, nominal average wage rates for Florida farm workers in January 1980 averaged $4.25 per hour, compared with nominal average wage rates of $5.94 per hour in January 1990, a 40 percent increase over a ten-year period (Table 5.6). This increase in nominal wage rates almost matches general inflation changes over the same period.

Wages paid for farm workers in Florida are consistently above wage rates paid for hired farm workers in the South and for the United States as a whole. However, the differentials between Florida and the United States have tended to narrow over time. Average wage rates in Florida, however, were almost $1 per hour over average rates in the South for 1990 (Table 5.6).

Table 5.5. Average Annual Earnings in Florida in Agriculture, Forestry, and Fisheries, Compared with Selected Other Industries, 1989 and 1990

Industry	1989 $	1990 $	% Change
FLORIDA	20,072	21,032	4.8
Agriculture, forestry, and fishing	12,309	13,040	5.9
Construction	21,001	21,907	4.3
Manufacturing	24,290	25,666	5.7
Retail trade	12,375	12,773	3.2
Services	19,907	20,906	5.0

Source: U. of Florida 1992.

Wage rates of workers employed by independent farm labor contractors tend to be slightly higher in Florida when compared with average wage rates paid by farm employers (Table 5.7). This is primarily due to the fact that most workers of labor contractors are paid by the piece-rate method. When translated to hourly wage rates, these rates tend to be higher than workers who receive their pay on the basis of hourly wage rates.

Demographic Characteristic

There have been dramatic changes in the demographic composition of the Florida farm work force over the past two decades. Based on studies at the University of Florida for the 1970–71 and 1987–88 seasons, the following ethnic/racial patterns have emerged for seasonal farm workers in Florida (Table 5.8):

1. The relative importance of both domestic white workers and black workers has dwindled compared to Hispanic workers,

2. The relative importance of Mexican and other Hispanic workers has increased tremendously, and

Table 5.6. Wage Rates in Florida, the South, and the United States for All Hired Farm Workers, by Season, 1980, 1985, and 1990

Region	January	April	July	October
\$/Hour				
1980				
Florida	4.25	4.33	3.97	4.13
South	3.51	3.47	3.29	3.51
United States	3.69	3.61	3.54	3.85
1985				
Florida	5.20	4.84	4.95	5.05
South	n.a	4.07	3.99	4.31
United States	n.a	4.51	4.24	4.56
1990				
Florida	5.94	5.91	5.87	6.28
South	5.00	4.98	4.85	5.12
United States	5.70	5.54	5.30	5.64

Source: USDA, NASS 1991.

3. Haitian and other foreign black workers now account for 8 percent of the total farm work force, compared to almost none 15 years ago.

There are currently some wide differences in the use of Mexican and black workers, depending on commodity. For example, in strawberries essentially all of the seasonal farm workers are Mexican or of other Hispanic origin. For oranges and grapefruit, however, roughly 30 percent of the total seasonal work force is black, including African Americans, Haitians, and other foreign black workers. In tomatoes, approximately 78 percent of the seasonal work force is Mexican or other Hispanic, with almost 18 percent African Americans or Haitians. Less than 5 percent of seasonal tomato workers are non-Hispanic domestic Caucasians (Polopolus 1989).

The proportion of women in the seasonal agricultural work force in Florida has dropped somewhat over the past two decades. In the early 1970s, 30 percent of the work force was female. It is believed that the female component of the seasonal work force had dropped to roughly 24 percent by the 1987–88 season. Women, however, now account for approximately one-third of the work force in foliage and strawberries (Table 5.9).

The farm work force in Florida has become relatively younger over the past two decades. In the 1970–71 season, 60 percent of agricultural workers were over 35 years of age. For the 1987–88 season, only 28 percent of Florida's seasonal horticultural workers were over 35 years of age (Table 5.10).

Table 5.7. Comparison of Wage Rates of Workers Paid by Farm Employers and Labor Contractors in Florida, 1980, 1985, and 1990

Type of Employer	January	April	July	October
	$/Hour			
	1980			
Farm Employers	4.25	4.33	3.97	4.13
Labor Subcontractors	4.56	4.72	4.61	4.65
	1985			
Farm Employers	5.20	4.84	4.95	5.05
Labor Subcontractors	5.54	5.49	5.19	5.42
	1990			
Farm Employers	5.94	5.91	5.87	6.28
Labor Subcontractors	6.23	5.93	6.11	6.24

Source: USDA, NASS 1991.

**Table 5.8. Ethnic/Racial Distribution of Farm Workers in Florida, 1970–71
and 1987–88 Seasons**

	Percentage	
	1970-71[a]	1987-88[b]
White	32.90	6.20
Black		
American Black	56.00	14.47
Foreign Black	0.00	8.54
Hispanic		
Mexican	6.40	64.30
Other Hispanic	3.10	5.83
Other	1.60	0.66
Total	100.00	100.00

Source: Polopolus and Emerson 1975; Polopolus 1989.

[a]Based on random sample of 2,500 farm worker interviews.
[b]Based on mail survey of 1,400 farms.

More recent surveys by Emerson, Chunkasut, and Polopolus of
the demographics of Florida orange pickers in 1990 and 1991 reveal
results similar to the 1987–88 surveys of Florida horticultural workers. Without providing the detailed data for the 1990–91 surveys, the
data can be summarized with the finding that 81 percent of Florida's
citrus pickers in this period were of Mexican origin or heritage. The
next largest ethnic groups were Haitian and African American workers, each representing 8 percent of the total number of workers. White
and Jamaican workers represented only 1½ percent each.

The 1990–91 orange picker surveys also revealed that very few
women are involved with harvesting oranges. Only 8 percent of the
total number of workers surveyed were female, with the remaining
92 percent being male. This may be due to the physically demanding
nature of orange harvesting, where ladders and bags are usually involved with the harvesting process.

Orange harvesters are also quite young, having an average age
of 32 years. The average picker has had an average of almost seven
years of citrus harvesting experience.

Table 5.9. Distribution of Agricultural Workers by Sex in Florida, 1970–71 and 1987–88 Seasons

Type of Farm	1970-71		1987-88	
	% Male	%Female	%Male	%Female
All Types	69.80	30.22	–	–
Horticultural			76.26	23.74
Oranges			80.69	19.31
Grapefruit			80.06	19.94
Strawberries			68.62	31.38
Tomatoes			80.90	19.03
Foliage			65.72	34.28
Fern			81.19	18.81

Source: Polopolus 1989.

Table 5.10. Age Distribution of Agricultural Workers in Florida, 1970–71 and 1987–88 Seasons

Age Group	1970-71	1987-88
	%	
Less than 20 years	10.7	15.0
21-35 years	29.3	56.9
36-64 years	55.8	26.0
65 & over	4.2	2.1
Total	100.0	100.0

Source: Polopolus 1989.

The educational level of Florida citrus harvesters is quite low. The average highest grade of school completed for respondents in the 1990–91 surveys was just four grades of school. This is a very low level of education with which to compete in the American economy (Emerson, Chunkasut, and Polopolus 1991).

IRCA and Florida Agriculture

The Immigration Reform and Control Act makes it unlawful for any employer in the United States to employ an alien not legally entitled to work in the United States. It requires that employers in Florida and elsewhere carefully monitor Form I-9 for completeness, making this form available to officers of the U.S. Immigration and Naturalization Service or U.S. Department of Labor for inspection upon advance (three days') notice.

In addition to recordkeeping and I-9 requirements, IRCA has a broad array of provisions, including the imposition of civil and criminal sanctions or penalties for knowingly hiring illegal aliens and a broader program (H-2A) for obtaining temporary foreign agricultural workers.

Impact of SAWs

There were over 120,000 Special Agricultural Worker (SAW) applicants in Florida representing at least 117 different foreign countries. Mexico led the list of foreign nations with 51,428 SAW applicants as of July 17–21, 1989. Haiti was a fairly close second with 41,087 SAW applicants. Thus, two countries, Mexico and Haiti, supplied Florida with 92,515 (or 75%) of the total of 122,045 SAW applicants. Other countries with at least 1,000 SAW applicants in Florida included the following: Jamaica (4,805), Colombia (2,986), Brazil (2,811), Guatemala (2,151), El Salvador (1,860), Honduras (1,524), Peru (1,414), Trinidad (1,129), Nigeria (1,092), and the Dominican Republic (1,070).

A relatively large number of SAW applicants in Florida were believed to be fraudulent. Fraud appeared to be more widespread among applicants from Asia, Africa, Europe, South America, and the Caribbean, compared with SAW applicants from Mexico and Central America. While fraud may have been much less prevalent among Haitians, once given Temporary Resident Alien status, these workers tended to quickly abandon seasonal farm work for nonseasonal nonagricultural employment more readily than Mexican SAW workers.

University of Florida surveys of horticultural employers in 1988 revealed increased competition for seasonal labor from the nonagricultural industries in Central and South Florida. This competition came primarily from the food service, hotel/motel, tourist, and construction industries (Polopolus 1989).

Despite a large influx of SAW workers in Florida, early indications were that a large percentage of them were never farm workers or workers who shifted rather quickly into nonfarm employment in view of the strong market for service/construction workers in the late 1980s in Florida. This "void" or loss of SAW workers has been filled by the entry of "documented illegals" following the passage of IRCA in 1986.

Adequacy of Farm Labor Supplies

Shortly after the passage of IRCA, surveys of horticultural employers indicated that IRCA would have the effect of reducing labor supplies and increasing wage rates (Polopolus, Moon, and Chunkasut 1990). The predominant view of farm employers in 1988 was that many of the legalized SAW workers would abandon seasonal agricultural employment for year-round nonagricultural work. Given the presumed enforcement provisions of IRCA and a smaller labor supply, labor supply problems were a reasonable and likely outcome.

There is mounting evidence since our 1988 study that SAW workers have indeed left seasonal agricultural work in Florida, particularly Haitian SAW applicants. While there were almost as many Haitian SAW applicants (41,087) as Mexican applicants (51,428), recent demographic studies of Florida farm workers report the dominant influence of Hispanic, primarily Mexican, workers.

In the recent study of winter vegetable workers in South Florida (Griffith and Camposeco 1993), Mexican/Chicano workers accounted for almost 50 percent of the total work force, with Guatemalan workers accounting for 20 percent. Haitian workers represented only 16 percent of the total. Moreover, a recent demographic study of Florida orange harvest workers revealed that the percentage of Haitian workers in Valencia orange harvests had dwindled from 11 percent of the total work force in 1990 to 6 percent in 1993 (Polopolus, Chunkasut, and Emerson 1993).

The potential labor supply problem previously expected in 1988 from IRCA did not materialize. Thus, wage rates for seasonal agricultural work did not increase at rates above the general rate of inflation. It is hypothesized that the adequacy of farm labor supplies in Florida since 1988 is due to the continued entry of illegal workers (mostly documented) and the softening of nonfarm labor markets.

Impact of New Alien Workers

The serious concern regarding seasonal farm labor supplies in Florida evaporated by the early 1990s, partly because of the laxity in enforcement of the IRCA law. While precise data are unavailable, there is a consensus among farm labor market experts that new waves of "documented" illegal aliens have entered Florida's seasonal agricultural labor markets in recent years. These documented illegals have replaced former domestic workers or SAW workers who have shifted out of seasonal farm work.

The Role of Independent Labor Contractors

Whether Florida's seasonal farm work force has a large presentation of legalized SAWs or documented illegals, we know from our various demographic studies of horticultural industries that most of these workers are now of Hispanic or Caribbean origin (Polopolus 1989; Polopolus, Chunkasut, and Emerson 1993). This particular situation provides independent labor contractors with a special advantage in labor recruitment. Labor contractors, defined here as employers of farm laborers, are advantageous where workers are likely to be foreign born, migrant, illegal, unskilled, uneducated, and unorganized. Without labor contractors, workers with these characteristics face difficulties in finding jobs; consequently, these workers rely on informal networks of friends, relatives, and contractors for employment information (Polopolus and Emerson 1991).

Labor contractors become indispensable to seasonal labor markets because of their extensive contacts with farm worker communities and migration networks. They also have bilingual skills that serve to bring non-English-speaking workers into seasonal job markets.

The level or intensity of contract labor usage is much higher for Florida when compared to the nation as a whole or even California and Texas. Labor contracting is most pronounced in the Florida citrus industry; in our 1990 Valencia orange survey, 42 of 55 responding employers, or 76 percent, were independent farm labor contractors (Polopolus and Emerson 1991).

Impact of IRCA Sanctions

Particularly for Florida citrus harvest labor markets, there is some indication that the threat of IRCA sanctions or fines for knowingly hiring illegal aliens has caused some reorganization of farm firms. This has included the creation of separate legal entities to produce, harvest, haul, and pack or process citrus fruit, where previously these functions occurred under one corporate name. Entrepreneurs have offset the risk of IRCA sanctions by shifting responsibilities to labor contractors or by creating subsidiary harvesting corporations. In either strategy, the primary producing firm has become insulated from the risk of IRCA sanctions (Polopolus and Emerson 1991).

Reliance on a Temporary Work Force

As already noted, the Florida farm labor market is highly seasonal. Peak employment in January can be at least three times the level of employment in August. Thus, the bulk of the work force is seasonal in the framework of a casual labor market. The number of jobs changes over the calendar year. There is also turnover, with different people filling the same seasonal jobs.

Supplemental Foreign Labor

If and when the current IRCA law is fully enforced, there could be labor shortages in Florida agriculture. In addition to the continuing but diminishing need for H-2A workers for harvesting Florida sugarcane, there could be requests for H-2A workers in the citrus, tomato, strawberry, and tobacco industries, among others.

Modern Labor Management Techniques

Our 1988 surveys of horticultural employers strongly suggested the need for education and training programs on modern labor management. More effective programs and benefit packages for workers could obviate the need for supplemental or temporary labor from non-Florida sources. The development and maintenance of labor management programs will require the coordinated efforts of labor, management, government, and university personnel in Florida.

The Potential Labor Problem in Southwest Florida

There is a clear and emerging agricultural labor/housing problem in southwest Florida (Hendry, Collier, Lee, Charlotte, and Glades counties). Large acreages of citrus have recently been planted, and considerable acreages of pasture land have been converted for vegetable production. Citrus processing plants are also under construction. Many of these agricultural changes have occurred without corresponding infrastructure development, particularly housing, roads, and schools. This portends problems with the adequacy of seasonal farm labor supplies in southwest Florida.

International Competition

Trade liberalization agreements in the form of GATT and NAFTA could have an adverse impact on Florida's agricultural industries, particularly its fruit, vegetable, and sugar industries. It is these industries that account for the large proportion of hired labor expenditures in Florida agriculture. Thus, any adverse effects of lower tariffs and the lessening of other U.S. barriers to foreign imports would probably reduce the number of workers needed in Florida for agricultural production and harvesting operations.

As the competitive game is played out in commodity markets, Florida farmers will probably turn to new cost-reducing technology, which will also have the effect of reducing the number of workers employed.

Conclusions

The ineffectiveness of IRCA enforcement has minimized the previously expected problem of inadequate labor supplies for Florida's seasonal agricultural industries. However, if and when IRCA enforcement arrives, it is possible that the problem of inadequate seasonal farm labor supplies could appear. Any future labor supply problems assume not only strict enforcement on entry of illegals, but also a strong nonfarm economy for Florida and continued growth in agricultural output.

Another key variable in Florida farm labor markets is the net effect of trade liberalization policies on Florida agriculture and its resulting impact on the demand for seasonal farm workers. The worst-case scenario from employee and worker perspectives would find Florida's citrus, vegetable, and sugar industries adversely affected by the free trade agreement with Mexico and the lower support prices for U.S. sugar under GATT. Thus, NAFTA and GATT could weaken the competitive position of Florida agriculture and dampen the demand for seasonal farm labor.

6

IRCA and Apples and Vegetables in New York

ENRIQUE E. FIGUEROA

Introduction

New York is a major producer of Seasonal Agricultural Service (SAS) crops and ranks among the top 5 producing states for 11 SAS crops.[1] In 1991, the farm-gate value of New York fruit and vegetable production was $442 million.[2] While some crops have declined since 1986, the production and value of many crops has increased and the author expects the increases to continue.

Farm workers in SAS crops in New York have historically migrated from the South, come from Puerto Rico, or have been brought under the H-2/H-2A program. Since 1987 the influx of Mexican, Mexican American, and Central American farm workers has increased significantly, particularly in central and western New York. Though farm labor contractors (FLCs) have also increased their presence in the farm labor market in New York, they are not as prevalent as in the rest of the country. Most of the labor hired through FLCs works in the apple industry or on the larger vegetable-producing farms.

As in most other states, information and data on farm labor in New York are lacking. To help ameliorate this lack, I am presenting data collected directly from New York state (NYS) vegetable producers.

A total of 348 surveys were sent to members of the NYS Vegetable Growers Association during January and February 1993. The return rate was 43 percent (150) and the vast majority of respondents completed the survey with ample details. Respondents were categorized into three groups: strictly fresh market producers; strictly processed market producers; and multimarket producers — producers for both markets.

The survey is the first attempt to collect data on how vegetable producers have adjusted to IRCA. Also, the experiences I had collecting data from NYS apple producers leads me to conclude that it is

very difficult to survey this industry, particularly on matters concerning farm labor. It would be safe to say that current data should be
viewed with skepticism and should certainly be scrutinized.

The policy arena surrounding farm labor in NYS has changed
since IRCA. Two reports commissioned by the governor's office have
evaluated and issued recommendations concerning farm labor policy.
The crux of the issue, as in other parts of the country, is the implementation of statutes covering farm workers. The right for collective
bargaining by farm workers as yet does not exist in New York, but a
Governor's Task Force Report recommended that farm workers be
given such a right. Conversely, many producers of SAS crops complain about entities such as Legal Services, Rural Opportunities, Inc.,
and the federal Department of Labor, to name a few. The complaints
fall under various categories, but many are a result of misunderstand-
ing or a lack of awareness of what the laws allows.

SAS Crop Production in New York

The production of SAS crops in New York has not declined since
IRCA. Factors such as the Canadian Free Trade Agreement, exchange
rates, the North American Free Trade Agreement (NAFTA), opening
export markets, and other issues have and continue to be as important to SAS crop producers as farm labor. Also, the structure and size
of the apple industry is considerably different than other commodity
industries. The same can be said of the processed vegetable industry.
The apple industry commands both political and economic power,
whereas the processed vegetable industry is closely associated with
processing cooperatives who are experiencing declining demand for
their products.

The Apple Industry

Table 6.1 presents both the volume and value of NYS apple production since 1977. Between 1977 and 1986, the average annual production of apples was 760,420 tons, while between 1987 and 1991 (the
last year data were available) production averaged 677,580 tons. However, 1992 was a high production year and therefore when incorpo-
rated into the post-IRCA figure, the average increases. More important is the distribution of the apple crop between the fresh and processed market. The average share going to the processed market pre-

IRCA was 38.9 percent, while the post-IRCA share is 45.4 percent. Therefore, since fresh market production requires more labor per ton of production, the total demand for harvest labor in the apple industry has most likely increased. In addition, new plantings of orchards are primarily semidwarf, dwarf, and trellis systems that have more trees per acre. In turn, these systems require more harvest labor per acre. The preceding description reflects how NYS apple producers have responded to increasing demand for fresh market apples.

Table 6.1. Utilized Production and Value of Production of Fruits in New York, 1977–91

Year	Utilized Production-Tons	Value of Production
1977	581,200	$111,217,000
1978	767,450	154,434,000
1979	721,250	160,581,000
1980	771,300	158,175,000
1981	576,050	150,111,000
1982	761,950	150,083,000
1983	783,000	165,327,000
1984	738,850	163,978,000
1985	717,750	114,510,000
1986	646,550	137,796,000
1987	652,850	131,058,000
1988	648,170	151,925,000
1989	666,880	153,669,000
1990	667,800	179,735,000
1991	752,200	200,599,000

Source: <u>New York Agricultural Statistics</u>, New York Agricultural Statistics Service, 1 Winners Circle, Albany, NY 12235. Various issues.

The apple industry has relied on H-2/H-2A workers for many years, with eastern NYS more dependent than western NYS. A farm labor cooperative exists in the Hudson Valley and it recruits many of the H-2A workers for the industry.[3] In western New York, a few FLCs (one large one) have emerged since IRCA, and they primarily, if not entirely, serve the apple industry. In addition, an organization named Agricultural Affiliates, Inc. has emerged in western New York to provide educational seminars or other services to the SAS crop industry — primarily apple producers.

The influx of Mexicans, Mexican Americans, and Central Americans into the apple harvest labor population has been significant post-IRCA. No shortages of labor have been documented by state agencies and the piece rate for harvesting apples has declined in real terms.[4] Many of the new entrants into the harvest labor pool have entered through farm labor contractors (FLCs), a crew leader system, or word of mouth. Growers, extension agents, and farm worker social service agency personnel indicate that many workers come just for the apple harvest and then return to Florida, south Texas, or Arizona. Though industry participants will not say it publicly, unofficially they assert that many apple harvest workers are illegal or "illegally" legal — in other words, possess fraudulent documents. However, the penetration of FLCs into the apple industry is not at the level found in other parts of the country: In California one in three SAS crop workers is hired through a farm labor contractor.

The apple industry has grown and the forecast is that it will continue to grow — the number of young and nonbearing trees is large. It is clear that the industry is shifting more production to the fresh market and that the eastern- and western-producing areas of the state are increasingly more integrated — packer/shippers now pack for growers from both regions of the state. The industry is relatively well organized and the political clout of the industry at the state and local level is significant. However, signs indicate that the industry is amenable to developing programs to encourage more work-related benefits for apple pickers — some growers and extension agents report that the more progressive growers provide for health coverage.

The Vegetable Industry

Table 6.2 presents harvested acreage and the value of production for both fresh and processed market vegetables. On average, 3,000 and 6,000 fewer acres were harvested of fresh and processed market vegetables since IRCA passed in 1986, 1984 and 1978 being peak years for fresh and processed vegetables, respectively. However, yields for many of the vegetables have increased since IRCA, and therefore production has not declined as much as one might suspect. The production value of onions, cabbage, fresh and processed sweet corn, fresh and processed green beans, tomatoes, strawberries, cauliflower, cucumbers, carrots, and processed green peas show a statistically significant value-of-production growth trend over the past 16 years (based on regression analysis). In fact, there is evidence to suggest that the crop mix within the vegetables category has changed to more labor-intensive rather than less labor-intensive crops, indicating a response to consumer demand.

Table 6.2. Harvested Acreage and the Value of Production of Vegetables in New York, 1977–91

YEAR	VEGETABLES			
	FRESH MARKET		PROCESSING MARKET	
	ACREAGE	VALUE	ACREAGE	VALUE
1977	65,300	$90,000,000	72,900	$25,000,000
1978	64,900	108,500,000	85,700	28,000,000
1979	67,900	107,300,000	80,700	30,300,000
1980	69,800	156,500,000	76,600	31,600,000
1981	72,200	155,900,000	74,500	33,700,000
1982	71,900	131,600,000	75,700	36,100,000
1983	71,600	179,900,000	73,800	31,700,000
1984	73,700	139,800,000	75,600	33,800,000
1985	72,900	135,600,000	80,500	37,600,000
1986	67,500	167,400,000	65,600	26,500,000
1987	67,700	166,800,000	71,000	30,900,000
1988	64,800	165,700,000	69,400	24,100,000
1989	65,400	176,300,000	70,200	32,300,000
1990	67,700	172,800,000	68,700	36,400,000
1991	67,900	208,400,000	71,100	33,000,000

Source: New York Agricultural Statistics, New York Agricultural Statistics Service, 1 Winners Circle, Albany, NY 12235. Various issues.

Table 6.3 is from a survey I conducted of NYS vegetable producers during the winter of 1992–93. Approximately 50 percent of respondents indicated that the level of mechanization in their farms changed since 1986. The change was greater for fresh market vegetable and multimarket (both fresh and processed) producers. More important is how the producers changed their level of mechanization. Table 6.4 indicates that the change has, in fact, been toward more mechanization, but the multimarket[5] producers were evenly split — nearly half increased their level of mechanization while the other half increased their use of labor. For the entire sample, one in five producers changed their level of mechanization toward a greater use of labor.

**Table 6.3. Vegetable Producers Indicating that Their Level of
Mechanization Has Changed since 1986**

Producers	Number of Affirmative Responses	% of Category Response
Fresh Market Producers	56	48.3%
Processing Market Producers	3	25%
Multi-Market Producers	11	55.0%
Total	70	47.3%

Source: Figueroa, E.E., and Curry, P. Department of Agricultural Economics, Cornell
University, March 1993.

**Table 6.4. New York State Vegetable Producers Indicating a Change in
Their Level of Mechanization**

Producers	More Labor	More Mechanization
Fresh Market Producers	14.3%	85.7%
Processing Market Producers	—	100%
Multi-Market Producers	45.5%	54.5%
Total	19%	81%

* — Of producers changing to more labor, the average increase was 30%.
** — Of producers changing to more mechanization, the average increase was 123%
Source: Figueroa, E.E., and Curry, P. Department of Agricultural Economics, Cornell
University, March 1993.

Table 6.5 presents information of how crop mix has changed for
NYS vegetable producers since 1986. First, nearly 60 percent — 87 of
the 150 respondents — of the sample indicated that they changed
their crop mix. The largest shift was to higher-value crops, 35 percent, followed by more labor-intensive crops, 27 percent, and less labor-intensive crops, 21 percent. The reader may think that there are
inconsistencies between tables 6.4 and 6.5, but what appears to have
occurred is that respondents increased production, thereby increasing the level of mechanization. However, many also shifted their crop
mix toward more higher-value and more labor-intensive crops, but
this rate of growth was probably not as large as the rate of growth
toward mechanization.

The vegetable industry in New York continued to grow after the
passage of IRCA. There was some crop-mix shifting within the vegetable category, but it was not entirely to less labor-intensive crops.

Table 6.5. How New York State Vegetable Producers Have Changed Crop Mix Since 1986

— 58.0%, or 87 of 150 respondents, indicated a change —	
To Higher Value Crops	35.1%
To More Labor Intensive Crops	27.0%
To Less Labor Intensive Crops	20.9%
To Less Perishable Crops	13.5%
Other	3.4%
Total	100%

Source: Figueroa, E.E., and Curry, P. Department of Agricultural Economics, Cornell University, March 1993.

The level of mechanization did increase, but the larger farms had a lower propensity to mechanize after IRCA. Though the largest shift was to higher-value crops, wage rates of farm workers did not commensurately increase with the increase to higher-value crops. Many of the state producers of vegetables are relatively small producers, hiring less than ten workers at peak season.

Environmental Horticultural Crops

Very little industrywide data exist specific to New York producers and workers. Some information regarding the floriculture sector exists, but this sector is not a large employer of workers within the state. The industry has grown since IRCA, but the recession in the Northeast over the past three years has played a much more important role in the industry as compared to labor cost or availability. Some of the larger nurseries in the western part of the state have progressively relied more on Mexican and Mexican American workers since IRCA — one large nursery's entire hourly labor force is comprised of Mexican and Mexican American workers.

Market Channels

The market channels for the products mentioned above follow traditional channels of marketing. However, "the nearness-to-market" syndrome has caused many producers not to develop or join marketing organizations to the extent found in places like California.

Also, the direct marketing share of total sales is relatively larger in New York (and the Northeast in general) as compared to other parts of the country. One positive outcome for farm workers is that many growers grow a diversified set of crops that extends the harvest season and thereby increases the value of a farm worker to an individual grower. Also, a farm worker does not need to move as much because a diversified grower has more weeks of work for the farm worker.

Profile of Farm Workers

As most researchers involved with farm labor know, data on the subject are at best spotty and at worst misleading. New York state is no exception. No one entity can confidently quote the number of migrant farm workers in the state. The figures for "hired farm workers" include such a large number of dairy industry workers that it is difficult to use the figure to discern the number of SAS crop workers. Social service agencies such as Rural Opportunities, Inc. and NYS Legal Services have no real handle of their service population. If absolute figures are not available, then the ethnic composition of SAS crop workers is certainly not available. Therefore, the following information needs to be viewed within the parameters of the situation.

Ethnic Composition

The ethnic composition of New York SAS crop workers has changed, but the extent of the change has not been quantified. It is clear, however, that the change has been similar to what has happened in other parts of the country: more Mexicans, Mexican Americans, and Central Americans, and fewer African Americans. Over 41,000 (3.4% of the national total) SAW applicants were submitted by "residents" of New York, but a large number were considered fraudulent. In addition, SAS crop producers indicate, in private, that many of their current employees are illegal or "illegally" legal. My survey of the NYS vegetable industry found that 25 percent of the workers were Mexicans, Mexican Americans, or Central Americans, but the figure is based on both full-time (mostly white) and seasonal workers. If one excludes H-2A workers, then the majority of the remaining apple pickers are Mexicans, Mexican Americans, or Central Americans. If this pattern of employment continues, then by the turn of the century the harvest labor in NYS will be made up almost entirely of

Mexicans, Mexican Americans, or Central Americans. In fact, enclaves of these groups can be found in Rochester, Buffalo, and other cities. Year-round living is significantly curtailed by the cold weather during winter months, and therefore the establishment of families will not be as rapid as in other parts of the country.

Migratory Patterns

The typical migratory pattern before IRCA was for individuals to begin the winter season in Florida and travel up the Southeast and mid-Atlantic states during spring and early summer and arrive in New York for vegetable work during August and September. Apple pickers generally arrived during late September and stayed until mid-November, but most just came for the apple harvest and did not work in vegetable harvesting in New York. Many SAS crop producers have been hiring the same migrant workers for over 20 years. Direct hiring was and is the most prevalent and dominant form of hiring SAS crop workers.

Now the pattern is somewhat similar, but the use of crew leaders and FLCs has emerged. Though direct hiring is still the dominant form of hiring — 68 percent in my vegetable industry survey — the use of farm labor contractors and crew leaders has increased in the apple industry. Also, more individuals migrate form Texas and Arizona than in the past. Table 6.6 provides the distribution of the various hiring categories by NYS vegetable producers during the 1992 season. Though it is still a relatively small share — 16.9 percent — multimarket producers are the only vegetable producers utilizing FLCs. Since multimarket producers are relatively larger producers than fresh or processed market producers, it is safe to conclude that economies of scale play a role in encouraging producers to hire FLCs.

Table 6.7 provides the ethnic distribution of workers hired by NYS vegetable producers during the 1992 season. The combined information of tables 6.6 and 6.7 indicates that multimarket producers hire relatively more Mexicans and Mexican Americans and therefore many of the workers contracted by FLCs are Mexicans and Mexican Americans. The majority of the Puerto Ricans were hired by vegetable producers in Long Island. One very large processed market producer skewed the sample because he alone hired 144 "other" workers. For the entire sample, Mexicans, Mexican Americans, and Central Americans represent one in four workers hired by NYS vegetable producers during the 1992 season. Of the nearly 2,900 different workers rep-

Table 6.6. How New York State Vegetable Producers Hired Their Labor Force in 1992

Hiring Arrangement	Fresh Market Producers	Processing Market Producers	Multi-Market Producers	All Producers
Direct Hires	58.6%	85.5%	59.5%	67.9%
Crew Leader Arrangements	7.2%	7.2%	15.5%	9.9%
Farm Labor Contractors	0.7%	—	16.9%	5.9%
Off-Shore Puerto-Ricans	5.0%	—	—	1.7%
H2-A	0.8%	—	—	0.3%
Family Members	19.0%	—	—	6.3%
Other	8.7%	7.3%	8.1%	8.0%
Sum	100.0%	100.0%	100.0%	100.0%

* — Total Sample: 148 of which 116 were Fresh Market Producers, 12 were Processing Market Producers, and 20 were Multi-Market Producers.

Source: Figueroa, E.E., and Curry, P. Department of Agricultural Economics, Cornell University, March 1993.

resented in the sample (Table 6.8), many are year-round employees while others are "local seasonal" workers. Most of these year-round and local seasonal workers are most likely Caucasian. Therefore, the "migrant seasonal" workers represented in the sample are mostly Mexicans, Mexican Americans, and Central Americans.

The NYS vegetable producers survey sample represents 59,000 total farm acres, of which nearly 38,000 were planted in vegetables in 1992. Of the 38,000 planted acres, 35,700 were actually harvested. Multimarket producers harvested 12,500 of the acreage in the sample and the fresh market component of the multimarket category was 17 percent — 2,160 acres. The assumption is that the 808 total workers listed in Table 6.7 under the multimarket category were distributed consistent with the distribution of the acreage. Therefore, 17 percent of the 808 — 137 workers — worked in producing fresh market vegetables, while 671 worked in producing processed market vegetables.[6] Continuing the process one step further reveals that of the 2,900 total workers in the sample, 2,030 worked in producing fresh market vegetables — 70 percent — while 780 worked in producing processed market vegetables.

**Table 6.7. Ethnic Distribution of Workers Hired by New York State
Vegetable Producers in 1992**

Ethnicity	Fresh Market Producers	Processing Market Producers	Multi-Market Producers	Totals
Caucasian	46.3% (877)*	18.4% (35)	29.7% (240)	39.8% (1,152)
African-American	18.1% (342)	0%	5.2% (42)	13.3% (384)
Mexican-American	8.0% (151)	2.6% (5)	12.9% (104)	9.0 (260)
Mexicans	6.1% (116)	3.2% (6)	32.2% (260)	13.2% (382)
Puerto Ricans	11.1% (211)	0%	7.9% (64)	9.5% (275)
Caribbean	3.4% (70)	0%	1.2% (10)	2.8% (80)
Central-Americans	4.0% (75)	0%	2.5% (20)	3.3% (95)
Other	3.0% (51)	75.8% (144)	8.4% (68)	9.1% (263)
Totals	100% (1,893)	100% (190)	100% (808)	100% (2,891)

* — Numbers of Workers in parenthesis.
Source: Figueroa, E.E., and Curry, P. Department of Agricultural Economics, Cornell
University, March 1993.

Approximately 18,455 harvested acres — nearly 52 percent — in
the sample were fresh market vegetables. For 1992, the sample represents 27.2 percent of the total fresh market vegetable acres harvested
in NYS and 24.3 percent of the processed market harvested acreage.
Therefore, based on both the worker and harvested acreage sampling
rates, I estimate that during 1992 the NYS vegetable fresh market industry hired 7,463 workers, while the processed market producers
hired 3,210, for a total of 10,673. This is a conservative estimate because it is strictly based on harvested acreage, not planted acreage,
and therefore planting labor is indirectly left out. Also, the sample
includes a disproportionate acreage share of less labor-intensive SAS
crops such as potatoes.

Table 6.9 presents figures for the relationship between workers
and harvested acres. The NYS Department of Labor estimates that at
peak season — September 1 to 15 — in 1991, the total number of agri-

**Table 6.8. Number of Different Workers Hired by New York State
Vegetable Producers in 1992**

Producer Type	Number
Fresh Market Producers	1,893
Processing Market Producers	190
Multi-Market Producers	808
Total	2,891

Source: Figueroa, E.E., and Curry, P. Department of Agricultural Economics, Cornell
    University, March 1993.

cultural seasonal hired workers was 13,195. The NYS Department of
Education estimated that 10,504 "true migrants" were in New York
during 1991 and that about half were working in the dairy industry.
The Migrant Health Program of New York estimated a total population of "Migrant and Seasonal Farm Workers" (including all dependents) of 30,811 for 1989. Farmworker Legal Services estimates that
there are approximately 60,000 migrant farm workers in NYS. With
the exception of the Department of Labor reports, I have not reviewed
the methodologies that arrived at these estimates.

**Table 6.9. New York State Vegetable Producers' Labor Use on a Per
Harvester Acreage Basis, 1992**

	Acres
Acres Harvested Per Year-Round Worker	104.033 (0.00958)*
Acres Harvested Per Seasonal Local Worker	86.07 (0.0116)
Acres Harvested Per Seasonal Migrant Worker	78.46 (0.0127)
Acres Harvested Per All Three Above Categories	28.71 (0.0348)

* — Workers Per Acre Harvested.
Source: Figueroa, E.E., and Curry, P. Department of Agricultural Economics, Cornell
    University, March 1993.

H-2A Workers

Over the past 20 years, the number of H-2/H-2A workers in New York has ranged between 2,000 to 2,700 workers per year and 90 percent of them pick apples in the Lake Champlain or Hudson valleys. Prior to IRCA, nearly 100 percent of the H-2 workers picked apples in the two valleys. There is some evidence to suggest that some western New York apple producers have switched to H-2A workers, but the numbers are still relatively small.

The important question is whether an H-2A worker program is needed. At this point, the effect of the H-2A program on the farm labor market in New York has been detrimental to farm workers in terms of wages and earnings. Since IRCA, the farm labor market in New York has more often than not had an oversupply of apple pickers. Consequently, real wage rates have declined and the H-2A program looms as a ready source of labor if domestic apple pickers "cause problems." For example, last year an apple producer wanted H-2A workers, but the NYS Department of Labor could not certify his application because domestic workers were available to do the picking. The domestic workers were sent to the producer's farm, but after a day the producer "fired" the workers (or workers resigned, depending on whom you ask) because their "efficiency of picking" did not meet the standards of the producer. After the producer's "evaluation" of the domestic workers' apple-picking efficiency, the NYS Department of Labor certified the H-2A application of the producer. This example is perhaps one of the more odious, but it illustrates how the "spirit" of the H-2A program can be used to the detriment of domestic workers.

It is clear that one of the intents of IRCA — to improve the working conditions of farm workers — is compromised by the use and abuse of the H-2A program. Producers claim that domestic workers just cannot pick fruit, but at the same time producers are reluctant to increase piece rates to attract "better" pickers. The H-2A program affords apple producers a ready and easy way to control the harvest labor force. The net effect is to signal to domestic workers that the standard by which they will be evaluated is based on foreign workers. Foreign workers have no real stake in this country, nor do they compare their wage rates to U.S. standards.

Labor Market Structure and Performance

Since IRCA, the farm labor market in New York is more linked (primarily through farm labor contractors) to the national labor market, though it is not yet an efficient market. Clearly, more information and sources of information are available to both the demand and supply sides of the market, but the influx of new non-English-speaking workers contributes to farm labor market imperfections. The increased role of FLCs and crew leaders has benefited the demand side more than the supply side because of the relative "unawareness" of the new entrants into the labor pool. The majority of SAS crop producers in NYS hire their labor through direct hiring, and many producers have a long working relationship with many of their workers.

The apple industry has moved more to crew leader and FLC hiring. To a lesser extent, producers who grow both fresh and processed market vegetables have also used relatively more crew leaders and FLCs as compared to vegetable producers that only grow fresh market or processed market vegetables. Eastern NYS apple producers rely very much on H-2A workers, while western NYS apple producers do not. Most, if not all, producers that also have direct market operations hire almost exclusively through direct hires. The larger the farm, the higher the probability the producer will use crew leaders or FLCs.

One rationale for the apple industry shifting more to crew leaders or FLCs is because the crop value has increased. The shift toward more fresh market apple production coupled with high-density plantings translates to a higher value per acre. To minimize the risk of not being able to pick the fruit when it is ready, producers have chosen to increase the probability of having adequate labor — crew leaders or FLCs can provide that increased probability. In addition, the ability of both crew leaders and FLCs to have illegals or illegally documented legals increases the supply of labor. Finally, the influx of Mexicans, Mexican Americans, and Central Americans who have limited English-speaking skills makes it more difficult for growers to communicate with them. The above rationale can also be applied to the larger multimarket vegetable producers.

Table 6.10 ranks the farm labor issues of most concern to NYS vegetable producers. Labor availability is at the top of the list, followed by regulations. The next tier of concerns are paperwork, worker productivity, and cost. Language, transportation, and legal status are the issues of least concern. Availability is interpreted to mean both willing and able workers, be they local or migrant. The fact that availability is the number-one concern bespeaks the real need for good communication between potential workers and potential employers. Nonetheless, the issue may be a real or perceived problem by NYS vegetable growers and needs to be addressed.

Table 6.10. Labor Issues of Greatest Concern to New York State Vegetable Producers

Issue	Concern Index*
Availability	3.73
Regulations	3.95
Paperwork	4.25
Productivity(Worker)	4.42
Cost	4.46
Housing	7.13
Legal Status	8.41
Transportation	8.56
Language	10.76

* — 1 = Greatest Concern, 11 = Least Concern.
Source: Figueroa, E.E., and Curry, P. Department of Agricultural Economics, Cornell University, March 1993.

Conclusions

The Immigration Reform and Control Act of 1986 clearly did not meet its objectives in New York. There are probably more illegals or "illegally" legal farm workers in New York than before IRCA. Wage rates have not improved, nor have living conditions for many farm workers. No real noticeable changes in crop mix, such as the lower production of labor-intensive crops, has taken place, and there is some evidence that more higher-value or labor-intensive crops are now grown. Based on the survey of NYS vegetable producers, they feel overburdened by the paperwork requirements of IRCA as well as other regulations. The use of FLCs in SAS crop production has increased. A higher percentage of the migrant labor force is made up of single non-English-speaking males. A higher percentage of migrant families are comprised of legal and illegal members of the same family. The number of H-2A workers has changed little after IRCA, even though labor shortages have not been documented and in some years labor oversupply has been the case. And producers or producer groups and farm labor advocacy groups have become more rather than less combative.

The increase in consumer demand for SAS crops during the latter part of the 1980s maintained favorable economic conditions for SAS growers in New York. Employer sanctions and INS enforcement have not been effective and therefore the pull factors have not been abated. The black market for fraudulent documents proved effective and profitable and FLCs became not only contractors of labor but also brokers of legalization.

What can be done now? First, the mechanisms of how to administer an SAS crop worker benefits package across state borders and across multiple employers need to be investigated. New York SAS crop producers might consider an employee benefits plan that would allow for a number of employers — say, a citrus grower in Florida, a nursery operator in Tennessee, and an apple grower in New York who would contribute to Pedro Gonzalez's health insurance program. Apple producers in New York recognize the need for stabilizing their supply of workers and they also recognize the stabilizing effects of health insurance and other benefits. The bottleneck is how to administer a program across states and across employers.

Second, the H2-A program may need to be phased out because it deters producers from investing in a stable domestic labor force. In addition, the H-2A program has exerted downward pressure on wage rates and no evidence of labor shortages has been documented in NYS — the provisions for Replenishment Agricultural Workers were never used.

The NYS Department of Labor needs to embark on an SAS worker data-gathering program that is comprehensive. The producer community is becoming more aware of the need for accurate data and therefore will be more cooperative with state statisticians.

Public, quasi-public, and private farm worker social service agencies need to attract and retain more bilingual Spanish-English staff. The "Latinization" of rural America has not taken place in New York as it has in other parts of the country, but the process has begun. The sooner government agencies are geared to communicate with their clientele, the better the social service delivery system will be. In addition, both state citizens and SAS crop producers will benefit from a healthier and more stable labor force.

Last, all interested parties in NYS need to arrive at the understanding and appreciation that SAS farm labor is different than dairy farm labor. Working conditions, length of employment, manner of compensation, number of different employers, existence (or nonexistence) of a benefits package, language and cultural differences, and other factors point to very different worker profiles and employee/

employer relationships. Programs that address hired farm workers will fail if they are not designed within the unique parameters of SAS crop employment.

NOTES

1. Because of a lack of data, the chapter excludes environmental horticultural crops.

2. Excluding potatoes.

3. Approximately 2,800 H-2A workers were certified in New York state in 1992. The Hudson Valley Cooperative applied for 1,475.

4. In 1982, the prevailing piece rate for 1⅛ bushel box for fresh market apples in western New York, Hudson Valley, and Lake Champlain were $.52, $.46, and $.55, respectively. In 1992, the comparable figures were: $.60, $.60, and $.70. These growth rates represent approximately half the growth rate of the U.S. Consumer Price Index (CPI).

5. It is evident from the survey that the largest producers in the sample are the multimarket producers.

6. Most likely, the assumed distribution is skewed toward the processed market — a higher proportion of the multimarket workers probably worked producing fresh market vegetables than the derived 137.

7

IRCA and Agriculture in Southwest Michigan and Central Washington

EDWARD KISSAM

Introduction

In contrast to the earlier literature on farm workers and the U.S. farm labor system, recent farm labor research has focused on the heterogeneity of the industry and presented a picture of U.S. agriculture as a "mosaic" — a wide range of specialized strategic adaptions to the "problem" of labor force recruitment and supervision, although some ubiquitous features can be observed in every labor market — such as seasonal employment of workers, reliance on labor intermediaries, payment on a piece-work basis, and lack of employee benefits. In this chapter, I analyze differences in labor recruitment, supervision, and farm labor migration patterns as unique adaptions within a common structural framework for addressing universal problems of farm labor management, such as matching labor supply to constantly changing labor demand.[1]

This discussion rests primarily on case studies of tree fruit and row crop production conducted for the Commission on Agricultural Workers (CAW) in Michigan and Washington from spring through fall of 1990. Some additional observations stem from 1989–90 field research in south Texas, south Florida, and central California as part of the Farm Labor Supply Study (FLSS), a study of factors affecting post-IRCA labor supply (Kissam and Griffith 1991).

We used similar methodologies in both our CAW case study areas, sampling workers at each production operation in proportion to the size of the labor force. All interviews were conducted in person. In Washington, we interviewed 18 employer/managers, 87 workers, and 9 field supervisors/crew leaders (since some small producers provided their own field supervision). In Michigan, we interviewed 14 employers, 4 crew leaders, and 98 workers in 24 different labor camps (including one public migrant housing project). In Michigan, the employers (who included pickle producers, apple producers, and peach, plum, tomato, pepper, fresh market cucumber, strawberry,

berry, and multicrop producers) were broadly representative of overall production in the area. In Washington, only apple and asparagus
producers were interviewed, but the study sample was representative of these major crops. Interviews were conducted during the peak
harvest season, so the sample best represents the harvest labor force,
as opposed to the off-season "core" labor force.

The CAW case-study regions in southwest Michigan (Berrien and
Van Buren counties) and central Washington (Yakima County) are
both production areas devoted primarily to labor-intensive agricultural production. Also, both rely heavily on migrant workers during
a long season stretching from April through September in Washington and May through November in Michigan. Both regions are leading apple production areas, providing a good opportunity to examine in detail the differences in labor utilization between areas and in
relation to market niche. Both regions have extensive row crop production. In central Washington, our case study of row crops was conducted in asparagus (a major crop in the area), and in southwest Michigan, our case study of row crops focused on pickle production but
also included interviews with a variety of vegetable harvesters
(Kissam, Garcia, and Runsten 1991).

Labor Recruitment

Central Washington and southwest Michigan are both "upstream"
labor markets relying heavily on migrants to meet peak farm labor
needs, and both have extensive and fairly well-understood histories
of relying on Mexican-origin migrant workers to meet their labor
needs. From the 1950s through the early 1970s both areas relied largely
on Texas-based migrants (who were, themselves, immigrants from
central and northeastern Mexico), but over the past 20 years the migration patterns of both areas have changed greatly as the farm labor
force participation of Texas-based "green card-era" migrants has
dwindled. In the 1990s, however, they contrast sharply in their reliance on labor market intermediaries for labor recruitment and supervision and in migration patterns.

The southwest Michigan labor market has, for at least the past
two decades, relied on "downstream" labor supply areas in south
Texas and south and central Florida to meet their labor needs. Prior
to IRCA, southwest Michigan's relative reliance on Florida-based
migrants arriving with Florida crew leaders had already begun to
increase, but this increase accelerated after IRCA. Many small farm-

ers in the area had established long-term relationships with Texas-based family migrants who had come north year after year, but these relationships were already weakening in the pre-IRCA period and were further weakened as a result of IRCA. An important factor in Michigan's historic reliance on family migrants was the availability of employer-provided free or low-cost housing for migrant workers.[2] However, year-round housing has been in chronic short supply and relatively few Texas migrants have settled in Michigan. The result is that the labor market has continued to rely extensively on crew leaders migrating from Texas and Florida for worker recruitment and supervision.

In contrast, central Washington, which had relied on the same labor force of Texas-based family migrants as had southwest Michigan during the 1960s and 1970s when asparagus production was rapidly increasing, came during the 1980s to rely more on settled local farm workers and on Michoacan-based transnational migration networks to meet their peak labor needs. During the mid- to late 1980s, networks of Mixtec migrants who had established themselves in several important California and Oregon labor markets began to extend their migrant circuits to include central Washington also. The substantial settled population of former Texas and Michoacan migrants living in the *colonias* of the Yakima Valley has provided an important labor pool for recruiting in-house foremen, supervisors, and managers, eliminating the need for Washington agricultural producers to rely on labor contractors.

Network Recruitment

As in all industries, production operations in labor-intensive agriculture, which appear on the surface to consist of many parallel and apparently autonomous worker activities, in fact require a substantial amount of coordination, worker interaction, and problem solving. The "quality" of work-place dynamics has, over the past several decades, been recognized to be a major component of industrial productivity. Managers, supervisors, and workers internalize many aspects of work-place organization, particularly in informal work environments such as agriculture, and make explicit only partial and often stylized pictures of the content of worker interactions (Darrah 1991). This does not mean that the work place is unstructured, but simply that the structural characteristics and dynamics of the work place are to a large extent latent — the rules governing worker-supervisor interactions are so deeply embedded in informal interactions that it may be difficult for participants to formulate them or for observers to recognize them at the surface level of daily interactions.

The traditional view that farm work consists simply of unskilled labor confuses the content of work with the activity in context. Setting a ladder in the fruit harvest is, at once, as simple and as complex as the typical bank officer's scrutiny of a loan application. The view that farm work is simple rests in part on the fact that traditional, informal approaches to worker recruitment and supervision so successfully incorporate into the labor management system of U.S. agriculture the fine-grained social dynamics of Mexican systems of mutualism and cooperation. These informal systems of social relationships, which evolved in the context of peasant agriculture, have been successfully transplanted into urban settings as well as rural ones, in both Mexico and in the United States. I refer here to social relationships related to *compadrazgo* and their role in facilitating the development of relatively large cooperative networks integrating work, housing, and social cooperation (Lomnitz 1977). In the U.S. agricultural context, reliance on this nonformal system of social dynamics has provided a crucial foundation for a broad spectrum of worker-manager interactions — negotiations regarding overtime or temporary furloughs, conflict resolution, assignment of employer-owned housing, and arrangements for regular seasonal workers to return or not.

In both southwest Michigan and central Washington, Mexican network dynamics have played a key role, not only in worker recruitment, as has been commonly recognized, but also in managing, coordinating, and problem solving related to production work. In the context of small family farms in Michigan, reliance on a work force composed primarily of migrant families has provided a self-organizing, self-directing labor force for well over 30 years, making possible the continued existence of the "family farm." In central Washington, reliance on Mexican village networks and the close ties within a relatively small number of "transnational communities" for labor recruitment and supervision has been a key element in the current rapid expansion of the apple industry to successfully establish a niche in the global market. The rapid expansion of central Washington asparagus production 30 years earlier relied similarly on south Texas migrant-sending villages, where traditional network arrangements prevailed among a population who then consisted primarily of recently arrived "green-card" immigrants from central and northeastern Mexico.

The contribution made by Mexican culture to the traditional system of U.S. farm labor management has been the extended family network, governed by an elaborate system of reciprocal obligations as the fundamental building blocks of work-place organization labor recruitment, task assignment, task coordination, supervision, and conflict resolution.[3] The American contribution has been an exploitative system of risk management and cost containment — payment by piece rate, reliance on contract or pseudo-contract labor in lieu of bona fide employer-employee relationships, and ambiguous supplemental motivational arrangements such as the end-of-harvest bonus and employer-provided housing (often for a rental fee). The exploitativeness of all these arrangements stems primarily from their deceptiveness, deceptiveness based in part on the fact that arrangements embed unfamiliar computational features or legal features (such as a subcontractor's legal obligation to withhold income tax from his employees) and that labor contractor-worker relationships mimic but do not duplicate traditional network relations of *compadrazgo*. At the intersection of the Mexican system of social organization in the work place and the American system for structuring economic relationships lies a common attachment to entrepreneurial output-based economic agreements (for example, piece-rate payment or sharecropping) in lieu of input-based ones (for example, hourly wages). However, in many cases such arrangements are attractive to prospective workers, either because they facilitate the employment of marginal workers (such as older workers and children) or they serve to maximize family income if not individual earnings. From the employer perspective, these arrangements assure a buffer against mismatches of labor supply and demand, since family workers who are not primary wage earners can be easily mobilized during peak periods of work and easily idled during labor demand troughs.

Piece-rate payment and assigning individual plots to migrant families in an actual sharecropping or pseudo-sharecropping arrangement achieve a number of important risk management objectives in Michigan, Washington, and many other farm labor markets. Piece-rate payment eliminates the need for effective worker screening, since the cost of labor per unit of production is fixed. In Michigan, the assignment of individual plots to pickle pickers motivates workers to increase levels of labor inputs (including reliance on children for some picking) and an easy rationale for employers not to pay workers for critical nonharvest tasks such as hoeing and training pickle cucumber vines. Asparagus production in Washington relies on the assignment of individual rows to migrant families or work teams of unrelated individuals for similar reasons. Such arrangements in asparagus also maximize labor inputs and rationalize the savings realized by not paying workers for nonharvest tasks (such as removing frozen or damaged asparagus spears).

In the areas we have studied, piece-rate payment system is seldom grossly abused but often tilted subtly in favor of the employer. For example, in the case of Washington asparagus, it is widely believed that weight adjustments in payment to workers, made to account for trimming spears, overcompensate workers; although trimming may remove 10 to 20 percent of weight, cutters' paychecks reflect a still greater downward adjustment for weight. In Michigan, at the variable-rate pickle producers' establishment in our study, workers were not paid for the oversized and deformed pickles they picked, but these pickles were, in fact, brined for relish and sold, a highly visible practice that gave rise to bitter accusations by workers that the employer was cheating them.

In traditional family networks, the close link between recruitment and supervision — functions typically separated in "modern" personnel management structures — serves to assure mutual accountability among workers and supervisors in the small farm setting. Workers are accountable to the relative who has recruited them, who is also their supervisor. In turn, the supervisor is accountable to his worker-relatives — both practically and in terms of social mores. Network ties of actual and fictive kinship convey both rights (as in supervision and mediation) and responsibilities (as in emergency support and negotiation), serving to dampen, if not eliminate, conflict. From a practical standpoint, in the small farm/migrant family network setting, core workers must also retain the allegiance of their workers because their informal role as de facto supervisor is the source of special benefits (such as the best available housing) and security (such as continued employment, even during periods when work slows down).

While the small family farmers we interviewed in Michigan, California, and Washington often pride themselves on their ability to "get along" with workers, the reality is that the limited-Spanish-speaking, small farmer-managers have only marginal organizational control over their work force. Management-labor relations are characteristically implemented, articulated, and tensions smoothed by the "invisible" supervisory structure consisting of heads of household and core workers interacting in the context of extended family networks.

In the work-place context of the small family farmer managing one or several migrant family units, the invisibility of the supervisory structure gives rise, in some cases, to the mistaken impression by managers that no management is needed for unskilled work or that current management techniques are tremendously effective. The manager's delegation of responsibility may, in fact, be the most effective management strategy, but the point is that the manager or supervisor's own interpersonal and communication skills are not the key to operational success. The primary responsibilities retained by the small farmers we interviewed in Michigan were: task assignment (including specifying particular quality standards of what to pick or not), assigning workers to housing, and firing.

In reality, small farmers' reliance on migrant families serves as the structural equivalent of "quality circles" in organizational management, fulfilling key functions of production coordination, supervision, quality control, problem solving, and support for worker morale. The link between quality and overall productivity is often recognized by farmers in differentiating between the speed of unrelated, unaccompanied male workers and the reliability of family workers.

There are both strengths and limitations in relying on extended family networks. Network ties of mutual obligation can only extend over a limited domain of social distance. The characteristic structure and functioning of farm labor networks is related, in part, to demographics because families with large numbers of siblings provide a substrate for more extensive networking than do smaller families. Other important factors affecting network functioning include the extent to which children from farm worker families intermarry, ease of residential clustering, and range of occupational alternatives in home-base areas. As a farm becomes larger, labor demand eventually outstrips the capacity of network recruitment. Recruitment and supervision, while conforming to the nominal patterns of extended family ties, reach a point at which mutual obligation has been attenuated to the extent that it no longer exists. The core worker then begins to function as a crew leader or farm labor contractor whose sphere of influence has now been expanded to the labor pool of a *colonia* or, at times, multiple networks in a village. However, as the sphere of influence widens, the stabilizing forces linking workers and their supervisors weaken.

Core Worker to Crew Leader to Labor Contractor

The functioning of labor market intermediaries — including *mayordomos*, crew leaders, and FLCs — rests on the establishment of "artificial networks" structured along the lines of traditional informal family-based networks. These networks, on the one hand, exert the control and provide the support services rendered in more traditional settings by extended family members, while on the other hand, they "exploit" workers by charging directly or indirectly for a variety of services: establishing the link between the worker and the employer (part of the "service package" offered by long-distance *raiteros*), providing support services (for example, selling beer, soda, and food in the field), and making loans to workers experiencing cash-flow crises. Whether or not such interactions are, in fact, benign or exploitative is not easy to determine; key issues are whether services "sold" to workers are fairly priced, whether financial arrangements are deceptive or not, and to what extent coercion is used in inducing workers to "buy" services. For example, charges made by *raiteros* in California for providing workers with transportation to work may or may not be considered exploitative. They are not deceptive and in terms of cost may be no more overpriced than public or private transportation once insurance charges are factored in, but coercion by supervisors to induce workers to pay for rides to work instead of driving their own cars or carpooling tips the scales in a clear-cut way toward exploitation.

One of our most interesting observations of the relationship between network recruitment, ties of mutual obligation, and supervision stems from discussions with a group of long-time "traditional" Texas-based crew leaders working in Washington asparagus, who found that as crop productivity declined and as employment options opened up in the lower Rio Grande Valley, they could no longer rely on family networks to recruit workers for their Washington employer. These crew leaders reported that some extended family members who were experienced workers no longer agreed to migrate north because they knew from experience that their earnings would not be as much as before, in that the production unit consisted primarily of overmature beds of asparagus. Because the labor force had been so stable over many years, experienced workers were aware of the decline in productivity of the fields. Other family members — sons, daughters, brothers-in-law — continued to migrate north because they could be assured that the crew leader would assign them the most productive asparagus beds.[4]

It was only at this point that crew leaders, who had formerly functioned in the role of extended family recruiters, took on the role of farm labor contractors. Given the difficulties the crew leaders experienced in recruiting family members with whom they were linked by bonds of mutual obligation, they turned to recruiting inexperienced workers from their hometowns (who agreed to migrate north because they did not know that asparagus productivity was falling) and to "instream" recruiting of recently arrived workers from Mexico to fill out their crews.

The "second tier" of "outside" workers who had been recruited informally, either by being "invited" to come north with the crew or hired in central Washington, were assigned the worst plots (since the best plots had been promised to close family members), had the most difficulty working (since they had the least experience in asparagus), and had the highest turnover (since they had the least obligations to the crew leader, the worst working conditions, and the worst earnings). The key to problem solving on the part of the core-worker crew leaders functioning as labor contractors rested on managing to recruit outside their own network, thereby assuming less accountability to disclose what working conditions actually were. Recruiting within the network would have required too much accountability — a danger constantly present within the flexible functioning of traditional networks (Lomnitz 1977; Lewis 1961). The success of this recruitment hinged, in part, on the ambiguous nature of the process of recruiting outside the network, a recruitment that tacitly implies but does not guarantee workers the protections afforded members of an extended family network.

The recruitment, supervision, and labor management processes of the agricultural work place all have elements of both altruism and exploitation. However, in part because individual labor market intermediaries are at different points on the continuum from network recruitment to direct hiring, there is a great deal of variation in the labor market intermediary's concept of his or her own role and specific mutual obligations. In Michigan, for example, a well-known labor contractor strongly professed traditional values despite extensive documentation of her abuse of workers, stating to us, "*Yo sin mi gente, no soy nadie*" ("Without my people, I am no one"), as she discussed with us the roles and responsibilities of a *troquera*. Yet, at the same time, several of her workers strongly criticized her — not for wage rates or the shacks she had arranged for them, but for refusing to give them a ride as they walked back to work in the rain. Her refusal symbolized to them, more than any of her other legal or personal shortcomings, indifference and "uncaring" attitudes not expected within a network that at least implied mutual support. Another very successful crew leader, who was fairly well regarded, discussed with us his reluctance to hire within his extended family network because, if he did, people would always "bother him" for favors — in other words, demand that he provide them with various types of support such as loans or rides. A key element in his personnel strategy was to avoid bonds of mutual obligation, providing only a minimal package of support services to his workers.

In Washington's Yakima Valley, a production area that has few labor contractors, in-house *mayordomos* fulfill the same structural roles in recruitment and supervision that crew leaders do in Michigan or that labor contractors do in the eastern migrant stream. However, there are important differences in both recruitment and supervision practices, in large part because traditional extended family networks still dominate the work place. An interesting observation from this area is the close linkage between network-based recruitment, the strong ongoing flows of recent immigrants from home-base villages in Mexico, and high-quality supervision. The *mayordomos* working for apple producers who were responsible for both recruitment and supervision relied on migrant-sending villages in Michoacan, Zacatecas, or another core-sending area of Mexico to meet labor demand. Because the linkages within these transnational communities are so powerful, both Mexican American and traditional Mexican *mayordomos* accepted a wide range of responsibilities relating to the welfare of their workers. The workers were actual or fictive kin and violation of network mores would decrease their effectiveness as recruiters.

Housing, Migration, and Supervision

Southwest Michigan and the Yakima Valley of central Washington represent two sharply contrasting patterns in terms of housing, migration, and supervision arrangement, although both are upstream labor demand areas relying heavily on migrants to meet peak labor needs. Our observations of Michigan and Washington have given us an excellent opportunity to examine different pathways by which migrant networks mature, due to the fact that both areas were important upstream labor-demand areas in the migrant circuit of post-bracero family farm workers in the lower Rio Grande Valley. After the viability of follow-the-crop migration peaked in the early to mid-

1960s, former migrants began to settle in local areas that were once part of their migrant itinerary. As labor demand dropped in key crops and labor markets, such as sugar beets in Colorado, cherries in Michigan, cotton in west Texas, and tomatoes in California, following the crop became an increasingly untenable economic strategy (Runsten and LaVeen 1981; Briody 1985). This was not so much because of low wages but because there were decreasing amounts of available work and employer-provided housing was allowed to decay (Kissam and Griffith 1991; Kissam, Garcia, and Runsten 1991).

In southwest Michigan, most farm worker housing is employer provided and there has been relatively little settling of Mexican migrants in the area. Because there is not a local settled farm labor force, agricultural producers who have outgrown their ability to supervise a work force of 15 to 20 workers must rely on labor contractors to meet their worker recruitment and supervision needs. While more and more employer-owned seasonal housing is being rented to migrants instead of being provided as a "fringe benefit," there is little year-round rental housing and house ownership is virtually impossible. One of the few migrant families that had settled in the town of Hartford, Michigan called our attention to an important reason for little settling: Banks are not willing to make house loans to farm workers and, in actuality, making monthly payments during the winter months is extremely difficult, since the winter trough of no agricultural work is even more daunting than the harsh climate. The workers we are aware of who have successfully settled in the area are those involved in nursery production, which is less seasonal than the major crops. The result is that there is a restricted labor pool from which to recruit in-house supervisors. The alternative is to rely on core workers, who are themselves migrants, to return regularly, having helped the employer to replace the labor force via network recruitment in their home base; to rely on crew leaders, who are not officially recruiters but who informally recruit and formally supervise workers; or to rely on farm labor contractors, who are officially responsible for recruiting, hiring, and supervising workers.

In Michigan, traditional migration patterns, housing arrangements, and labor force supervision have changed relatively little from the 1960s to the 1990s. The most innovative changes in labor market dynamics stem from the successful efforts of pickle producers to lengthen their growing season by producing pickles in the southern U.S. and to establish a "migrant itinerary" to extend the work season of a core of experienced and favored workers. Successful strategies for decreasing worker turnover and concomitantly reinforcing the "standing waves" of migration patterns have included the provision of improved housing for peak-season migrant workers, reliance on complementary cropping to maintain a relatively steady flow of work and assure that migrants will not leave in search of better opportunities, and structured arrangements to pool labor demand and labor by "lending" workers to neighbors. The "transplantation" of networks of green card-era Texans to Florida, at the same time that traditional Texas *troqueros* were evolving into modern farm labor contractors, has made possible southwest Michigan's continued access to ongoing flows of new immigrants to replace departing workers.

In contrast to the southwest Michigan situation, settlement of Mexican and Texas-based migrants in the Yakima Valley burgeoned in the 1970s. The Yakima Valley has a relatively long work season due to early season demand for pruning labor, the early season asparagus crop, and a late season apple harvest. Also, as settlement began to be a significant factor, local employers abandoned their privately operated labor camps, in part due to regulatory pressures, in part because they shifted away from hiring Texas-based migrant families, thereby accelerating a settlement process that was already underway. Ironically, employers' refusal to continue to provide housing for migrant workers played a key role in facilitating Mexican settlement in the Yakima Valley, because it made it possible for settled migrants to supplement their shaky income for housing payments with rental income from arriving migrants who would share housing with them. The inadvertent privatization of the farm labor housing market in the Yakima Valley has played an important role in establishing viable Mexican communities in the valley.

In this labor market there is extensive reliance on in-house *mayordomos* for recruitment and supervision. The settled population provides an important pool for recruitment of harvest workers but, even more importantly, for semiskilled workers needed for off-season operations such as pruning and packing-house work, and for supervisory personnel. At the same time, the Mexican settlements in the Yakima Valley provide opportunities for international migrants to find low-cost shelter in crowded, shared housing. The availability of family and village networks and "artificial support networks" in local communities obviates the need to rely on farm labor contractors to provide such services. The participation of younger, locally raised Mexican or Mexican American supervisors in the farm labor market is an important factor in the functioning of the central Washington labor market. On the one hand, these supervisors can rely on their own (or their parents') hometown villages in Mexico for a supply of farm labor and can communicate effectively in supervising them because they speak Spanish and function within the social system of networks. On the other hand, their relatively high levels of education and ability to speak English make it possible to integrate them into the overall operations of large and rapidly growing corporate agricultural operations.

These differing patterns of adaption to the constant problem of farm labor recruitment can be expected to drive the ongoing evolution of labor management interactions in the post-IRCA era. In contrast to models of labor management interaction that relate the quality of working conditions simply to the balance of labor supply and demand, our comparative analysis of central Washington and southwest Michigan indicates that working conditions are likely to remain better in labor markets in which labor is supplied by traditional transnational migration networks than in labor markets relying more on semimodernized labor managers: farm labor contractors. The many labor abuses of farm workers reported regularly by newspapers, legal service providers, and other observers are not the result of Mexico-U.S. migration and labor surpluses, per se, but rather of the dynamic of labor brokers' current roles. With neither the constraints of traditional mutual obligations under *compadrazgo* systems nor the accountability of the genuinely modern labor manager, farm labor contractors cannot be expected to respond either to the interests of their employees or the interests of the agricultural producers who hire them.

Work-place Organization and Productivity

A key measure of quality work-place organization — from the perspectives of farm workers, labor market intermediaries, and producers alike — is effective coordination to smooth out fluctuations in the match between labor supply and demand. While some aspects of harvest coordination are decided at high levels of management in large firms, in crops such as apples important decisions are also made by field supervisors and site managers working on an hourly basis. Critical problem-solving tasks relate to commitments made to workers about temporary layoffs, reassignment to alternative tasks when the harvest in a key crop slows down, and negotiations about when "the end of the season" occurs (triggering bonuses). Coordination can also include arrangements for longer-than-average hours during peaks of labor demand. Retaining workers during periods of low labor demand is a chronic problem — addressed in part by the provision of free housing in southwest Michigan, labor-exchange arrangements among neighbors in southwest Michigan, and assignment to nonharvest tasks in both Michigan and Washington.

In both Michigan and Washington, workers at establishments judged to be well managed had higher weekly and annual earnings than workers at poorly managed establishments, because work was steadier. Among multicrop producers in southwest Michigan, the provision of free housing and coordination with neighbors to arrange work for workers idled from one crop (say, strawberries) until the next crop (say, pickles) was ready to harvest was a key element in attracting and retaining a stable and productive work force. A common element in the strategies described by producers in both Washington and Michigan to improve labor management included efforts to downsize the harvest labor force and to restructure the organization of work to the extent possible to provide an extended season of work for core workers. Family-oriented employment plans were also a key element of employer strategies, since worker assessment of work-place quality influences the availability of packing-house work done by wives and teenage children in the family.

Labor stabilization by reliance on complementary cropping, task reassignment, careful coordination of harvesting among differently maturing blocks of trees, and work-place restructuring such as downsizing the labor force used for pruning and thinning provides one of the most clear-cut "win-win" situations in contemporary labor-intensive agriculture — in part because the economic benefits accruing to farm workers from steady employment can (and must) be assessed in relation to the opportunity costs of underemployment or temporary unemployment, as well as in terms of payment received per unit of work (hourly wage versus piece-rate payment). The real cost and worker-perceived risks associated with underemployment also take into account the problem of cash-flow crises experienced by these (or any other) very low-income families. Because farm workers have very poor access to credit, fluctuations in cash flow reap an especially high toll. Because peak-season temporary unemployment is not insured by the unemployment insurance system, the "real cost" of peak season unemployment is higher than that of off-season unemployment when living costs are lower in a home base and when opportunities for employment in the informal economy are more easily available.

The Washington apple harvest provides a good example of how weekly earnings (the key measure of farm worker economic welfare) can diverge significantly, due to overall work-place productivity (including crop yield, efficiency with which tasks are assigned, arrangements for picking up boxes), as well as from the piece rate paid. Table 7.1 shows the relationship between piece rates, hours worked, hourly earnings, and weekly earnings for our sample of Washington apple producers.

Table 7.1. Variations in Yakima County Apple Workers' Weekly Earnings by Employer, 1990

Employer Site	Prior Week Mean Earnings	Prior Week Mean Hours	Prior Week Mean Hourly Earnings	Piece Rate
Employer A	$362.23	34.4 hrs.	$10.53	$13.00/box
Employer B	$144.63	21.3 hrs.	$6.79	$9.00/box
Employer C	$494.48	53.4 hrs.	$9.26	$10.25/box
Employer D	$205.80	39.5 hrs.	$5.21	$10.50/box
Employer F	$332.11	52.3 hrs.	$6.35	$10.00/box
Employer G	$266.95	38.3 hrs.	$6.97	$11.00/box
Employer H	$328.68	36.0 hrs.	$9.13	$9.00/box
Employer I	$189.73	22.4 hrs.	$8.47	$9.00/box

Source: Kissam, Garcia, and Runsten 1993.

As Table 7.1 shows, the number of hours of work per week available per worker based on orchard productivity plays as large a role in determining workers' earnings as do the piece rates offered. While piece rates in theory adjust for variations in productivity of individual units (and are sometimes negotiated on this basis, as in the case of differential piece rates for pruning), this adjustment is only partial and represents just one aspect of an employment offer package. In terms of worker decision making, expected weekly earnings — a function of orchard productivity and work-place organization — have, for experienced farm workers, a weight equal to or greater than that of the actual piece rate offered.

Employer C, whose workers averaged almost $500 per week during the peak of the apple season, is considered to be one of the most successful producers in the Yakima Valley. The midlevel piece rate, $10.25 per box, paid by Employer C yields a high hourly wage — $9.26 per hour, as well as high weekly and seasonal earnings. This stems from Employer C's commitment to downsizing harvest crews as a means of stabilizing his labor force. By utilizing fewer harvest workers, Employer C provides more work for each individual worker, increasing that worker's earnings during the harvest season and strengthening his or her attachment to him. The other key component in this win-win strategy is that Employer C employs a high proportion of his harvest crews as core workers in a year-round program of intensive orchard management, giving them higher annual earnings. This program also increases per acre yields, which increases worker earnings when harvesting on piece rates. In contrast, Employer D's workers, getting a per box piece rate that is 25 cents higher, are earning less. Employer C's intensive orchard management strategy increases off-season labor demand and serves to further stabilize the labor force. Although Employer C's labor-intensive orchard management strategy increases total per acre production costs, it is likely that these cost increases are offset fully or partially by the improved efficiency in the use of the land.

Overall, the patterns of piece rates, weekly earnings, and hourly earnings in the apple harvest provide impressive evidence that with the expectation of improved yields, medium-sized producers with 100 to 499 acres of orchard in production (such as employers C and H), who undertake especially intensive programs of orchard management, can control their harvest costs while providing workers with better-than-average earnings on an hourly, weekly, and annual basis. The larger producers in our Washington sample — those operations with 500 or more acres in production (employers B, G, and I) — appear to lose some efficiency as size increases, resulting in slightly lower hourly wages for their workers. However, some of the managers of larger operations are emphasizing a variety of staff training programs for orchard managers and harvest supervisors, which they hope will increase both production yields and the efficiency with which the harvest labor force is used.

Where the labor market adaptions of Michigan and Washington producers and differing migration patterns give rise to different dynamics, the key factor appears to be supervision. For labor intermediaries such as *troqueros*, crew leaders, and farm labor contractors in southwest Michigan, the opportunities and incentives to upgrade supervision, improving the quality of work-place organization and overall productivity, are few. Crew leaders and labor contractors who have easy access to the depersonalized labor markets in labor-surplus immigrant-receiving areas of south Florida and central California have little motivation to increase the quality of the supervision they provide. They speak Spanish, understand the social processes of the labor market, and can replace departing workers easily with newly arrived workers. In contrast, the settled labor market intermediaries of the Yakima Valley must pay attention to the quality of supervision in order to continue to attract workers. Their adaption to network-based recruitment and the benefits in terms of reduced turnover, greater work-force flexibility, and overall productivity stemming from these supervisors' traditional relationships with workers outweigh their increased accountability to provide workers with stable employment and a modicum of personal support.

Impacts of IRCA

Contrasting the cases of central Washington and southwest Michigan provides insight into the degree to which the impacts of IRCA have been context-sensitive — modulated by the distinctive circumstances of each local and regional labor market.

The major impacts of IRCA in southwest Michigan have, in many respects, been similar to the impacts in other leading areas of labor-intensive fruit, vegetable, and horticultural production in the United States. These impacts have been: increased migration flows, a shift toward reliance on unaccompanied male transnational migrants, and decreasing worker earnings as larger numbers of workers competed for the same amount of work. However, our research in southwest Michigan shows that the mechanisms by which these changes have taken place are more complex than has been generally recognized.

IRCA's impacts in southwest Michigan stem mostly from transformations in migration patterns to south Florida, labor market functioning in south Florida, and employers' secondary adaptions to these changes. By spurring migration to south Florida at the same time that agricultural employers' apprehension about adequate labor supply was increasing, IRCA strengthened southwest Michigan's reliance on Florida-based farm labor contractors and weakened its reliance on Texas-based family migrants. IRCA also effected indirect but long-lasting changes in southwest Michigan labor practices by making it advantageous to employ unaccompanied male migrants recruited by employment packages combining work in Michigan and Florida pickles and who were relatively indifferent to upstream housing conditions, as opposed to Texas family migrants whose concerns included the quality and availability of employer-provided housing. These experienced, long-term family workers were also less motivated by the availability of winter- and spring-season work in Florida — a year-round migrant itinerary — facilitated by a major southwest Michigan employer retaining a labor force dominated by recent immigrants, who had difficulty finding work without the help of a labor contractor or special arrangements for employment in complementary crop activities.

IRCA destabilized the labor market in many regions of the United States by spurring migration in the period before the legalization window closed and by increasing cyclical migration in the long run (in part because Special Agricultural Workers [SAWs] could, with legal documentation, more easily and cheaply cross the border each year to return home to their villages of origin). However, in central Washington, IRCA's impact has been to stabilize the farm labor market. This stems from the special circumstances of the Yakima Valley — mature migrant networks, high rates of worker settlement, a strong and growing high-wage agricultural production base, a relatively long season of labor demand in apples, and the availability of packinghouse work for farm worker family members. Because the Yakima Valley is a popular settlement area and worker recruitment relies extensively on extended family and village networks and little on farm labor contractors, the SAW provisions of IRCA have strengthened existing stable arrangements based on family and village networks rather than disrupting them. The international family and village-based networks, by providing moderate levels of replenishment workers without generating large labor surpluses, as in California or Florida, have decreased worker turnover.

There is a structural commonality underlying the regional variations in the impact of IRCA. Changes in the legal context of migration versus settlement, which have given rise to changes in the costs of migration and the demographic and social composition of the farm labor force, are modulated by the structural characteristics of the local labor market. These characteristics include factors such as the availability and quality of housing, the historical reliance on labor contractors versus network recruitment, the seasonal profile of labor demand, and the processes of specific migration networks dominating the area. The prevalence of family recruitment and the predominance of one regional network — the Michoacanos — in the Yakima Valley has "controlled" migration by fostering patterns in which most (but not all) arriving migrants are "invited" by relatives to come north to work in a "guaranteed" job. This pattern of international migration contrasts sharply with mass migration to downstream labor markets such as south Florida and parts of California, famous for their strong off-season labor demand, where there is extensive reliance on farm labor contractors, and where well-developed artificial networks facilitate the short-term employment of even workers with few connections, making it possible for large surpluses of severely underemployed workers to accumulate.

In summary, the impacts of IRCA have varied not only on upstream and downstream farm labor markets, but also in relation to other key structural features of the farm labor market, including those affecting the dynamics of labor recruitment and supply (extended family and village networks, prevalence and practices of labor market intermediaries) and labor demand (seasonal profile of labor need, availability of complementary employment for family members, wages, availability of work, and availability of housing).

Conclusions

The ability to distinguish the structural characteristics of a farm labor market within an integrated analytic framework is important both methodologically and in practical terms for farm labor policy analysis. Recent studies of the post-IRCA farm labor market sponsored by the CAW, the Department of Labor, and California's Employment Development Department provide a wealth of descriptive detail on the current farm labor market. What must now be done is to gain improved understanding of the dynamic processes that determine the social ecology of the farm labor market — the syntax of interrelationships between workers, intermediaries, and their employers.

Structured observations of verbal interactions among coworkers, workers and their supervisors, and actual behavioral routines in the work place must be linked together in an explanatory framework that allows us to understand how and why the work place has evolved into its present configuration. This sort of ethnographic analysis is critical as a basis for developing innovative policy responses to the challenges of farm labor and immigration policy that confront us in the post-IRCA era. It is also critical for developing innovative and effective strategies to structure the supervision and organization of work in large production units to maintain work quality and maximize productivity. Such efforts in job restructuring, provisions for quality supervision, and more smoothly flowing movement from one crop task to another will need to incorporate some facets of small farm network recruitment and supervision but, at the same time, recognize that this traditional system cannot simply be transplanted to larger production units.

Traditional methods of informal network-based recruitment and supervision of workers, despite their effectiveness, are not easily extended beyond medium-sized farming operations. This implies that as agricultural production in the United States is concentrated into fewer, larger production units, the agricultural work place will inevitably experience a transition from an informal structuring of the work place and production operations into a labor system of more formally articulated interrelationships. The fundamental changes brought about by this structural transition in work-place organization, as the small family farm gives way to the large "mega-factory in the field," will seriously threaten the labor-intensive sector of agriculture unless careful and explicit attention is given to innovative methods for structuring work activities, deploying labor supply in response to constantly shifting labor demand, and assuring the effective management and supervision of agricultural operations.

Overreliance on analyses of macrolevel data, the mainstay of economic policy formation, provides inadequate guidance in developing truly effective policy responses to problems such as immigration, employment policy, and education policy (Kissam 1993; Kissam 1991). In examining immigration policy, for example, careful attention to the work of researchers such as Massey, Mines, Alarcon, and Kearney would very likely have predicted the failure of IRCA's efforts to control migration from Mexico, a "discovery" that was only recognized after passage of a major piece of legislation that is now almost unanimously recognized to have been ineffective. In developing policy for educational services and employment training strategies for immigrants there has been a similar lack of attention to existing research on learning theory (Wrigley and Guth 1992) or actual skills demand (Carnevale, Gainer, and Meltzer 1989; USDOL, SCANS 1992).

A parallel reliance on intuitive approaches to labor management provides the private sector with only a shaky foundation for improvements in work-place organization and for increasing farm labor force productivity. Fixation on the image of the family farm has obscured recognition by agricultural producers that agricultural production units are really business organizations and that those organizations require investments in organizational development in order to maximize productivity. In this regard, agricultural advocates who have so commonly evoked the romantic goal of "saving America's family farmers" may have made it virtually impossible for many producers in the labor-intensive sector to compete in a global produce market. Several small producers we interviewed in southwest Michigan bitterly observed that easy access to government-subsidized farm credit allowed marginal producers to continue in business, lowering prices and decreasing the profitability of their production sector in general.

As a case in point in examining the empirical foundations of public-sector policy, it is useful to look at the relation between unauthorized immigration and labor market characteristics. The labor forces in both of our CAW case-study areas have very similar proportions of post-IRCA arrivals — 17 percent of the total labor force in central Washington and 18 percent in southwest Michigan (Kissam and Garcia 1993; Kissam, Garcia, and Runsten 1993). While apparently similar at the macrolevel of analysis, the two labor markets differ significantly at a microlevel of analysis — the work place — and with respect to migration history. Labor force turnover is much higher in southwest Michigan (where 53 percent of the labor force are working with the current employer for the first time) than in central Washington (where only 26 percent of the labor force is working for the first time with the current employer). The proposition that uncontrolled migration is the main determinant of inferior wages and working conditions is not viable unless it is modified to take into account the unique dynamics of each labor market. These structural characteristics — prevailing labor recruitment strategies, labor demand profile, practices concerning labor supervision, and housing arrangements — are key factors in differentiating otherwise similar labor markets. Driving all adaptions are the employer-worker game strategies to minimize the risk of a shortage of work (on the worker side) or minimize the risk of a shortage of workers (on the employer side).

In a practical vein, our observations of the central Washington apple industry provide strong evidence that at least some sectors of labor-intensive agriculture can address labor management issues in a systematic and effective way within a strategic planning framework. Clearly, Washington's strong position for apples in a national and global market has been an important element in corporate decisions to address issues of work-place organization in a systematic way through staff development efforts, incentives and career ladders for supervisory personnel, and efforts to decrease worker turnover. Yet our Michigan study shows that quite large and profitable corporate producers can be oblivious to or inept in addressing the difficulties of rapidly growing production. It is widely believed that there are no incentives for agricultural producers to improve their labor recruitment and management practices in the post-IRCA period of access to virtually unlimited supplies of labor. Our research in the CAW case studies shows, instead, that the incentives for better management, while weak, do exist, even if they may not be recognized or appreciated by many producers.

The absence of much observable impact from Agricultural Extension Service efforts to improve farm labor supervision practices is particularly significant. Producers' main sources of information regarding labor management issues have been their neighbors or competitors. The lack of Extension Service impacts on employers' beliefs about labor management, supervision, and work-place organization stems partly from the minimal attention given by these university-based programs to strategic planning, organizational development issues, or linkage between investments in work-force improvement and productivity. The outmoded strategy of directing outreach efforts to the minimally educated, unsophisticated small family farmer has resulted in the Extension Service being unable to make much of a contribution to the labor practices of corporate managers, while isolating this technical assistance delivery system from the mainstream of labor-intensive agriculture.

Yet the analysis presented here suggests that a key element in the Agricultural Extension Service's inability to play an effective role in leveraging positive change in farm labor management is also inadequate appreciation of the extent to which farm labor market processes and work-place interactions are governed by Mexican labor market intermediaries, an audience that has not been targeted by any outreach efforts. Farm labor contractors, unlike corporate managers but like old-fashioned family farmers, often combine extensive practical experience with minimal educational background, making them an ideal audience for educational outreach. Yet, even in this realm, it may be that two-year community colleges are more institutionally suited to the marketing challenge than the state college system.

Agricultural managers — both FLCs and corporate supervisors — are intellectually isolated with respect to analyzing and developing new labor relations strategies. This unfortunate isolation seems to be related in part to the fact that agricultural producers and farm worker advocates have, in the course of a long history of adversarial dueling on the battlefield of labor management relations, created an analytic smokescreen regarding the realities of the agricultural work place. This smokescreen obscures a fundamental observation about agricultural labor relations — the degree to which the agricultural work place is, in fact, an environment rich in interpersonal interaction, complex in structure, and evolved to address, if not successfully resolve, a number of formal problems (for example, matching labor demand and supply).

Although agriculture, like many natural resource industries, has staunchly refused to address labor management issues proactively, this does not mean that union organizers' equally short-sighted vision of the ideal agricultural work place provides a much better blueprint for the future. Effective resolution of management-labor conflict requires careful attention to innovative strategies for maximizing workers' earnings and employment security, and for improving working conditions on the basis of increased productivity — developing win-win scenarios for workers and their employers. Effective approaches will require increasing attention to producers' and workers' respective strategies for managing economic risk in the face of uncertainties stemming from fluctuation in labor demand. Effective strategies for change will require recognition that agricultural producers' reliance on labor contractors or their functional equivalents is part of an inevitable trend away from monolithic production structures and toward "articulated" networks assembled to make up "virtual corporations."

Clearly, the United States will not be able to compete with Mexico and other third-world countries in a global market for agricultural commodities if the proper goal of agricultural policy appears to consist simply of assuring a surplus of low-wage labor and our strategic planning is based on competition for cost containment. As noted in a recent series of research studies on the relative competitiveness of California and Mexican agricultural producers, infrastructure plays a key role in success ("Special Report: International Trade" 1991). What is needed now is to expand agriculture's concept of infrastructure to include, in addition to pumping facilities, irrigation canals, and marketing mechanisms, a serious strategy for investing in human capital.

Recasting farm labor policy issues so that they are more closely linked to mainstream issues of industrial productivity and industrial policy in a global economy is an important element in a well-conceived move to remove "agricultural exceptionalism" from the landscape of U.S. government policy. Federal, state, and local efforts to address farm labor policy issues will benefit significantly from increased emphasis on substantial restructuring of the dynamics of the agricultural work place, much more so than efforts to maintain the precarious balance of the status quo — official but ineffective control of immigration coupled with official but ineffective efforts to maintain minimum labor standards, and continued indirect subsidies to agriculture by tax-supported services that provide too little help for too few farm workers and their families.

NOTES

1. See Mines, Gabbard, and Samardick (1993) for macrolevel data on pioneering into new receiving areas.

2. IRCA's impacts in this regard stem, in part, from the fact that southwest Michigan's stock of migrant housing is aging and has not been replaced as rapidly as necessary to accommodate the peak-season labor force.

3. Recruitment in mature family-based networks tends to be conservative, in that it most commonly encourages migration when there is an actual job opening to be filled, and sometimes discourages potential migrants from coming by conveying information to the sending village that it is difficult to find work or that there is a large surplus of workers. In contrast, the information flow in newer networks is worse, and it is more likely for potential migrants to optimistically travel in search of work in a labor demand area, even if there is already a surplus of workers. For further details, refer to Kissam and Griffith 1991.

4. Different market niches induce producers to set very different standards for quality work. Washington producers mostly agreed that pruning and thinning were key tasks in determining produce quality and establishment profitability in a market in which there are very large price differentials based on produce quality.

8

Starting a Migrant Stream in Kentucky

GIL ROSENBERG

Introduction

When the first Kentucky tobacco farmer hired Latino migrant farm workers, he did it in order to harvest his crop, not to start a new migration. However, since 1988, an increasing number of Latino/Latina migrant farm workers (LMFs) have been filling harvest jobs previously occupied exclusively by a more "American" labor force. The population of LMFs has increased from a marginal presence in the fruit and vegetable and tobacco work forces in the mid-1980s to around 6,000 tobacco workers alone in 1993.

Kentucky is one of the few states to see the composition of its work force change since the passage of the Immigration Reform and Control Act of 1986 (IRCA). The impact of IRCA is found in its (1) direct shaping of the harvest labor force, (2) influencing attitudes about the status of this labor force, and (3) triggering effects on the enforcement of other labor regulations, such as the Migrant and Seasonal Agricultural Workers Protection Act (MSPA) and the Occupational Safety and Health Administration (OSHA).

IRCA's effects can also be measured in efforts to avoid or ignore it. In Kentucky this avoidance is symbolized by farmers' refusals to complete I-9 forms and in attempts to frame LMFs in economic instead of legalistic terms. To understand how IRCA can have such varied effects, it is important to see how different interest groups have framed issues such as labor shortages, reliability, legality of residence, and empowerment to meet their needs. This study attempts to develop a context-sensitive understanding of the farm labor market in Kentucky. To do this I examine the key transitional issues of harvest labor shortages and the implications of what has happened and what may follow in Kentucky's use of hired labor.

This study was shaped by three research questions:

1. What role did IRCA play in the introduction and employ-
    ment of Latino/Latina migrant farm workers?

2. How did IRCA and other factors facilitate or hinder the ac-
    ceptance of LMFs?

3. How did IRCA change the structure and functioning of the
    labor market?

The study used several sources of data to address these questions. Information was gathered from interviews conducted over the
course of two growing seasons (1989–90) with farmers, farm workers, farm labor contractors, and government and industry representatives.

More than 92 persons were contacted initially, most of them located in the central region of Kentucky (the tobacco belt). Key informants were recontacted as conditions changed. Taped interviews were
done with a small sample of the key informants (N=15).

Other sources of information were integrated into the overall
analysis. These included:

1. Mail survey responses by farmers to labor attitude questions
    recorded during a previous stage of the fruit and vegetable
    study (N=340; see Thigpen and Coughenour 1988 for meth-
    odology).

2. Data from two statewide telephone surveys of public atti-
    tudes toward migrants.[1]

3. Impressions from my participant observation as a harvest
    worker and member of the Cooperative Extension Service
    Farm Labor Committee.

The Tobacco Harvest Labor Force[2]

Kentucky is a state characterized by small farms. It ranks fourth
in the number of farms — 91,000 — which average 152 acres and
$22,450 each in sales. Two-thirds of Kentucky farms rely on tobacco
as their main cash crop. Tobacco is primarily raised on small plots
(averaging 3.25 acres). The farm price is $1.80 per pound, so that gross
sales of over $4000 per acre gave the state tobacco sales of $863 million in 1992. Federal quotas allocate production at the farm level. In a
time when tobacco production is being threatened from a variety of

directions, it is ironic that another threatened crop (marijuana) is the second leading cash crop in the state.

The hired and seasonal labor market in Kentucky reflects the dominance of the tobacco harvest. Seasonal labor provides 134 hours per acre for the harvest of tobacco and 40 percent of total variable costs. Although Kentucky has been ranked second nationally in the number of farm jobs reported by farmers — over 200,000 in 1982 — many of these jobs last only a short time, so the state is only twenty-eighth in hired labor expenses — $180 million in 1992. Wages are low: The tobacco harvest can be paid as piece work or by the hour; hourly earnings average $5.50 an hour. Women do more of the less strenuous postharvest work of stripping the leaves off the stalk, which can pay $2.00 per hour less than harvest work.

The post-IRCA employment of LMFs in Kentucky represents a significant shift in the harvest labor force, but Latinos/Latinas are not the state's first migrants. In the past it was estimated that 10 percent of farm workers were migrants, with established supply streams coming from areas within the state such as impoverished eastern Kentucky, which has less tobacco production. Many return migrants come from surrounding states, with many farmers also relying on both in- and out-of-state relatives for help. While African Americans have traditionally had an impact on the harvest (especially in counties surrounding urban centers), LMFs are seen as newcomers. The shift toward employing LMFs in tobacco began in 1988, rose quickly to 4,300 workers in 1992, and was estimated to be 6,000 workers in 1993. These workers reflect a change in supply sources, recruiting systems, housing demand, worker sponsorship, community relations, and work structure.

Harvest Labor Shortages

A major analytical focus for the study of farm labor force transitions concerns claims of harvest labor shortages. Harvest labor shortages are hard to document because of the seasonality of the work, the marginality of the labor force, the variety of labor and organizational strategies of farmers, the lack of a mandate for government documentation, and the difficulty in knowing producers' labor needs from year to year. These same factors help to determine the supply and demand for agricultural labor as they interact within the agricultural labor market. A labor market involves: (1) a myriad of employer and prospective employee contracts and wage bargains; (2) a system of networks of employer-employer, employee-employee, and employer-employee relationships; and (3) ideologies that rationalize and justify conditions for the exchange of labor power for wage income.

Because of IRCA, the government has a specific interest in determining harvest labor shortages. The government's techniques to certify these shortages are questionable, because the determination of a labor shortage is made long before domestic workers are interested in the jobs or are even in the area. The decision is based on information from agency officials who ignore the existing informal methods of recruitment. They rely on state employment services or agencies to certify labor shortages, yet in a study on the effects of IRCA, Heppel and Amendola (1991, p. 11) found that these agencies are rarely utilized by either growers or workers. Besides the inefficiency and inability to assemble sufficient numbers of workers, growers feel that workers going through an employment service are generally lazy and not worth hiring. With the prevalence of informal recruitment methods, growers contend there is no reason workers should go through a State Employment Service unless they simply lack initiative. Workers, of course, recognize that the Employment Service is not where recruitment is done.

Because of such governmental oversights, Heppel and Amendola claim that the unintended consequences of IRCA have eclipsed the anticipated effects. These unintended consequences serve as a preview for what is happening in Kentucky. In eight labor regions across the United States these researchers found no labor shortages, even in areas being certified for H-2A workers. Also unmet was the goal of a widespread stabilization and rationalization of the farm labor market (Heppel and Amendola 1991). The saga of the undocumented worker continues as, counter to the intent of IRCA, growers have avoided sanctions against hiring these workers by using a loophole in the law that allows them to accept false documentation. With a fraudulent document industry in place, it is easier for illegal aliens to enter agriculture in the post-IRCA era than before (Heppel and Amendola 1991). While these types of illegal documents have appeared in Kentucky, at this stage their sales and production still only occur in the more dense labor supply regions such as Florida, Texas, and Mexico.

Although another goal of IRCA was to reduce the number of farm labor contractors (FLCs) or middlemen, the increased paperwork and regulation have produced the opposite result. In Kentucky, FLCs dominate the labor market; they capitalize on geographical and cultural differences between farmers and workers and leverage their superior knowledge of labor regulations and networks. In many ways it seems that it is "business as usual" since IRCA. Heppel and Amendola (1991, p. 2) summarize their study by concluding that "it is incorrect to assume that IRCA hasn't happened yet to agriculture, [and] it is safe to say that neither the anticipated benefits nor the feared repercussions have yet to occur." While Heppel and Amendola make these generalizations about IRCA, they stress that there is enormous regional variation in the ways IRCA is perceived and acted upon. The lack of integration of farm labor markets on a national scale is shown by this contradiction: Certain locations are described as having shortages at a time when there is a national farm worker surplus. This surplus exists because there are both adequate numbers of workers and constraints to their upward mobility. There are more than enough workers because, contrary to IRCA's intention, the harvest labor supply has continued to grow (Heppel and Amendola 1991).

The response to IRCA in Kentucky differs from the other states that have been studied because it has not been part of an existing stream of foreign migrant farm workers. Instead of one foreign migrant population replacing another in Kentucky, it is locals being replaced by foreign migrants. With little or no previous experience, Kentucky farmers are being introduced into the foreign migrant system and a new set of regulations. At the same time there are concerns about how to deal with public opinion against the "migrant invasion."

The first problem that Kentucky farmers face in making this transition to a new employment network is trying to understand how it works. In many ways it is more formal than the old network because it introduces labor contractors, regulations, paperwork, and advance commitments of time and money. These things may seem as foreign as the language and culture of the new workers. In contrast to the local worker situation, where farmers claim that there are not enough workers willing to work, farmers are faced with the dilemma of knowing that there are plenty of willing migrant workers but not knowing what the risks are. The lack of information and the inefficiencies within the migrant recruitment system help to create feelings of vulnerability for both the farmers and the LMFs.

Farmers' Attitudes about Labor

Worker Reliability

The first wave of LMFs came into Kentucky in 1987 to harvest in the state's minor production of fruits and vegetables (F&V). A mail survey (N=340) of farmers' attitudes about labor just previous to the beginning of this transition gives insight into the extent that harvest labor shortages are self-defined. The labor problem in raising F&V is not strictly confined to a numerical lack of workers. The cost of labor and farmers' orientations toward using labor in farming are also critical. Most farmers who are not growing fruits and vegetables indicate that the cost of hired labor constrains their starting F&V production. The cost of hired labor has been a common complaint of farmers, perhaps for generations.

Consequently, farmers believe that their main recourse is to rely on family labor. However, most (53%) believe that there is not sufficient family labor to raise F&V either. The prevailing attitude among farmers generally is that growing F&V is not feasible due to both the lack of reliable hired labor and the shortage of family labor. The bulk of growers (85%) believe that it is hard to find reliable labor. More than half (56%) believe that hiring labor takes the profit out of farming, and 64 percent do not intend to expand their farming operations beyond what they can handle with family labor. Like many farmers, they take the point of view that the costs of land and capital are sunk costs. The cash income made from the farm is primarily earned by one's labor and paying cash for hired labor simply reduces the cash income for the family. The point of view is simplistic but is widely held. One farmer expressed it this way:

"If you pay minimum wage then that's an enormous amount of money . . . and you don't know for sure what you're going to get out of it anyway. . . . If you've done a lot of work through the summer and you come up and you don't think you've got the money you ought to have, well you'll soon get disgusted. . . . Like I said, there's easier ways to kill *yourself*" (A Wayne County tomato grower, March 1989).

Attitudes about hiring migrant labor are related to this cluster of attitudes. Growers who believe that hiring labor takes the profit out of growing F&V and do not intend to expand beyond what they can handle with family labor are not inclined to hire migrants (Table 8.1). By contrast, those who are willing to hire migrants also are among the minority who do *not* believe that hiring labor takes farm profits and do *not* limit farming operations to what can be handled with family labor. Farmers who are willing to hire migrants also tend to believe that finding reliable labor is difficult. Since these farmers participate more in the labor market, their evaluations of the labor market should be more valid. However, this advantage may be negated because farmers using both labor strategies measure reliability against their past personal experiences, which intertwine labor management and recruitment practices with their assessment of the labor market. This theme came through in comments similar to this one:

Table 8.1. Fruit and Vegetable Growers' Attitudes about Labor (N=340)

Item	Strongly Agree, or Agree %	Strongly Disagree, or Disagree %
Easier to work with those you know	73	10
Know workers well for reliability	60	17
Hard to find reliable labor	85	6
Managing labor is hard	46	26
Hired labor takes F&V profits	56	26
Expand only with family labor	64	20
Would hire migrant workers	20	53

Source: Coughenour, Zilverberg, and Hannum 1991s p.90.

"Then in the thirties when I took over the farm, I could get all the kids . . . I could go downtown in the truck and just round up pickers, easiest thing in the world . . . and now these kids work for McDonald's and Arby's and these fast-food places" (A Boone County apple grower, December 1988).

From the farmer's viewpoint, reliability can be divided into labor supply (obtaining) and labor control (retaining) problems. Although they usually mention both problems in the same sentence, obtaining and retaining can be isolated problems. With regard to problems in *obtaining* labor, a pickle company executive complained that *"even the migrants* won't pick the smallest pickles," while a bean grower remarked:

"It's not quite manly enough to pick beans. I don't know. People just don't seem like they're at all interested in that kind of work, whatever the price is. *Money can't buy labor sometimes*" (A Grant County bean grower, December 1988).

On the other hand, there are situations in which the focus is on *retaining* workers, such as in this strawberry grower's predicament:

"What I would do if I needed 15 pople, I would take up to 35, 45, even 50. And most of the time there would be enough of them to get me *through the day with a very nasty headache*" (A Russel County tomato grower, March 1989).

Most F&V growers not only believe that it is easier to work with people that you know (73%), as indeed do most people, but also that the better they know their workers the more reliable the workers will become (60%) (Table 8.1). In other words, most farmers try first to recruit local workers whom they know, with nine out of ten farmers primarily hiring local workers in 1986. They believe that better working relationships can be developed with the local workers, whom they already know to some extent, and that local workers are an important source of additional workers, if needed. These attitudes reflect first-hand experience with local workers, along with fear of the unknown (FLCs, LMFs, contracts, IRCA and other regulations) and resistance to change. Farmers interested in employing LMFs face the predicament of sacrificing both their locally based tests of reliability and systems of authority, while further complicating the overall farming operation.

The authority system for the supply and control of local labor rests on kinship ties and word-of-mouth recruitment. Few farmers have recruited through formal means (such as newspapers or employment services) and even fewer have succeeded. These are often seen as measures of last resort. Whether recruiting in the local pool room, through a cousin, or out of a classroom, it is important that there is a "previous account" of the worker via a social network already tied to the farm. Because of this networking, reliability is often assessed on the basis of group performance. Many farmers are pleased with a kin group that brings a truckload of workers every day, without questioning whether the same workers are returning each day. The traditional payment in cash wages helps these groups remain "faceless." This system is in effect an internal labor market, in that hiring is not based on individual skills and groups without such ties (for example, LMFs) are excluded from the labor market even prior to the employment process (thus avoiding IRCA's antidiscrimination policies). Eventually LMFs have been able to establish ties, and these internal labor markets have been duplicated, with the recruitment emphasis on numbers of workers and not faces or documentation.

Managing Labor

Even though they are working with people they know, a substantial proportion of farmers (46%) believe that managing labor is a difficult task, and those who feel this way are also inclined to think that finding reliable labor is difficult and that labor takes the profits out of growing F&V (Table 8.1). Perhaps, it is more accurate to say that farmers who have difficulty finding reliable labor and believe that hiring labor is unprofitable are predisposed to thinking that managing labor is hard. The relationships among the attitudes about the difficulty of finding reliable labor, hiring migrant labor, of managing labor, and the profitability of labor can be expressed technically in terms of correlation coefficients (Table 8.2). For example, the belief that it is hard to find reliable labor is correlated with the strategy of confining farming operations to what can be handled with family labor (r=.30), with the belief that hiring labor takes the profit out of F&V (r=.24), and with the belief that management of labor is hard (r=.30). The three attitudes — family labor strategy, profitability of hired labor, and difficulty of managing it — can be regarded as the justification or rationale for the belief that finding reliable labor is difficult. Because of this cluster of attitudes farmers can claim that there is a labor shortage, even when there is an excess of workers.

Table 8.2. Correlation Coefficients among F&V Growers' Attitudes toward Labor (N=340)

Attitude	1	2	3	4
Hard to get reliable labor				
Use only family labor	.30			
Labor unprofitable	.24	.41		
Managing labor hard	.30	.15	.24	
Would hire migrants	.22	-.27	-.18	.01[a]

[a]All the coefficients except this one are statically significant: P ≤ .05.
Source: Coughenour, Zilverberg, and Hannum 1991s p.93.

Favorable attitudes toward hired labor also are clearly revealed by the correlations. That is, growers who reject a family labor strategy and who do not believe that it is unprofitable to hire workers are most inclined to hire migrants. (See the negative correlations between readiness to hire migrants and use of family labor and unprofitability of hired labor in Table 8.2.) At first, this willingness to hire migrants has not excluded the hiring of local labor, as well.

The positive correlation between the belief that it is "hard to get reliable labor" and each of the other four attitudes (Table 8.2) indicates support for this belief coming from each of the groups with contrasting beliefs about hiring labor. In other words, those who believe that family labor is the only successful strategy for farmers *and* those who are willing to hire labor but have difficulty in finding workers locally tend to agree that it is hard to get reliable labor. Thus, when a minority of farmers first chose to hire LMFs, a majority of farmers still supported their right to search for this labor based on their common claim for reliability.

The aura of reliability of LMFs is rooted in IRCA, both in terms of labor supply and labor control. In this post-IRCA era there have been enough LMFs (labor supply) to allow Kentucky farmers to select the "reliable" workers among these new migrants. Because of the costs, distance, and time involved in recruiting LMFs, this weeding-out process often occurs between and not within harvest seasons. One common scenario is that a farmer asks a "reliable" or "proven" migrant worker on his or her farm to be an informal labor contractor for the following year, thus increasing labor control through this worker's sponsorship and at the same time eliminating (FLC) fees. Labor control also exists in the balkanization of LMFs within the harvest labor force.

Farmers' Labor Alternatives

Kentucky farmers with a hired labor strategy and a labor scarcity problem have chosen four options (separate or combined): (1) mechanize, (2) reduce acreage or change commodities, (3) decasualize seasonal labor, or (4) employ LMFs. Mechanization exists for some very minor crops (for example, canning tomatoes or chipping potatoes), but it has rarely been used to replace existing labor-intensive operations. This is because the labor supply has not been problematic and economically the technology is not feasible or appropriate for the economy of scale of most operations. In the case of the top cash crop of tobacco, where mechanization would have tremendous impact, the predominant small plots and uneven topography do not fit the size or price of the current prototype.

Another strategy toward labor scarcity is to reduce the acreage of labor-intensive crops or switch these crops to less demanding ones (such as field corn or beef cattle). This may allow the farmer to come closer to a more manageable family labor strategy. Besides facing a possible reduction in income, other obstacles to change include the permanence of systems like orchards or the need to invest in fencing; dependence on allotment programs, such as in tobacco; and consumer constraints on the number of fresh market producers for specific commodities.

Decasualization and the employment of LMFs are options for farmers who want to maintain a hired labor strategy. Included in the first option are the attempts to capture local labor; this includes the decasualization of labor and other work modifications to fit the job to the level of the workers' attachment to the job. Decasualization means converting seasonal work positions into year-round employment, thus securing a work force, stabilizing worker incomes, and increasing opportunities for benefits and advancement. Although this is typically done by combining harvest work with livestock work and general farm maintenance, it can also be done in combination with seasonal off-farm work. A variation on this theme is how some factory workers and school children (where allowed) take tobacco harvest "vacations."

While the labor structure of most farms does not allow for the maintenance of enough full-time workers to cover their harvest needs, there have been other attempts to deal with the variable levels of attachment of local workers. These include efforts to (1) work shifts of housewives in fruits and vegetables during school hours, followed by high school students after school; (2) hire senior citizens, especially around roadside stand operations; (3) recruit workers from areas or groups with great economic need; and (4) use a daily payroll. The farmers who practice these methods see them more as limited acts of desperation than innovation. Yet for the group of farmers that wants to hire labor and refuses to hire LMFs, supply shortages make labor retention the key to survival.

Many farmers who choose the final option of LMFs see it as a choice dictated by lack of other options. In interviews they were concerned about community reaction to their hiring LMFs and about increased levels of regulation and enforcement. Still, many farmers are not in a financial position to reduce the scale of their operations or to risk expansion into the few commodities that are mechanized. For them the move toward LMFs reflects the irreversible trend toward the lower attachment of local workers to farm jobs. It is not in terms of costs, but on the issues of labor supply and reliability that the idea of using LMFs is sold.

LMFs also bring farmers extra responsibilities, regulations, and paperwork. One responsibility is facing concerns of community members about social and economic changes. Because of all these extra obligations most of the interested farmers have taken a "wait-and-see" attitude and let a few "pioneers" bring in the first group of LMFs. Overall, the high level of initial interest about LMFs was indicated by the large turnout of farmers at statewide county Extension meetings focusing on farm labor. At these meetings economic questions predominated over legalistic ones. The leading questiong were: (1) How much does it cost to employ these workers? (2) Can they do tobacco work? and (3) Will they stay the season?

Labor Network Transitions

For the initial transition to occur, farmers' responses expressed a need to be convinced that these unknown and untested LMFs would be affordable, reliable, and adaptable to Kentucky commodity production systems. Before the LMFs arrived, farm labor contractors (FLCs) played a key role in portraying the farmer and the worker to each other. The FLC's attendant myths about the new worker system are used to make the transition seem unproblematic.

The methodology of the FLCs (Rosenberg 1992) connects three interrelated themes of commodification, systemization, and dependency to make this labor transition process seem unproblematic. The first theme was the commodification of LMFs, which is to depersonalize them and then to make them an article of commerce. To this end were found comments about (1) product performance, saying that LMFs were *adaptable, specialized, and efficient*; (2) product movement, by saying that LMFs were *disposable, replaceable, and professionally mobile*; in addition to (3) product characteristics, by saying LMFs were *economic, invisible, and guaranteed*. That this commodification was a conscience strategy came through FLC statements like this one:

"And the farmer can order workers that are bilingual. . . . No extra charge, its kinda like brake lights, it comes with the vehicle" (A western Kentucky FLC, February 1990).

Once LMFs were commodified, they fit the second theme, which is systematization. FLCs presented a description of a homeostatic system that screens, ships, and replaces a virtually homogenized product on the command of the farmer. Computers only help to enhance this image, as one FLC told us:

"We have tons of people on the computer . . . by the job description. And you call down there and you tell them what particular category of person you want . . . all they have to do is . . . ship them off to ya" (A western Kentucy FLC, February 1990).

Commodification and systematization serve the third theme, which is the dependency of LMFs on the farmers. Besides having the farmer as instigator of all transactions, dependency comes through in terms of citizenship, culture (language), economics, and mobility. Before they arrive farmers are reminded by the FLC that the LMFs

"don't have the money sittin' down there in Mexico to buy a bus ticket up to western Kentucky. And they don't have the money to fund themselves. . . . The grower fronts them that money by sending it to the company" (A central Kentucky FLC, March 1990).

Once here, the dependency even extends beyond the farm:

"There's only two or three of them that were taken in to do the shopping for the whole crew. So, there was always the farmer or the farmer's wife with them, and showed them around" (An H-2A association representative, March 1990).

There is also dependency based on institutional systems such as IRCA. For example, labor legislation dictates what farmers must provide for workers, while often putting the farmers in the gatekeeper or sponsorhip position for making or recording these provisions. For example, when farmers are asked if they have H-2A or SAW workers, the standard protective answer is "I have Mexicans."

The second ideological shift that has to take place for the transition to a new labor network is for the community to accept the new work force. With the large amount of seasonal workers employed in Kentucky, there is potential in the long run for a large influx of LMFs into many rural communities and clashes stemming from cultural, racial, and class differences fueled by the LMFs "outsider" status. The public perception of the local farm labor force is central to opinions about the hiring of migrants. The results of a statewide telephone poll, done early in the transition, showed that the public did not identify with farmers on the issues of labor scarcity and reliability. While local workers were considered to be valuable and more productive, migrants were seen as job stealers and a burden to the community. This tied into the belief that migrants would work for lower wages.

Overall, the majority of respondents preferred that farmers not hire migrants, previous to this being a widespread practice. Similiar to the farmers, community members carried attitudes and beliefs that shaped their opinions about the harvest labor markets. The gap in opinions between these two groups illustrates that to a great extent labor shortages are social constructions or issues that groups try to frame (Snow et al. 1986).

It is mainly up to the farmers to establish a social, economic, and political network to defuse community members' objections. The importance of the strength of this networking can be seen when observations about the successful community resistance to migrant vegetable workers in 1987 are compared with the relative ease of acceptance of migrant tobacco workers only two years later. In 1987, an out-of-state produce company brought migrant harvesters into their new Kentucky operation as part of their normal operating procedures. They underestimated the local reaction to this move and found themselves two steps behind in a losing public relations battle over issues such as migrant camp zoning, job loss, and aerial spraying. Since they represented a marginal commodity (vegetables) in the state, there was no last-minute reprieve from the avalanche of attacks found in letters to the editors, lawsuits, public petitions, and local hearings. Two years later, LMFs were brought into Kentucky without any organized protest. This is because the leading edge of the second wave was composed of respected, local producers of the state's top cash crop of tobacco, and the threat of economic loss was believed to be an economic loss to a larger segment of the whole community. This change represents the removal of a major barrier for the occurrence of a farm labor force transition.

Lag Time

Farmer hegemony in advancing ideologies about the new composition of the farm labor force is facilitated by the "lag time" that exists between the time farmers start negotiating for migrants and when their impact is felt off the farm. Even though antitransition ideologies are dominant among the public, the off-farm reaction to migrants is slow and uneven, compounded by a system of compartmentalized and bureaucratized agencies (and regulations). The effect of the "lag time" on institutional response to the migrants' arrival can be seen in the following ways:

1. Regulations exist but enforcement is lacking or uneven. One enforcement official put it this way.

   > "I have 120 counties to cover by myself on these migrants, and I'm supposed to do this on only 10 percent of my time. And now I don't even have an interpreter. . . . You figure it out" (A labor enforcement agent, June 1990).

   Even though IRCA was passed seven years ago, County Cooperative Extension agents report that the I-9 form is not in use on most Kentucky farms (Rosenberg 1992). In response to uneven regulation, the farmer may assume that it is "business as usual" on the farm, only to be "stung" later as enforcement officials respond to the increased awareness of the LMFs' presence. But that awareness is slow to develop, for many reasons, including the following points.

2. Farm worker advocacy groups may be slow to respond or never materialize. Whether or not these are worker-based groups, early awareness of farm-based activities may be lacking. Empowerment is slowed by the low concentration and thus low visibility of LMFs, their citizenship status, and the seasonality of the work. Organizing efforts may also be slowed by the need to mobilize community members with a less vested interest than the farmers.

3. Reinstitutionalization of farmer bias, especially in agencies such as the Cooperative Extension Service and the Farm Bureau, which have historically seen farmers as their clients more than farm workers. In this study, Extension (with some funding from a tobacco company) has used meetings, publications, videos, and satellite television to educate farmers, while none of these early efforts have been translated into Spanish or directed toward the farm workers.

4. Privatization of public information as a hedge toward success for those selling labor. This stems from the "information lag" occurring in the private sector. By making this public information seem private, these firms are able to enhance the image of services that they alone can offer. For example, one labor brochure attempts to privatize publicly available mate-

rial to farmers, such as information on regulations, (especially IRCA), housing inspections, and tax exemptions. Even more interesting is the issue of the privatization of public employees within this same association. The chairman of the board served concurrently as a state representative with strong ties to both tobacco and Extension, and the executive director was recently the top aide to the state commissioner of agriculture. He may have gained information and an overall economic advantage, in that part of this aide's government job was to talk to farmers about migrant farm workers and now many farmers mistakenly consider this to be a government organization. It is no coincidence that this group has become the largest supercontractor in the state or that a state law was passed allowing for a portion of every tobacco allotment to be used to support farm labor recruiting.

With the increase in LMF activity, information and advocacy will become more prevalent, causing some of these lag situations to diminish. Still, those groups on the front end of the lag may derive benefits from this initial advantage in information and activity. This is especially true when these relationships are embedded in bureaucracy. The H-2A program in Kentucky is a classic example. Representing only 1 percent of seasonal farm workers (277 H-2As in 1992), it is a widespread consensus that they were never needed and could easily be replaced (by SAW). This more expensive and bureaucratic program is sold based on sponsorhsip of the H-2A workers by farmers, which means that they are legally restricted to work for that farmer. It is also sold through the sponsorship of farm leaders who have networks to other states where H-2A workers are employed in tobacco (for example, Virginia). Sponsorship is also an important factor on the local level. The counties where H-2A workers are heavily employed coincide with the residences of the board of directors of the H-2A association who also serve as its main recruiters. There are other indicators that the H-2A program in Kentucky is not administered as a direct response to labor shortages or as a "last resort." It seems that the employment of H-2As is a luxury instead of the intended effect of a last resort. They are sold to farmers who don't realize there are other options or to others who see "reliability" in the only classification of workers that are anchored to their farm. To cover the recent trend away from H-2A workers, the same association has developed a privatized advantage for bringing in SAW by exclusively using the Employment Service to recruit workers, allowing them to charge farmers one-third the price charged by other FLCs.

Conclusions

Kentucky farmers and educators went to Atlanta in 1987 to visit various regional bureaucracies involved with LMFs. They left confused. Two years later when they invited these bureaucrats to Lexington, they found out that these people had never met each other. More confusion. In the meantime, farmers within the state sought LMFs using the path of least resistance. A path paved by fast-talking FLCs who have quickly monopolized the market, and surrounded by weak enforcement of regulations. In this labor market, SAW and H-2A workers have been defined to farmers more in economic than in legalistic terms, yet it is the legal aspects of IRCA that continue to frame issues such as labor supply and control, which are then used to define economic differences between sources of workers.

It can be said that Kentucky farmers came late to IRCA or that IRCA came late to Kentucky farmers (or that it is yet to come). These farmers were not invididually involved in the legalization process that verified the SAW who are the predominant category of LMFs in Kentukcy. While this process happened elsewhere, it helped stabilize the "push" of LMFs into new harvest regions like Kentucky. In Kentucky, this push remained dormant until there was a "pull" from Kentucky farmers, based on their perception of the reduction in reliability of the local farm labor force. The result of this process is that worker reliability is being linked to a new work force based on different restrictions to worker mobility, restrictions aided by the intended and unintended consequences of IRCA.

As the 1993 harvest season comes to an end, a new restriction on worker mobility has developed. My recent field trips through Kentucky have found that a surplus of mostly undocumented and uncontracted workers have ended up in urban homeless shelters and abandoned barns. For now, the main problems are (1) more jobs available than housing and (2) a mismatch of worker arrival time and location to harvest needs. Soon there will be many more workers than jobs. The workers not housed by farmers fill the needs of other farmers who cannot afford housing. This also removes a level of worker sponsorship, which increases friction within the communities where LMFs are settling. Meanwhile, the housing issue has some community members asking how to stop the new stream of undocumented workers without threatening the tobacco harvest. One angry citizen proclaimed: "There ought to be a law!" Looking back at IRCA, the best answer to this may be that there is — and yet there isn't.

Table 8.3. Public Opinion about Farm Labor (N=597)

Opinion	1989 (%)	1990 (%)
Enough local workers available for harvesting		
Yes	65	40
No	35	45
(Don't know)	(15)	15
		(4)
Should farmers hire migrant workers if labor is needed		
Yes	39	
No	47	
Uncertain, depends	14	
(Don't know)	(6)	
Would migrant workers be a benefit or a burden to the community		
Benefit	13	
Burden	37	
Makes no difference	50	
(Don't know)	(12)	

Migrant workers take the jobs of local workers

Strongly agree	27
Somewhat agree	33
Somewhat disagree	27
Strongly disagree	13
(Don't know)	(10)

Migrant workers are more productive than local workers

Strongly agree	48
Somewhat agree	27
Somewhat disagree	13
Strongly disagree	12
(Don't know)	(4)

Compare to 1989 the number of migrant workers in Kentucky is

Greater	38
About the same	49
Less	13
(Don't know)	(55)

Source: Rosenberg 1992, p.80.

Table 8.4. Factors Influencing Community Unrest about Migrant Labor (Participant Observation)

Unrest About Migrant Labor	
1987 - Protest Situation	1990 - No Protests
Minor commodity (vegetable)	Major commodity (tobacco)
Out of state firm	Local firms
Larger farms	Medium farms
Quick worker influx concentrated	Slow influx/dispersed
Near metro area	Most in rural counties
Mobile workers	Farmbound workers
No advance public relations	Relations networks in place
Government non-commitment	Government and Extension support

Source: Rosenbery 1992, p.82.

NOTES

1. Complete replies were received from 62 percent of the adults contacted, resulting in a total of 597 (1989) and 661 (1990) respondents. The margin of error for the survey is slightly less than 4 percentage points at the 95 percent confidence level.

2. The data for this summary of trends were compiled from the *Census of Agriculture* for 1982 and 1987 (USCensus 1983 and 1989); *Kentucky Agricultural Statistics, 1991–1992* Kentucky Ag. Stat. Serv. 1992); and tobacco cost and return estimates (Rosenburg 1992). Data on LMFs are based on surveys of Cooperative Extension agents in 1991 and 1992.

9

IRCA and Seasonal Farm Labor in North Carolina

STEVEN G. ALLEN AND DANIEL A. SUMNER

Over the last half century the North Carolina seasonal farm labor force has changed substantially. Two decades ago sharecroppers were replaced by hired workers from the local area, supplemented by migrant labor. In more recent years significant numbers of immigrant workers have become an important source of farm labor supply.

Issues surrounding seasonal agricultural labor have long been a concern to farmers and worker advocates, as well as to national policy makers and policy analysts (Fisher 1952). In North Carolina, flue-cured tobacco producers, the major employers of seasonal farm labor, have had labor concerns for many years. Recent issues of *The Flue-cured Tobacco Farmer* have documented that concern.

In this chapter new evidence is presented on the recent patterns in the major seasonal farm labor markets in the state. The research has provided new data on seasonal farm labor markets and focused analysis on the impact of immigration reform and other factors affecting the farm labor market. We conducted four annual mail surveys of several hundred North Carolina flue-cured tobacco growers with questions related to the 1986 through 1990 crop years. This chapter provides new information about the use of different types of labor; the compensation practices used by growers; and how employment, pay, and agricultural production have changed following the Immigration Reform and Control Act of 1986 (IRCA). We previously reported partial results through 1989 (Allen and Sumner 1991).

In North Carolina the immediate impact of IRCA was to raise concern among both growers and workers over the Seasonal Agricultural Worker (SAW) program, the Replenishment Agricultural Worker (RAW) program, and the H-2A program. At the same time a variety of worker sanitation, housing, and safety regulations were either modified or enforced differently, raising additional anxiety in the labor markets. Thus, the North Carolina seasonal agricultural labor market has faced the similar stress prevalent in other major states. However, there are unique factors, as well, including the major crops grown (tobacco versus fruits and vegetables) and the size and variety of the agricultural industries and labor market. In this chapter, I also explore the reaction in seasonal farm labor markets to regulatory and other pressures.

Methodology

This study is based on four successive mail surveys of growers from 18 eastern piedmont and coastal plain counties that produce well over half of the North Carolina flue-cured tobacco crop and most of the major vegetable crops. The first survey, covering 1986 and 1987, was mailed in January and February 1988. The results examined for 1986 and 1987 are based initially on 597 responses to that survey that were mostly complete.

A second survey, this one pertaining to 1988, was mailed in February 1989 to those who responded to the previous survey, plus about 330 additional names and addresses obtained from the respondents to the first survey. The survey covering 1988 received 408 responses that were complete enough to be used in the analysis. A third survey covering 1989 was mailed in February 1990. The analysis is based on 354 complete or nearly complete responses. In February 1991 we mailed a fourth survey covering 1990, and proceeded as in the previous three years. In this case, 357 usable responses were received from active producers.

In each year the questionnaires were designed to obtain information on a wide range of subjects, including the cross-section/time series patterns of employee and job characteristics, wages, capital-labor substitution, crop mix and acreage, time allocation of farmers and their families, and farm costs and returns. The survey also contained questions to gauge the extent to which growers relied on illegal immigrants as a source of labor supply before the immigration law became effective.

Sample Characteristics

Our surveys have yielded five years of information for repeated cross sections of North Carolina producers in which most of those surveyed responded to more than one survey and a substantial number responded in all five years. There are two basic approaches to analyzing data with these characteristics. The first is simply to treat the data as a set of repeated cross-section samples. The second is to create subsamples of panel data by including in this approach only those who responded to all the surveys or at least more than one. In our data, as in most other cases, the descriptive characteristics of the samples differ with these two approaches. Representative respondent characteristics for each of two samples for each year are in Table 9.1. The first five columns report on all responses in each year that were mostly complete; the second is a matched sample of 146 farms that responded to all four surveys.

Table 9.1 also reports data from the *1987 Census of Agriculture* (USCensus 1989) for the 18 counties in our survey area. Our survey was not designed to represent the average crop farm in the region, but instead to reflect farms that use or may use seasonal labor in tobacco or other crop enterprises. Our sample underrepresents small or part-time farms and those that produce mainly grains and other crops that are not labor intensive relative to the census. Alternatively, the census includes many farms that are not active participants in the labor market.

The respondents to our survey represent a sizable percentage of crop production in the 18-county area under study, accounting for 7 percent of total acreage and more than 10 percent of tobacco acreage in 1987 (the census year). The average farm in our full sample grew 263 acres of crops in 1987 and had a gross revenue of $126,000. It was operated by someone who was 49 years old, had a high school education, and had operated that farm for 21 years. Most of these farms (85 percent) grew tobacco and the average tobacco grower harvested 27 acres. Average farm size increased by 1990 to 362 acres, gross revenue rose to $197,000, and average tobacco area almost doubled. These results are partly attributable to changes in the composition of the sample between the surveys.

Smaller farms were more likely to leave the sample between the first and last surveys. Within the set of respondents who answered all surveys, the percentage of farms with less than $10,000 of output fell from 7 to 4 percent. This decline was much more sizable (18 to 6 percent) when the sample was expanded to include those who did not respond to all of the surveys. The greater exit rate of small farms from the sample follows because smaller farms are simply less likely to respond to another survey. Further, average acreage reported by respondents at the new addresses added to the mailing lists on the second and third surveys was larger than the initial sample average.

Table 9.1. Characteristics of Survey Respondents Compared to Census of Agriculture

	Entire sample					Respondents to all surveys					1987 Census of Agriculture (18 Counties)
	1986	1987	1988	1989	1990	1986	1987	1988	1989	1990	
Number of farms growing crops	597	587	408	354	357	146	146	146	146	146	15,326
Total acres (thousands)	153	154	130	125	129	41	42	44	46	44	2,069
Average acres	256	263	318	353	362	278	290	299	313	305	135
Average output ($ thousands)	120	126	168	183	197	136	147	160	183	154	93
Percentage of farms, by output											
Less than $10,000	18	18	7	7	6	7	6	6	4	4	40
$10,000 - $49,999	31	30	24	24	23	31	29	23	25	27	29
$50,000 - $99,999	20	20	19	21	17	25	25	22	25	21	11
$100,000 or more	32	33	51	48	54	37	40	49	46	48	19
Tobacco farms	502	497	379	329	317	136	136	136	134	132	9,107
Average tobacco acres	25	27	36	41	49	27	29	32	38	39	15
Age of operator	-	49	49	49	50	-	48	47	48	49	51
Experience of operator	-	21	22	23	24	-	22	21	23	23	-
Schooling of operator	-	12	12	12	12	-	12	12	12	12	-

The average farm size of all respondents was smaller than those in the panel data for 1986 — $120,000 in sales compared to $136,000. But by 1990 the average size in the full sample was larger — $197,000 compared to $154,000. So, while there was growth in the panel of continuing farms, from $136,000 to $154,000, it was much less than the growth in size of all farms in each sample. Clearly, small farms were more likely to leave our sample and larger farms were added.

The average size of farms in our sample was considerably larger than in the *1987 Census of Agriculture*. Average acreage was about twice as high and the dollar value of output was 35 percent greater. A likely explanation again is that small farms are less likely to respond to a survey focusing on the farm labor issue. Simple averages across all farms in this sample give more weight to the larger establishments than an average for all North Carolina crop farms. However, large establishments are more likely to use hired labor and should receive greater weight, since the focus of the analysis is on hired farm workers.

The Farm Labor Market

The next several subsections describe the basic patterns we found in the North Carolina seasonal farm labor market for the years 1986–90. Because our focus is on change over time, all results relate to the matched panel data of 146 farms over 5 years.

Workers

Worker characteristics were developed from questions about the individual respondents' farms and about the local county labor market. We first asked respondents to report the fraction of seasonal farm workers in their county that they estimated were immigrants. The results in Table 9.2 show that the percentage was 25 percent in 1986 and 29 percent in 1987. After IRCA became effective, the estimated share of immigrants in the seasonal farm labor market was 37 percent in 1988, 41 percent in 1989, and 44 percent in 1990.

In another question respondents were asked to identify the race or ethnicity of most seasonal workers on their farm. The percentage of growers who reported that most of their seasonal workers were Mexican (or other Spanish speaking) was 19 percent in 1986 and 1987, but jumped to 32 percent in 1988, 35 percent in 1989, and 40 percent in 1990.

This shift toward reporting more immigrant labor was true across our sample, even though the level of immigrant activity varied substantially across counties (Table 9.3). By 1990 more than half of the seasonal farm work force in the relatively urban and industrialized piedmont region was comprised of immigrants. In all our survey years immigrants were a relatively less important source of seasonal farm labor in the counties at the South Carolina border. In these counties the estimated percentage of immigrants was only 13 percent in 1986 and 1987 and 30 percent in 1990. This South Carolina border region had the fastest growth in number of immigrants over the period. There was relatively slow growth in the piedmont region from 1986 to 1990. This could reflect a dampening of the trend or an often observed tendency for measurement error to reflect a regression toward the mean in panel data.

The question of how many immigrants were in the country illegally is much more difficult to gauge. The surveys asked respondents to estimate the fraction of seasonal farm workers in their county that they thought were illegal immigrants. These estimates declined slightly from 20 percent in 1986 to 15 percent in 1990 (Table 9.2). In the first survey we asked, "Before the [immigration reform] law became effective, roughly what fraction of the seasonal workers on your farm do you think may have been illegal aliens?" The survey was mailed out after the passage of the law, so it is not surprising that respondents reported lower figures for themselves (12 percent in 1986) than for their neighbors (20 percent in 1986). We varied the question somewhat in the next survey. Noting first that "special provisions in the immigration reform law allowed farmers to hire illegal aliens through November of last year [1988]," the second survey also asked growers about use of illegal immigrants on their own farm in 1988. This yielded an even smaller estimate of 8 percent relative to 16 percent for the county average.

A final question designed to measure the importance of illegal immigrants as a source of seasonal farm labor in North Carolina dealt with the SAW program. Our survey for 1988 found that 29 percent of the growers knew of a current or former employee who had filed for legal resident status under that program.

All of the estimates presented so far are simple averages across survey respondents. However, as expected, use of immigrants as a share of seasonal hired labor varies by farm size (other labor market variables also show this pattern). Table 9.4 illustrates that growers' estimates of the percentage of immigrants in their county increased with the size of the respondents' farm. The estimates of the percent-

Table 9.2. Use of Immigrants as Seasonal Farm Workers, by Year

	Year				
	1986	1987	1988	1989	1990
Percentage of seasonal farm workers in county thought to be immigrants	25	29	37	41	44
Percentage of growers with most seasonal workers Mexican or Spanish-speaking	19	19	32	35	40
Percentage of seasonal farm workers in county thought to be illegal immigrants	20	19	16	20	15
Percentage of seasonal workers thought to be illegal immigrants on respondent's farm	12	-	8	-	-
Percentage of growers with SAWs as current or former employees	-	-	29	-	-

Table 9.3. Estimates of Percentage Immigrant and Average Hourly Earnings, by Region and Year

Region	1986	1987	Year 1988	1989	1990
Piedmont (Edgecombe, Franklin, Harnett, Nash, Johnston, Wake, Wilson)					
Percent Immigrant	36	41	48	50	52
Hourly Earnings	4.09	4.12	4.22	4.53	4.54
Coastal Plain (Duplin, Greene, Jones, Lenoir, Pitt, Wayne)					
Percent Immigrant	16	20	31	39	41
Hourly Earnings	3.45	3.45	3.66	3.84	4.13
South Carolina Border (Bladen, Columbus, Cumberland, Robeson, Sampson)					
Percent Immigrant	13	14	24	27	30
Hourly Earnings	3.69	3.84	4.02	4.11	4.34

Note: Corresponding tobacco production regions are Middle Belt, Eastern Belt and Border Belt.

Table 9.4. Use of Immigrants as Seasonal Farm Workers, by Farm Size and Year

Gross value of output	Percentage immigrant in county					Percentage of farms where most seasonal workers are Mexican or Spanish-speaking					
	1986	1987	1988	1989	1990	1986	1987	1988	1989	1990
Less than $50,000	21	26	27	34	37	10	11	21	22	29
$50,000 to $99,999	29	29	37	43	44	17	17	24	32	32
$100,000 to $249,999	26	30	41	42	52	27	20	35	36	49
$250,000 or more	29	38	55	54	54	42	47	65	59	54

age of immigrant labor used by farmers with a gross output of $250,000 or more were almost 50 percent larger than the estimates by farms with an output of less than $50,000. In 1987 the estimate was 26 percent for those with sales of less than $50,000 and 38 percent for the large farms. This difference in perception may be attributable to the greater use of immigrants by large farms themselves. In Table 9.4, large farms were more likely than small farms to have a seasonal labor force that was mostly Mexican or Spanish speaking. In 1987, 11 percent of the smaller farms used primarily Mexican workers, compared to 47 percent for the large farms. The use of Mexican seasonal labor expanded in all size categories, but remained higher on farms with sales of more than $250,000.

Note from Table 9.4 that those with smaller farms (less than $50,000 in sales) perceived that the share of labor by immigrant workers in the county was higher than the share of Mexican-origin workers on their own farms. But for the respondents with farms of more than $250,000 in sales, the perception was the same or a lower proportion of immigrants in their county than Mexican workers on their farms.

In summary, based on our sample, immigrants were a large share of this labor force before IRCA; our best estimate is that they accounted for 20 to 30 percent of the labor force in 1987. The use of immigrants varied widely across the major flue-cured tobacco-growing areas, with growers in counties on or near the South Carolina border being much less likely to hire immigrants than growers from the other counties. The use of immigrants grew substantially between 1986 and 1990. In 1990 they accounted for nearly half of the seasonal farm labor force and there continues to be more use of immigrant labor on large farms and in the relatively urban counties in the central part of the state.

Jobs

Seasonal labor was used by almost all growers who responded to our survey. In 1986 and 1987 our survey found about 80 percent of the farms in the matched panel sample of 146 farms were hiring some seasonal labor. For the final three years that share was about 90 percent. By 1990 about 42 percent of the farms in our sample also hired full-time workers or part-time workers who were employed all year, and 8 percent of the farms had full-time managers.

Growers are using a larger variety of recruiting methods to find seasonal workers. The most commonly used method to hire seasonal farm workers is to hire someone who has worked for the grower previously, as shown in Table 9.5. This contradicts the widely held view that seasonal farm labor is a prime example of a casual, impersonal employment relationship. The use of this recruiting method held steady throughout our sample period. In every year the growers in our sample were least likely to use the Employment Security Commission to find workers. The percentage of growers who found seasonal workers through references (by farmers and by employees), labor contractors, and "walk-in" applications all rose by at least 15 percentage points. The percentage who hired workers through the Employment Security Commission also went up but remained relatively uncommon.

Starting in 1987, IRCA has required employers to document the legal residence of all new hires. In that year 13 percent of the growers in our sample reported that they were unable to hire someone because they lacked the proper documents. This percentage increased to 23 percent in 1990.

The tasks done by seasonal farm workers expanded in the years covered by our study. The main jobs done by seasonal workers in each year were harvesting, topping and suckering, and planting (Table 9.5). Fewer farms used seasonal workers for irrigation or applying fertilizer or pesticides, but the share of farms that used seasonal labor for these jobs also increased.

As they did these tasks, farm workers became more likely to have access to sanitation facilities. Nine out of ten growers provided access to drinking water in 1989 and 1990, up from 73 percent in 1986. The fraction of farmers providing other facilities more than doubled, to two-thirds for hand-washing facilities and 43 percent for toilets in the field.

Wages

In all years covered by our survey, by far the most widely used pay method was an hourly rate (Table 9.6). This method was used by about 80 percent of the growers in 1986–88 and increased to 93 percent by 1990 for those who responded in all years of the survey. The next most frequently adopted method was to pay a fixed amount for

Table 9.5. Percentage of Growers Using Various Recruiting Methods, Tasks, and Facilities, for Seasonal Farm Workers, by Year

			Percentage of Growers		
	1986	1987	1988	1989	1990
Recruiting					
Referred by other farmers	17	14	22	36	35
Referred by current or past employees	24	21	39	39	43
Worked for me previously	75	79	77	78	79
Employment Security Commission	3	3	6	8	8
Came to me looking for work	35	35	46	50	57
Through a contractor or crew chief	13	13	23	28	24
Tasks					
Harvesting	95	94	95	94	99
Planting	51	53	53	60	66
Irrigation	12	11	14	9	17
Spraying pesticides or applying fertilizer	5	6	12	12	14
Hoeing	34	35	41	51	47
Topping and suckering	74	72	74	80	86
Sanitation facilities					
Toilets	22		38	42	43
Access to drinking water	75		99	91	92
Handwashing facilities	38		65	70	67

Note: Columns sum to more than 100 because the categories are not mutually exclusive.

Table 9.6. Pay Methods, Earnings, and Benefits for Seasonal Workers, by Year

	1986	1987	1988	1989	1990
Method – Percentage of growers using:					
Piece rate	12	12	17	24	20
Hourly wage	83	82	85	86	93
Weekly or monthly salary	5	5	8	8	7
Fixed amount for a set job (e.g., harvesting a particular field or crop)	25	26	31	38	34
Hourly earnings[a] – Percentage of growers paying:					
Less than $3	6	5	2	1	0
$3 to $3.99	69	67	55	49	31
$4 to $4.99	21	24	33	36	54
$5 to $6.99	4	4	10	9	15
$7 or more	0	0	0	6	0
Estimated average hourly earnings	3.69	3.73	4.03	4.25	4.37
Benefits – Percentage of growers providing:					
Housing	37	36	41	34	43
Meals	16	15	20	18	16

[a] The sample is restricted to growers who reported average hourly earnings in each year.

a set job (such as harvesting a particular field or crop). Piece rates may be a common method for compensating farm workers for other crops or in other parts of the country, but in our North Carolina sample only 12 to 24 percent of growers used this pay method. Very few growers (less than 10 percent in any year) paid weekly or monthly salaries to their seasonal help.

Because many growers do not have records of the rate of pay for their seasonal workers, we did not ask respondents to report a wage rate for their workers. To assess wage levels we included this question: "On average how much do you think seasonal workers on your farm made per hour?" There were five possible categorical responses: less than $3.00, $3.00 to $3.99, $4.00 to $4.99, $5.00 to $6.99, and $7.00 or more. The percentage distribution for these responses, along with our calculation of the average wage, is reported in Table 9.6. The sample was limited to growers who responded to the question for all years.

Wages for seasonal farm labor increased during the period covered by our survey. This is shown most clearly by the reductions in the percentages of growers paying below $3.00 and $3.00 to $3.99 between 1987 and 1990 and the increased percentage of growers paying $4.00 to $4.99 and $5.00 to $6.99. Our estimate of the average wage increased by 6.6 percent from 1987 to 1988, 4.7 percent from 1988 to 1989, and 2.8 percent from 1989 to 1990. The total increase was about 17 percent from 1987 to 1990. In contrast, average wages for manufacturing workers in North Carolina rose by only 3.7 percent from 1987 to 1988 and again from 1988 to 1989, and by 4.3 percent from 1989 to 1990. For the whole period farm wages grew more rapidly, although for the final year manufacturing wages grew by an extra 1.5 percentage points.

These estimates are simple averages across growers. However, wages were higher on larger farms that hired more workers. Because average farm size in the sample grew during this period, the simple averages are likely to understate the true growth in the wage earned by the average worker. To provide estimates adjusted for changing farm size, we weighted the wage-rate estimates reported by farmers who responded to our first three surveys according to gross dollar value of output. Under this procedure, our wage estimate increased from 3.73 in 1987 to 4.06 in 1988, 4.49 in 1989, and 4.51 in 1990, representing an increase of 9.5 percent for 1988, 10.5 percent for 1989, and 0.5 percent in 1990.

Compensation for seasonal farm workers often includes housing and meals. The percentage of growers providing housing and meals to their workers shows little trend over the period of our surveys, with about 40 percent of growers providing housing and 15 to 20 percent providing meals (Table 9.6).

Estimates of wage rates by region indicate considerable geographical dispersion, as shown in Table 9.3. In 1990 wages were highest in the piedmont region and lowest in the coastal plain eastern region. At the county level, the sample size was small, but there is an indication that wages tended to be highest in the counties where immigrant labor was used most intensively.

Production

If immigration reform created labor shortages, this could manifest in several ways. In the short run, farmers could lose crops that had already been planted if adequate labor supplies were available only at wage rates that make the cost of tending or harvesting a crop greater than the expected market value of the crop. Over a longer period of time, farmers could change their mix of crops and production techniques to economize on labor, as well as make contractual arrangements in the labor market to assure that labor is available when needed.

The percentage of growers responding to all surveys who were "unable to plant, cultivate, or harvest some of their crops because [they] could not find enough seasonal labor" grew very slightly from 10 to 12 percent between 1986 and 1988, but then dropped to 4 percent in 1990. However, the percentage of growers reporting such labor shortages was quite high in certain areas in some years. In Franklin County (in the piedmont) in 1986 and 1987, 20 percent of the sample reported labor shortages. In 1990 there were no reported losses in Franklin County, but 13 percent of producers in Johnston County reported a loss.

It is difficult to infer much from our data about the impact of immigration reform on the production of specific crops because crop-planting decisions hinge more on expected prices and government program decisions than on the availability of labor. Also, since the immigration law was not in full effect until 1988, there has been little chance for long-term planting decisions to be observed. In our data on the acres and value of crops produced in each year by each producer, no clear shift toward less or more labor-intensive crops was evident. Finally, it is important to note that flue-cured tobacco production, with gross returns of some $5000 per acre and a poundage quota worth between $.40 and $.70 per pound, was produced fully up to the legal limit on any farm that had a quota. Further, the years of our survey were years in which the industry was stabilizing and expanding after a turbulent period in the early part of the decade.

Another potential reaction to the possibility of increased labor costs is the adoption of labor-saving cultural practices or harvesting techniques. Mechanical tobacco harvesting, a labor-saving method, was developed more than 20 years ago, but its use has never taken hold as the major harvesting method and actually decreased earlier in the decade. Harvesting methods vary across farms but also across years in response to weather and economic conditions. Long-term investment in mechanical harvesting depends on assessments of the profitability of tobacco production as well as on labor availability.

In our first two surveys (1986–87 and 1988) we asked growers to report "the technique used to harvest tobacco." Growers were then asked to select among walking primers, riding primers, and mechanical harvesting (some volunteered that they used combinations of these methods). For the last two years we asked a somewhat different question to reflect the fact that some growers use more than one harvesting method. For 1989 and 1990 we asked, "Did you use any of the following methods . . ." and allowed producers to indicate one or more among the same three alternatives.

Table 9.7 summarizes the pattern of harvesting method across the years of our survey. There was an increase in the primary use of mechanical harvesting, from 17 percent of the producers in 1986 to 24 percent in 1988. In 1989, 24 percent of the growers used mechanical harvesting exclusively and another 17 percent used mechanical harvesting along with one of the other methods. In 1990, the exclusive use of mechanical harvesting rose to 26 percent, but now only 11 percent used this method in combination. The largest decline was in the use of mechanical harvesting jointly with walking primers. Overall, the data indicate some reversal in the trend toward more mechanical harvesting in 1990. This is consistent with the observed slowing of wage-rate growth and fewer labor shortages. The conditions in the hired labor market may well be a major factor in the choice of harvest method. However, without further information, it is not possible to determine the relative role of labor, tobacco market factors, or weather.

Table 9.7. Percentage of Tobacco Growers Using Selected Harvesting Methods, by Year

	1986	1987	1988	1989	1990
Exclusive or primary use:[a]					
Walking primers	42	40	38	31	36
Riding primers	40	36	37	25	23
Mechanical harvesting	17	23	24	24	26
Combinations of the above	1	1	1		
Specific combinations:					
All Three Methods				3	2
Walking and Riding				4	5
Walking and Mechanical				10	4
Riding and Mechanical				4	4

[a] For 1986-1988, primary use is implied. For 1989-1990, exclusive use of the method indicated is implied.

Conclusions

The impact of immigration reform in North Carolina through 1990 has been neither apocalyptic or insignificant. The use of immigrant labor has increased over the period covered by this study, at the same time that wage rates have risen more rapidly for seasonal farm workers than for other types of labor. Working conditions seem to have improved slightly and more farmers are using mechanical harvesting. The adjustments that have not been observed, such as shifts from seasonal to year-round labor and shifts to less labor-intensive crops, may require more study, or perhaps they are just not feasible responses for these producers.

We have described some unique data and presented a picture of some features of one regional farm labor market. We have raised a number of research questions and hope to have generated interest in these issues. Econometric analysis of the hiring patterns, wage functions, farm production responses, and other labor market issues will be developed in subsequent research. More in-depth study is required to understand the patterns we have described.

Part II, Section B.
IRCA's Effects in Western States

Studies in this section include the western states of California, New Mexico, Oregon, and Washington. The western region is the dominant region for the production of fruits, vegetables, and horticultural products in the United States. Within this region (as well as the nation), California is clearly the dominant state. California produces a massive array of fruits and nuts (F), vegetables and melons (V), and nursery and horticultural (H) products, while other states in the region tend to have significant production in more selective commodities. For example, Washington is dominated by apple production, Oregon is notable for pear, berry, and Christmas tree production, and New Mexico has chile and onions as its labor-intensive crops.

The western region has a long history of employing Mexican immigrant labor in agriculture. While the *bracero* program provided a formal legal mechanism for Mexican workers to enter the United States for agricultural work between 1942 and 1964, the termination of the program did not end the migration of workers. With the dramatic differences in labor compensation in the two countries and the extensive shared border in the western United States, the flow of workers has continued despite the absence of legal documentation.

Agricultural employers from the western states, unlike the eastern states, have rarely utilized the H-2A program, or its predecessor the H-2 program, arguing that its provisions were too rigid for their short-term, specialized employment needs. They were consequently ardent supporters of the SAW and RAW provisions for IRCA that were to assist them in adjusting to a legal work force. The chapters in this section review the experiences of the included states in that adjustment process.

A dominant theme in the chapters is the expansion in production of FVH products throughout the region following IRCA. The presumption is that if labor were in limited supply, such expansion would not have been feasible; just the opposite appears to have happened. Martin and Taylor note the expansion of FVH production in California, including the movement into areas not previously in labor-intensive production. Oregon was characterized by a dramatic increase in the nursery component of FVH production (Chapter 11); the main labor-intensive crops in New Mexico — chile and onions — increased in production (Chapter 12); and in Washington, the main FVH crop — apples — increased in production by 39 percent from 1986 to 1991 (Chapter 14). The uniformity in this phenomenon is indeed striking.

Despite the increased production in labor-intensive FVH agriculture across the region, none of the authors report problems of a labor shortage in the post-IRCA period. Rather, the indication is just the opposite, namely that there have continued to be ample supplies of labor for FVH agriculture. In conjunction with the ample supply of labor are indications that real wages have not improved over the period. Rosenberg (Chapter 15) notes a reduction in real earnings in California. In Oregon and Washington, while modest increases in nominal earnings are noted, they are attributed to other labor law changes, such as an increased minimum wage rather than IRCA (Chapters 11 and 14).

Emphasis was also given to the demographic changes in the farm work force. The most common change noted was the increased Latinization of the work force. This has been particularly true in California where the workers are mostly Hispanic, often immigrant and non-English speaking (Chapter 10). The same characteristic is noted for New Mexico, as expected for a border state (Chapter 12). However, Cross, Mason, and Caraballo (Chapter 11) discuss the increasing presence of Hispanic workers in Oregon, clearly some distance from the border. Associated with this change, they note the increasing presence of family members who are often without documents. Martin and Taylor (chapter 10) also talk about the increased presence of young immigrant men, who are more productive than the more diverse work force they are replacing.

Rochin and Castillo (Chapter 13) give a major emphasis to the demographic changes in the work force of rural areas. Using Census Bureau data, they have characterized the development of *colonias* in California rural areas. They argue that the *colonias* have become impoverished communities of Spanish-speaking laborers. They find little evidence for the establishment of ethnic enclaves in these areas, but rather observe a tendency toward immiscibility.

Considerable attention is given to the role of farm labor contractors (FLCs) in this section, as in the other sections of the book. Martin and Taylor (Chapter 10) and Rosenberg (Chapter 15) suggest that the role of FLCs in the farm labor market has increased in California. Cross, Mason, and Caraballo (Chapter 11) find that among commodities traditionally using FLCs in Oregon, FLC use has increased. Eastman (Chapter 12) observes that in New Mexico FLCs were important in chile and onions both before and after IRCA. While the tendency is to ascribe the growth in labor contracting to IRCA, Rosenberg is somewhat more cautious. Although he argues that IRCA is a factor, he suggests that it is only one of a number of factors. Moreover, he points out that separation between the employer and the firm for whom the work is being done has been occurring in the rest of the economy, as well. Most importantly, farm labor contractors are not a homogeneous group, but are varied in size and associated business activities.

10

IRCA's Effects in California Agriculture

PHILIP L. MARTIN AND J. EDWARD TAYLOR

Introduction

The Immigration Reform and Control Act of 1986 (IRCA) was a case of good intentions gone awry throughout U.S. agriculture, in the sense that IRCA did not result in gradual farmer adaptation to a legal work force. The failure of IRCA to achieve its goals is particularly striking in California, the nation's leading producer of agricultural commodities. After two decades of profit-minded expansion accompanied by wage and benefit improvements for many farm workers, IRCA seems to have signaled a new era in farm labor: expansion accompanied by lower wages and worsening conditions for recently arrived immigrant farm workers who find jobs with the help of farm labor contractors (FLCs).

California has about 35,000 "employers" or reporting units whose primary activity is agricultural,[1] and these reporting units report about 1.2 million unique social security numbers to state tax authorities sometime during a typical year. Agricultural operations are almost 5 percent of the state's reporting units, and persons employed sometime during the year in agriculture include about 7 percent of the persons employed in the state (Martin, Taylor, and Hardiman 1988).

IRCA was expected to increase government involvement in the farm labor market. In particular, the anticipated reduction in illegal alien entrants was expected to encourage farm employers to improve wages and working conditions in order to retain newly legalized Special Agricultural Workers (SAWs), and farmers were expected to work more closely with agencies such as the Employment Service (ES) to recruit U.S. workers or demonstrate that U.S. workers were not available in order to receive certification to admit H-2A or Replenishment Agricultural Workers (RAW). These expected adjustments have generally not occurred, largely because illegal aliens continue to arrive and find jobs, often with fraudulent documents.

[181]

California Agriculture

California has the nation's largest and most diverse agriculture. California farmers produce milk, wheat, cotton, and other crops that are produced elsewhere, but the distinguishing feature of the state's agriculture is the production of specialty crops: fruits and nuts, vegetables and melons, and horticultural specialties such as flowers and mushrooms. These so-called FVH crops represent half of California's farm sales, and the orchards, vineyards, and fields in which they are produced are where most farm workers find jobs.

There are many ways to describe a California farm labor market in which 900,000 individuals find at least temporary jobs with 25,000 employers. One way is to focus on 3 Cs: concentration, contractors, and conflict. *Concentration* refers to the fact that most farm workers are hired by a relatively small number of large farms and that the relatively few year-round farm workers get most of the farm wages that are paid. *Contractors* refers to the presence in many farm labor markets of "intermediaries" who match seasonal workers with jobs. *Conflict* refers to the stormy history in which workers protested low wages or poor working conditions, but then individually got out of agriculture when their collective efforts to obtain improvements were rebuffed.

Agriculture is California's largest employer of adult immigrants. According to tax records, 24,500 California farm employers paid $4.6 billion in wages to 900,000 workers in 1990 (Martin and Miller 1993, p. 22). But these farm labor data conceal as much as they reveal. Most of these farm "reporting units" are small; half pay less than $10,000 in farm wages. The largest 1,250 farm employers — just 5 percent — pay about two-thirds of California's farm wages.

Concentration — in the sense that a few farm workers get most of the farm wages paid — is also the rule among farm workers. A "farm worker" is anyone who works for wages on a farm. Since farm work is seasonal, most farm workers do nonfarm work or, since many are immigrants, return to their country of origin when there are no farm jobs for them in California. Most farm workers are in the farm work force for only a short time. Over half of them have less than $1000 in annual farm earnings, which, at $5 per hour, means that they did 200 or fewer hours of farm work, or the equivalent of 5 weeks.

Concentration means that the largest 5 percent of employers and the 10 percent of farm workers who are employed the year round account for most of the farm wages paid and earned. The "average" farm employer is very small, but the farm employers who hire most of the state's farm workers may employ a peak of 5,000 workers and have a weekly payroll of $1 million, which makes them large employers by any definition. The "average" farm worker earns about $3000 annually, but there are in fact few workers with such earnings; instead, workers tend to fall into one of three groups. Just over half are very low earners, earning $500 or less. About 40 percent are in the seasonal category, earning $4000 to $6000 for half a year's farm work. The third group is the year-round workers, who earn $12,000 or more annually. For both farm employers and farm workers, importance increases as numbers decline.

Who are these large farm employers who hire most of the state's farm workers? Most are family-run corporations. An example is the Zaninovich table grape farm in California's San Joaquin Valley. This table grape grower sells about 5 million 25-pound boxes of grapes annually at an average of $10 per box, and hires 1,000 farm workers to help generate $50 million in grape sales. If these farm workers move from one area to another, they can find work for 1,000 hours between May and October and, at $5 to $7 per hour, earn $5000 to $7000.

The Bud subsidiary of Dole Fresh Foods, part of the Castle and Cook Corporation, is probably the largest California farm employer. This corporation hires some 7,000 farm workers to harvest lettuce and other vegetables. Some of these workers move among the company's far-flung operations in Arizona and California, and may be employed for 2,000 hours at an average of $7 to $8 an hour, earning $15,000 annually. Other Bud farm workers are employed 500 to 1,000 hours in the areas where they live, giving them annual farm earnings of $3500 to $8000.

Farm workers usually prefer to work for these large corporate farms, since they tend to pay higher wages and offer more benefits. The more typical employment situation in California agriculture is found on Fresno-area raisin farms. For 6 to 8 weeks each August and September, 50,000 farm workers are in the vineyards cutting bunches of grapes and laying them on paper trays to dry into raisins; this is the most labor-intensive activity in U.S. agriculture. Even if they work 10 hours daily for 8 weeks, raisin workers find just 560 hours of employment, which, at $5 per hour, yields $2800 in farm earnings, not enough to live on.

Workers employed in short harvest crops such as raisins can earn more if they can find other farm jobs, such as harvesting peaches and plums before the raisin harvest and olives and citrus after the raisin harvest. Many farm workers are able to string together a series of farm jobs, but it is not easy for many of them to go from farm to farm looking for work. Instead, many rely on labor market "intermediaries" — FLCs, crew bosses, or friends and relatives — to tell them about the next farm jobs.

The intermediaries who find jobs for farm workers and supervise them while they work are the second C of the California farm labor market: Most farm workers find a succession of farm jobs with the help of a contractor-like intermediary. This is not a new phenomenon; Chinese farm workers in the 1870s designated one person in the crew who could speak English to search for the next farm job for the crew. Over time, contractors evolved from one of the crew to independent businesses that seek to profit from the difference between what a farmer pays to have a job done and what the worker gets, and a major farm labor story for over half a century is that contractors who know the circumstances of the immigrant workers they employ have in some cases exploited immigrant farm workers.

Contractors are a major farm labor story of the 1990s. There were only 935 FLCs in California in 1990 who were registered to pay taxes on the wages they paid to workers, but they employed one-third of all farm workers, and their role in matching farm workers and jobs rose sharply in the late 1980s. Critics assert that there are far more FLCs who are not registered, and that farmers have encouraged the rise of labor contracting by switching from hiring workers directly to hiring them through intermediaries. According to farm worker advocates, farmers know that these "merchants of labor" must be cheating either the government or the workers, since the FLCs in some cases agree to do farm work for what appears to be a money-losing fee.

Farm worker advocates would like to eliminate this second C of the farm labor market by making farmers responsible for the activities of any FLC who brings workers to their farms. Farmers oppose this added liability. Why should farmers, they ask, be responsible for the independent businesses who supply them with the resources needed to farm? This would be analogous, they argue, to a homeowner who hires a painter or gardener to be responsible for that person's treatment of his workers: If the homeowner paid the painter and the painter failed to pay his employees, then, under the joint liability status that farm worker advocates want and farmers oppose, the farmer would have to pay the unpaid workers a second time directly. Farm worker advocates assert that this is the only way to ensure that farmers really investigate FLCs and use only those who are law abiding. Farmers counter that trying to eliminate the second C of the farm labor market would unduly burden them with the job of policing labor middlemen that the government has failed to regulate.

The debate over what to do about contractors illustrates the third labor market C: conflict. There is an inevitable conflict between employers and workers over the "fair" wage and the necessary level of effort a worker must expend while working. This conflict can be resolved in three major ways, one, by employers unilaterally, with the dictum either to work according to the employers' rules or leave. Second, a fair wage can be determined through collective bargaining. Third, work-place conflict can be minimized by government rules that stipulate minimum wages and work-place safety standards, as well as mandatory benefits for workers.

Conflict in the California farm labor market has rarely involved unions. Instead, workers have protested low wages and poor conditions in often-unrecorded disputes with individual employers, and the usual response to work-place dissatisfaction has been to exit the farm work force. There are examples of workers banding together and exercising a collective voice to improve farm labor conditions, but conflict in farm labor markets has often assumed unusual dimensions, for several reasons. Farm employers often band together and set a "standard" wage, so that there is peer pressure to resist worker demands for higher wages. Farm worker leaders have in many instances been radical outsiders in conservative rural communities, so that the entire community, not just farmers, often has been willing to tolerate violations of individual rights to get the troublemakers out of town, so that temporary migrant workers will go back to work. The workers who are dissatisfied with the wages they are offered have sometimes faced eviction from their temporary homes, making their choice about whether to strike a decision that has affected both their income and their housing.

Farm worker conflict has often been bloody because few devices have evolved to negotiate solutions to farm labor problems. In many U.S. work places, handbooks spell out work rules and describe how wage increases, promotions, and layoffs are handled. If there are disagreements, union representatives and personnel officers sometimes intervene, so that aggrieved workers can continue to do their jobs while their complaints are considered.

Such modern personnel practices are rare in agriculture. In many cases, farm workers are hired on the edge of a field by the only "employer" they will get to know. This foreman or crew boss is usually responsible for 20 to 40 workers, and his decisions about which workers get the easiest and hardest tasks cannot usually be appealed. Instead of protesting, aggrieved farm workers more often quit working for one crew boss and go to work for another. The resultant high worker turnover helps to explain why farm workers are often seen as interchangeable: Few farmers know where their workers worked before they arrived and where they will go after they depart.

Labor in FVH Agriculture

California has led the nation in farm sales since 1950. California has only 2 percent of U.S. farmland; it accounts for 10 percent of the nation's farm sales because it produces high-value FVH crops. This means that California farmers obtain far more revenue per acre than do most U.S. farmers. For example, California harvested 1.2 million acres of vegetables and melons in 1990, and they had a farm value of $3.5 billion. Nebraska farmers, by contrast, farmed 15 times more land, but they had only about the same level of farm sales.

California is the nation's largest farm state because its farms produce 40 percent of the nation's high-value FVH commodities. California began producing labor-intensive fruits and vegetables and shipping them long distances a century ago, but much of the expansion of California's FVH agriculture has occurred since World War II. In some cases there have been truly dramatic production increases. Broccoli production increased over sixfold between 1960 and 1990, almond and strawberry tonnage rose over fivefold, and nectarine and wine grape production more than tripled.

California produces over 250 commodities commercially, but most of these are various kinds of fruits and vegetables. Among the almost 200 FVH crops grown in California, a few account for most of the value of FVH crops. For example, raisin, wine, and table grapes accounted for one-third of the $5 billion in fruit and nut sales in 1990, and lettuce accounted for 20 percent of the $3.5 billion in vegetable and melon sales.

The need for farm workers depends on the physical volume of fruits and vegetables produced, and this is where rising yields have increased the demand for labor, even in crops whose acreage has not expanded. For example, there have been about 70,000 acres of table grapes since 1970, but production from these acres has doubled. The acreage of wine grapes has doubled since 1970 to 300,000, but the tonnage of grapes picked has jumped fivefold. Similarly, the acreages of strawberries and broccoli have more than doubled since 1970, but the amount of these crops produced has risen two- to threefold.

The major asset needed to produce crops is land, and the land used to produce fruits and vegetables is valuable. In 1990, for example, land in the Monterey area used to produce vegetables was worth an average of $10,000 per acre, and most vineyards and orchards in the Fresno area sold for $5000 per acre. A crude indicator of the amount of money tied up in California's FVH agriculture is the value of 1 million acres of vegetable land ($10 billion) and 2 million acres of fruit and nut land ($10 billion).

Fruits and vegetables are harvested and sold, and most of them generate farm sales of $2500 to $3000 per acre. The production of fruits and vegetables is considered "labor intensive," an adjective that is rarely defined but suggests that the costs of hired workers may be the single largest production expense. Labor costs in FVH production range from 20 to 50 percent, more than the 20 percent average in manufacturing but less than the 70 to 80 percent in many service industries.

A farmer grossing $3500 per acre from the sale of table grapes pays about $1500 to the workers who tend and harvest them. Is this too much or too little? From the point of view of consumers, labor costs are only a small fraction of supermarket prices. Farmers get about $.40 per pound of grapes and $.13 represents the cost of labor. However, the retail price of grapes averages $1 per pound, so that farm worker costs are just 13 percent of the retail price. If farm worker wages doubled, if farmers continued to use the same amount of labor as before, and if all of the increased cost of farm labor was passed on to consumers, then the typical wage of grape workers would be $12 hourly and the cost of grapes would be $1.13.

Labor costs are important to farmers but much less important to consumers, because farmers obtain only about one-third of the typical retail price of a fruit or vegetable. This means that whether farm worker wages go up or down, they affect only one-ninth of the retail price of a fruit or vegetable.

Seasonality and Farm Work

California fruits and vegetables do not ripen uniformly, so the peak demand for labor shifts around the state in a manner that mirrors harvest activities. Harvest activity occurs the year round, beginning with the winter vegetable harvest in southern California and the winter citrus harvest in the San Joaquin Valley. There are probably more hours devoted to pruning trees and vines during the winter months than are required to harvest tree fruits in the summer, but, since pruning occurs over several months, fewer workers are involved. During the rainy months between November and April, employment on farms is only half of its peak September levels.

Harvesting activity moves northward and toward the coastal plains in March. Workers harvest lemons and oranges in the south coastal area, and they are hired to work in flower and nursery crops, as well as to thin and weed vegetable crops in the Salinas area. By May, workers are picking strawberries and vegetables in coastal areas, and these harvesting activities continue to employ them throughout the summer.

In May and June, harvest activities move inland to the San Joaquin Valley. The thinning and harvesting of cherries, apricots, peaches, plums, and nectarines during the late spring, as well as harvesting activities in the Coachella Valley, produce a June mini-peak demand for labor. Harvest activities continue to require large numbers of workers throughout July and August, and other workers are employed to irrigate crops and weed cotton.

September is usually the month in which farm worker employment reaches its peak. A series of short but labor-intensive harvests, perhaps best symbolized by the employment of 40,000 to 50,000 workers to harvest the state's 300,000 acres of raisin grapes, keeps employers whose harvests are ending in August worrying about whether their workers will remain to finish the harvest of peaches or melons, and raisin employers worry that too few workers will show up in September. By October, only a few late harvests remain, including olives and kiwi fruit.

Workers willing to follow the ripening crops can find eight to ten months of harvest work. However, relatively few workers follow the ripening crops within California. A 1965 statewide survey of farm workers found that 30 percent migrated from one of California's farming regions to another, and a 1981 survey of Tulare County farm workers found that only 20 percent had to establish a temporary residence away from their usual homes because a farm job took them beyond commuting distance (Calif. Assembly Committee on Agriculture 1969; Mines and Kearney 1982).

There is some migration between farming regions that is organized by employers, such as the lettuce circuit from Salinas to Imperial to Huron, but such employer-organized migration seems to be decreasing. Instead, a lack of housing in other regions and surpluses of labor that can make finding a job difficult have discouraged the migration of workers from one farming region to another, unless they have firm housing and employment arrangements. Farm labor contractors continue to move crews of solo men from area to area, but there seem to be far fewer "free-wheeling" families who travel the state looking for farm jobs without prearranged jobs and housing than there were in the 1960s and 1970s.

Workers tend to stay in one area of California, both because follow-the-crop migration is less attractive in light of the general shortage of temporary housing and because the harvesting of some crops has been stretched out for marketing and processing reasons and the availability of service programs makes migration less necessary. Most California fruits and vegetables are grown for the U.S. or foreign fresh markets in order to obtain the highest prices; "surplus" production is directed to the lower-priced processing market or not harvested. To maximize the period during which fruits and vegetables can be sent to the fresh market, growers plant early-, mid-, and late-season varieties of fruits, or they plant more acres of a vegetable such as lettuce each week. Such efforts to "stretch out" production have labor market ramifications. Stretching out production for marketing reasons means that workers can remain in one region and harvest a crop for a longer period of time, such as in the strawberry harvest. Even when the crop is to be processed, the owners of processing plants have an incentive to get their growers to stretch out the harvest, so that fewer processing plants — which are idle most of the year — must be built to handle peak inflows of the crop at harvest time.

Many workers still migrate, but they tend to shuttle into the state from Mexico or Texas rather than to follow the crops after their arrival in California. It should be emphasized that the reduction in follow-the-crop migration does not mean that there is no migration; it means that the nature of migration has changed. In theory, migrant camps, open for six months annually, should experience considerable turnover as families move on to the next harvest. The fact that they do not experience such turnover highlights the importance of housing in explaining migration behavior. There is so little decent and affordable housing in rural California that once a worker finds housing, he is reluctant to move on unless housing will be available at the next destination.

The processing and packing work force is similar to the harvest work force, in that many of the seasonal workers employed in it are Hispanic, but it includes more U.S. citizens. Until the 1940s, it was common for the wives of field workers to be employed in the packing houses, but the success of unions in the packing houses increased wages there to twice field-worker levels and made them a first rung up the American job ladder for many field workers. Technology and trade are eliminating this packing rung up the ladder for field workers. Some (nonfarm) California packing jobs have been lost by the movement toward field packing, the practice of picking broccoli or melons and then putting them into cartons in the field. Other processing jobs have migrated abroad, such as frozen cauliflower and broccoli jobs to Mexico, where wages are about one-eighth U.S. levels.

Mechanization and Expansion

California farmers employ about 900,000 individuals for about 350,000 farm jobs, which means that, on average, three workers share one year-round-equivalent job. A year-round-equivalent job is a statistical artifact — it is a shorthand way of saying that if one person works four months on farms, another works four months, and a third works four months, a year-round job is created, even though all three of these seasonal workers were probably employed between June and September. For this reason, a seasonal industry such as agriculture will always have a "work force" that has larger-than-average employment; the issues are how large this reserve army of seasonal workers should be and who should pay for the time of the workers who wait for work to begin. The best (though inadequate) data indicate that there has been a 3 to 1 ratio between individual workers and average jobs in California agriculture, and neither changes in the concentration of farm employment, nor mechanization, nor the increased demand for fruits and vegetables are likely to change these work force and average employment numbers in the 1990s.

Mechanization — picking a crop by machine rather than by hand — has reduced employment in many commodities over the past 25 years, but the production of other commodities has expanded enough to create new jobs and to stabilize farm employment. Processing tomatoes is the outstanding example of what happens to farm worker jobs as a result of mechanization. In 1960, a peak of 45,000 workers — 80 percent of whom were *braceros* — hand picked 2.5 million tons of the processing tomatoes used to make catsup. In 1990, about 5,500 mostly female farm workers were employed to sort four times more tomatoes.[2] In this case, mechanization reduced the number of jobs for farm workers and changed the harvesting task from picking tomatoes into boxes to riding on a machine and sorting machine-picked tomatoes.

Farmers have often feared that they would not survive the loss of immigrant workers: Tomato growers in 1961 asserted that "the use of *braceros* is absolutely essential to the survival of the tomato industry" (Calif. Senate 1961). During the 1960s, many agricultural economists agreed; they thought that the choice was to mechanize or stop growing hand-harvested crops. For example, a major 1970 study asserted that "California farmers will continue the intensive search for labor solutions, particularly mechanical harvesting" (Dean et al. 1970), while another 1970 study concluded that "the door to employment is rapidly closing for those persons whose only qualifications for employment are a will to work and enough muscle to compete" (Cargill and Rossmiller 1970).

The expectation that there would soon be no need for those who could bring only strong backs to farm work was so widespread that the federal government began programs to help farm workers adjust to the nonfarm jobs it was expected they would soon have to seek. The justification for special farm worker assistance programs evolved from several concerns. First, there were farm labor reformers who had hoped to extend labor relations and protective labor laws to farm workers, and they saw programs that provided farm workers with health and education services as less controversial first steps. Second, there was national concern about the people left behind and a recognition that state and local governments can and do discriminate against the powerless, including migrant farm workers. Third, it was known that many of the children of blacks and whites who had participated in the "Great Migration" from small farms in the South to Chicago and Detroit during the 1950s were ill prepared for urban life. With mechanization threatening a new round of labor displacement in agriculture, it was hoped that federal programs could start to prepare farm workers and their children for the nonfarm lives they would be forced to live.

The mechanization that was expected to be the causal factor behind these changes did not occur. The mechanization of the tomato harvest proved to be the exceptional type of labor-displacing change in California agriculture, not the rule. The publicity generated by lawsuits against the University of California in the late 1970s for its part in using tax dollars to develop plants and machines that displaced farm workers led some people to believe that such sudden mechanizations are the rule and that more of such labor-displacing machines would soon drastically reduce the employment on California farms. This is not the case. The surprise is how little mechanization there was in California during the 1970s and 1980s, not how much.

There have been important labor savings in California agriculture, but they are usually less visible than machines replacing hand harvesters. Changes in production practices for perennial crops have saved labor, such as drip irrigation (which saves irrigator labor), dwarf trees and vines trained for easier hand or mechanical pruning, and precision planting and improved herbicides, which save thinning and hoeing labor.

There is also an important counter trend to displacing field workers. In the past, most crops were picked by field workers into bins or boxes and then hauled to a packing shed, where nonfarm workers sorted and graded the peaches and broccoli and put them into cartons. During the 1970s and 1980s, many crops began to be picked and packed in the field. Picking and packing grapes, vegetables, and melons in the field increases "farm" and decreases "nonfarm" employment. Several factors encourage field packing: Flexible conveyor belts and portable technologies make it easier to pick, pack, and cool cartons of produce in the fields or get them quickly to a cooling facility; there is less damage to the commodity because it is handled fewer times; and field worker wages are usually only half the levels of packing-shed wages. Today most table grapes and many melons and vegetables are sent directly from the field to the supermarket.

Field packing increases farm worker employment, but a more important reason for there being as many farm workers in the 1990s as there were in the 1960s, despite the mechanization of the tomato harvest and other labor-saving changes, is because there is so much more fruit and vegetable production today. It is sometimes hard to appreciate how much the expanded production of labor-intensive crops — reflected in both larger acreages and higher yields — creates additional jobs for farm workers.

Broccoli production provides an example. The average American's consumption of fresh vegetables rose 23 percent, to 136 pounds per person during the 1980s, but the increase was sharpest for broccoli: The per capita consumption of fresh broccoli almost tripled, from 1.6 to 4.5 pounds during the decade. Broccoli is hand harvested, and its production in California required an average of 52 hours of hired labor per acre in 1990 (Mamer and Wilkie 1990). There was a 50 percent or 40,000-acre increase in U.S. broccoli acreage — most of it in California — during the 1980s, so 2.1 million additional hours of hired labor were needed to produce broccoli. Even though broccoli is harvested over a long season, enabling workers to average 500 to 1,000 hours annually, the increased production of broccoli, a commodity worth just 1 percent of the total value of FVH commodities, required 2,000 to 4,000 additional seasonal workers just to handle the increase in production in the 1980s.

Americans are expected to continue to consume more fruits and vegetables, and the sharpest increases in consumption are expected for commodities that are eaten fresh and are thus most likely to be hand harvested. The U.S. consumption of most farm commodities increases about 1 percent annually due to population growth, but the demand for FVH commodities increases another 2 or 3 percent annually, or about as much as personal incomes typically go up, because Americans tend to spend about the same percentage more on fresh fruits and vegetables as their incomes rise. For example, if the population increases by 1 percent and personal incomes rise 2 percent, then expenditures on fresh broccoli and strawberries also rise by 3 percent.

Americans can continue to increase their consumption of fresh fruits and vegetables that farm workers harvest without increasing farm worker employment if more fruits and vegetables are imported. Imported fruits and vegetables today account for less than 10 percent of the typical American's consumption, and most of these imports are fruits such as bananas that are not grown in the United States. Tomatoes, sometimes considered a symbol of the ability to produce a fresh vegetable far from where it will be consumed, are mostly produced in the United States — only 11 percent of the fresh tomatoes in 1992 were imported, mostly from Mexico during the winter months.

Will the North American Free Trade Agreement (NAFTA) shift the production of fruits and vegetables to Mexico? Perhaps after the year 2000, but not during the 1990s (Martin 1993). Mexico's primary competitive advantage is climate; Mexico can produce fresh vegetables during the winter months when most U.S. production areas except Florida are not producing. But even if Mexico were to completely displace production in Florida, most fruit and vegetable production would remain in the United States because two-thirds of the production occurs in the summer and fall, when Mexico is not producing significant quantities. Mexico is not even likely to replace Florida as the source of most winter fruits and vegetables because, in many cases, low yields and low labor productivity make production there more expensive than in California.

Mexican reforms and NAFTA are unlikely to spur a quick shift of production to Mexico. NAFTA would eliminate tariffs on fruits and vegetables grown in Mexico, but these tariffs are already low — the average U.S. tariff on the Mexican fruits and vegetables exported to the United States in 1990 was 8 percent, so that, as a result of NAFTA, the price of Mexican produce in the United States may fall by 8 percent. But these tariff reductions will be phased in over a 15-year period, and that may not be enough of an incentive to quickly expand fruit and vegetable production in Mexico. Growers complain of many problems associated with producing in Mexico for the U.S. market: higher transportation costs, less public research on disease and other factors that reduce yields, and lower worker productivity. These Mexican disadvantages could be overcome with foreign investment and time, but Luis Tellez Kuenkler, undersecretary in the Mexican Ministry of Agriculture and Water Resources, assures that Mexico's fruit and vegetable agricultural exports will not skyrocket overnight, displacing U.S. farm workers. He may have been correct when he predicted in 1991 that Mexican FVH exports to the United States would rise at most by 40 percent within five years after NAFTA was signed, or from $900 million in 1990 to $1.3 billion by 1998. U.S. FVH production by then is likely to be worth $35 to $40 billion annually, demonstrating that even increased trade with the country best suited to produce for the U.S. market is unlikely to eliminate the need for farm workers in California agriculture.

The Farm Labor Market

How do 900,000 Mexican immigrant farm workers find jobs with the 25,000 California employers who hire them? The farm labor market is like all others, in the sense that it matches workers and jobs. However, it handles the three essential functions of a labor market in unique ways. First, farmers rarely come in direct contact with the workers they hire; job information is usually communicated in Spanish, which most farm operators do not speak. Second, agriculture has dealt with the problem of training and motivating workers to work by paying piece-rate wages for many jobs — for example, $12 per bin of oranges picked. Third, farmers assure themselves that there will be enough seasonal workers available by working collectively to maximize the supply of labor, rather than trying individually to identify and keep the best workers on their farms.

Contractors or intermediaries such as foremen and crew bosses handle the recruitment of new workers. Recruitment occurs in several ways. In some towns, such as those along the U.S.–Mexican border, there is a so-called day-haul labor market: Workers begin to congregate at 3:00 or 4:00 a.m., and contractors arrive with buses, tell the workers the task and the wage, and the workers then board the bus that seems to offer the best job. Some workers board the same bus every day, while others switch from bus to bus.

The most common way in which farm workers are recruited occurs when the crew boss tells the crew that more workers are needed, and the workers currently in the crew inform their friends and relatives that a job is available. Such "network recruiting" is very helpful to employers. There is no need to spend money on help-wanted ads, and workers who are often grateful for jobs tend to bring to work only "good" workers. Once hired, the friend or relative who brought the new worker to the work place is usually responsible for him or her: The experienced worker teaches the new hire how to work, the work rules, and other job-related information. In this decentralized hiring system, a worker may pay the crew boss to get a job and the crew boss may allow a worker to bring his or her children to work.

Crew bosses are often more than just employers. Especially when the workers are recent immigrants, the boss may be the workers' banker, landlord, transportation service, restaurant, and check-cashing service. Crew bosses provide such services to workers to make money from them and because newly arrived workers often need such services. Federal and state governments have enacted an ever-growing body of laws and regulations that regulate these sideline activities of farm employers, but they have failed to change the reality that, in the farm labor market, immigrant farm workers often pay premium prices for inferior services. Regulations requiring that farm worker housing be inspected, for example, have led to the destruction of housing on farms. Most workers today rent housing in nearby towns and pay rent as much as $4 to $5 daily for rides to work in the vans that have become the transportation system in rural California.

Day-haul and networking recruitment best describe the decentralized hiring systems most common in California agriculture. There are other recruitment systems, but they are not common. A few large farms do all of their hiring from a central site and require prospective workers to go to that site to fill out a job application.[3] Some of these farm employers hire workers in a nonfarm manner, with application forms, preemployment medical and drug screening, and occasional testing of prospective workers and checking of references. However, few workers are hired in such a manner. For example, fewer than 5,000 workers are sent to farm jobs annually by a union hiring hall, and fewer than 10,000 are sent to field worker jobs by the public Employment Service — less than 1 percent of annual hires.

If the seasonal farm labor market were organized to minimize the number of workers needed, there would be one central clearing house to which farmers would go for workers and workers would report for jobs. Reality is exactly the opposite; there are not even reliable data on the demand for and supply of workers, much less a central clearinghouse for jobs and workers. Decentralized hiring means that there are many farm labor markets. These markets may appear to be similar, in the sense that many of the jobs involve harvesting, most of the workers are Mexican immigrants, and the piece-rate wages tend to be similar from field to field, but take-home pay nonetheless can vary from worker to worker. Much of this variation in take-home pay is linked to how workers are recruited. Take-home pay is less if a worker must pay to get the job; if a worker needs or is required to accept housing and rides to work from the crew boss; or if a worker must pay for work equipment, check cashing, or many of the other needed services. Pay is also affected by hours of work; some crew bosses tend to overhire, so that each worker gets only five or six hours of work, even on days when weather or marketing do not stop the work.

Once hired, workers must be motivated to work. Two major wage systems are used to pay seasonal farm workers. According to the National Agricultural Workers Survey, 70 percent of the jobs held by California farm workers in the early 1990s were paid hourly wages and 30 percent were paid piece-rate wages or a combination of hourly and piece-rate wages (USDOL 1991). The most common hourly wage for entry-level workers is the minimum wage, which has been $4.25 in California since July 1, 1988. Some employers pay more — $4.50 or even $5 — in order to be able to select the best workers. But reports that hourly wages average, for example, $6.06, the figure reported for California field workers by the USDA for mid-July 1993 — can be misleading because such average wages are weighted by the hours that the various subgroups of hourly workers work. This means that the average hourly wage of a tractor driver paid $7 an hour for 10 hours of work daily, and two hoers who each work 35 hours a week for minimum wage are reported to have an average hourly wage of $5.62, even though neither the tractor driver nor the hoers are making a wage close to this average.

Piece-rate wages are straightforward: Workers are paid about $12 per bin of oranges picked or $.16 per tray of raisins cut. A combination wage is often paid to harvest table grapes: A typical wage in 1993 was $5.25 per hour, plus $.28 per 25-pound box of grapes packed.

The purpose of the wage system is to motivate workers. Generally, employers pay hourly wages when they want slow and careful work, such as to prune trees and vines, and when employers can easily control the pace of the work; field-packing broccoli by walking behind a machine or the traditional tasks of early-season picking and thinning and hoeing are examples. Combination wages are paid when the employer wants careful but fast work, such as harvesting and packing table grapes in the field. Piece rates are paid when the pace of work is difficult to regulate, when quality is not of great importance, and when an employer wants to keep labor costs constant with a diverse work force. The average hourly earnings of workers paid by the piece are typically higher than earnings paid by the hour, but this advantage tends to disappear when hours of work are considered. For example, according to the USDA, California farm workers paid hourly wages in July 1993 received about $6 an hour, while piece-rate workers had average hourly earnings of almost $7. But hourly workers who average 40 hours per week have about the same weekly wage as piece-rate workers who average 35 hours per week.

Recruitment gives California farm workers a Mexican immigrant work force and these workers are paid hourly or piece-rate wages. The third dimension of the farm labor market is retention — what do farm employers do to identify and persuade the best workers to return? The answer is: Not much. Perhaps the easiest way to describe the attitude toward worker retention is in an analogy to irrigation. Water is vital to produce crops in California, and there are two major ways to supply water to the crops. A field can be flooded with so much water that at least some trickles to each plant. At the other extreme, water pipes can be laid through the fields, and then only the water needed by each plant can be allowed to drip to each one.

The choice of irrigation technique depends largely on the cost and availability of water. Where water is scarce and expensive, farmers tend to invest in drip irrigation systems; where water is cheap, farmers tend to flood their fields with water. The choice for farmers is to work collectively so that there is plenty of water for all or to invest on a farm-by-farm basis to use available water as efficiently as possible. Farmers acknowledge that they view labor issues in the same manner: Should they work collectively to maximize the number of workers available or treat labor as scarce and try to select and retain the fewest workers necessary?

As the historical discussion makes clear, California farmers have usually chosen the flood-the-labor-market alternative. The best source of additional water is to dam up rivers; the best way to get additional seasonal workers is to open immigration doors, explaining why farmers have such a keen interest in immigration policy.

FLCs and Farm Workers

Immigrants are a majority of all farm workers and an even larger share of new entrants. The dominance of immigrants affects the structure and functioning of the farm labor market; for example, non-English-speaking immigrant workers often require intermediary bilingual FLCs or foremen to help them find housing and other social services; they are usually more willing to accept variation in days worked and piece-rate earnings than U.S.-born workers; and immigrants are more likely to pay without question charges for housing, transportation, and eating arrangements in areas that have farm jobs. Such behavior makes immigrants preferable to U.S. citizens in the eyes of most farm employers.

FLC workers tend to be newly arrived and often unauthorized immigrants. They tend to be less skilled, younger, and less educated and experienced than non-FLC workers. A unionized FLC worker is an oxymoron, as is the notion of a permanent FLC worker. Once employed by an FLC, a new immigrant quickly moves on to other employment inside or outside of agriculture (Taylor and Thilmany 1992).

There are striking differences between FLC and non-FLC workers and their employers. FLC workers have lower mean earnings, lower mean experience with the same employer, less consecutive employment with the same employer, and less regional experience than non-FLC workers. Employers of non-FLC workers have significantly larger average payrolls than employers of FLC workers.

The differences in socioeconomic characteristics between FLC and non-FLC workers were investigated in more detail with data from a 1983 UCD-EDD survey of farm workers (Mines and Martin 1986; Taylor 1992; Taylor and Espenshade 1987). FLC and non-FLC workers are similar in terms of age (36–38 years), sex (73–74 percent male), and education (5 years of formal schooling). Both FLC and non-FLC workers are predominantly foreign born (82 percent and 78 percent, respectively). Nevertheless, there are significant differences between the two groups with regard to immigration status and employment characteristics. Underreporting of illegal immigrants is an inherent problem in surveying the farm worker population. FLC workers, however, were more than 50 percent more likely to report themselves as illegal immigrants (29 percent of the FLC work force surveyed) than were non-FLC workers (19 percent).

FLC workers are concentrated disproportionately in labor-intensive and seasonal crops (for example, fruits) and not in mechanized crops (nuts) or year-round production (livestock). They are found disproportionately in low-skilled tasks (harvest) and not in higher-slated areas (machine operation, sorting). Because supervisors are required by all types of employees, FLC workers are as likely as non-FLC workers to be foremen and supervisors.

Farm labor contractors have a great deal in common with the workers they hire. A recent EDD (1992) survey found that most FLCs are Hispanic (83%) males (86%) who were born outside the United States (54%). Half were Mexican migrants. The average age of these FLCs was 48 years, with ages of individual FLCs ranging from 22 to 86. One-third of all respondents in the survey became FLCs after 1980.

Most FLCs (74%) were agricultural workers immediately before becoming FLCs. Eighty-one percent had been field workers for an average of seven years sometime before becoming FLCs. Nevertheless, they have important characteristics that set them apart from other workers. More than three-quarters (78%) were agricultural foremen. Moving up from field work into supervisory work seems to be the most important requisite for eventually becoming an FLC. Unlike the workers they supervise, most FLCs (54%) speak English well; only 18% speak little or no English. Average formal schooling of FLCs in Mexico is comparable to that of farm workers as a whole (about three years). FLCs, however, also have an average of six years of formal schooling in the United States. Schooling, language skills, and supervisorial experience set FLCs apart from the rest of the farm work force and are no doubt the key to FLCs' success at achieving self-employment status. (The reason that FLCs cited the most often for going into contracting work was to be self-employed.)

As ex-farm workers and supervisors, and as either Mexican immigrants or children of immigrants, FLCs possess an understanding of work processes on U.S. farm operations and have close ties with immigrant streams and migrant-sending places in Mexico. Knowledge of English and some formal schooling in the United States enables them to turn this knowledge and these contacts to their advantage, in a sector whose functioning and expansion are predicated on having continued access to new immigrant workers able to work efficiently in the U.S. farm labor system. Although immigration status is difficult to ascertain and was not explored in the EDD survey, status as a legal immigrant probably is also important in the transition from farm worker to FLC. Legalization of large numbers of farm workers under the SAW and amnesty provisions of IRCA may have helped to increase the number of FLCs.

Most FLCs have established relationships with several farmer-clients, but they complain that farmers have little loyalty. According to FLCs, competitors are constantly underbidding them in an attempt to steal away their farmer-clients. This means that an FLC whose costs increase, say because worker accidents raise workers compensation premiums, has a hard time raising the overhead or commission he charges because the farmer-client is aware that other FLCs will do the job more cheaply. Harvesting and other farm tasks usually involve a large number of units, and both FLCs and farmers acknowledge that, for example, a $.01 per vine cost difference on 2 million vines is $20,000 or the price of a new car, and farmers can and do switch between contractors for a $.01 per unit cost savings.

The personal nature of FLC-employer relationships is reflected in the absence of both an FLC organization and a secondary market in FLC businesses. There have been numerous attempts to organize an FLC organization to establish a code of conduct and minimum overheads and standards, but most of these associations have lasted only two or three years before collapsing.[4] There is no secondary market in FLC businesses; many FLC businesses are family affairs because it is difficult to maintain farmer-clients if the FLC business is transferred to a stranger.

FLCs and Worker Earnings

FLCs pay piece-rate wages for most jobs, so the continual movement of workers between employers, social networks that exchange information, and migrancy should tend toward equality in the piece rate per unit of work done. However, there is a surprising variety of piece-rate wages paid. One survey of Fresno-area FLCs, who picked green tomatoes that are then ripened before selling them, found that three of six FLCs paid $.37 or $.38 per 25-pound bucket in 1989; two paid different rates for first and second pickings ($.33 and $.38); and one paid a group or crew piece rate (Vaupel 1991).

This means that workers doing the same task in the same area will be offered slightly different piece-rate wages, but conversations with workers reveal that the announced piece-rate wage is just one determinant of their earnings, albeit an important one. Tomatoes are picked while they are green into buckets that weigh about 25 pounds when full, and each worker fills his two buckets, carries them to a tractor-trailer truck that moves slowly through the field, and then gets his individual picking card punched for each bucket picked by a checker who rides on the truck. Earnings vary between crews because some FLCs and foremen require workers to heap up tomatoes on top of the bucket or do not credit workers for buckets with "too many" pink tomatoes, so that workers who pick 10 or 20 percent more or fewer tomatoes may wind up with the same units of work credit at the end of the day.

Earnings vary not just with piece rates and the system for crediting work done; they also vary with conditions in the field and the efficiency with which the work is organized. The weather, pests, and farming practices affect yields, the ease of picking, and thus how fast a worker can fill buckets with tomatoes. Work organization refers, for example, to whether a worker who fills his or her buckets must walk a short or long distance to dump them and, once at the truck, whether there is a long line of workers with buckets ahead of him or her.

Just as a worker's earnings are affected by more than his or her own efforts and the piece-rate wage, a worker's take-home pay is affected by circumstances that vary from FLC to FLC. As recent immigrants, many FLC workers need everything from work authorization documents to housing to transportation to and from the work site. FLCs provide these services at various costs, and so a worker's take-home pay can vary between FLCs, even if earnings are equal.

The 1980's switch from employers hiring workers directly to hiring workers through FLCs has generally reduced take-home pay because farmers hiring workers directly are more likely to provide on-farm housing at little or no charge, thus affecting worker housing and transportation costs. FLCs sometimes provide these services, but more often they arrange with someone else to provide them because, if the FLC provides services to workers, they are covered by fairly strict regulations governing FLC services to workers.[5] In some cases, FLCs require their workers to buy some of these services in order to have a job; for example, local workers have complained that FLC foremen sometimes require them to ride to the field in the foreman's truck at a cost of $4 per worker per day, even if they have their own transportation. FLCs and their foremen have justified mandatory transportation charges by asserting that farmers do not want too many cars in the field or that workers driving their own cars may steal oranges or grapes.[6]

Despite the diversity inherent in agriculture that makes each farm task unique, farmers can protect themselves against FLCs bargaining for higher overheads by individually conforming to the normal overhead and working collectively to ensure that the labor market is awash with recent immigrants so that FLCs can flourish. Workers, on the other hand, usually have access to less information about the factors that will ultimately determine their take-home pay and, even if they wish to forego a FLC-supplied service, they may have to accept it in order to have a job. The FLC labor market is geared to the recent immigrants who replenish it, and they often need the services that FLCs offer for a fee. The fact that these services are made mandatory is simply another reason for workers with other U.S. job options to get out of the FLC labor market as soon as possible.

FLCs and IRCA

By offering farmers a less risky substitute for direct-hired workers at low cost, FLCs provide an almost ideal mechanism through which farmers can shed the risks and many of the costs associated with hiring and recruiting seasonal workers that result from IRCA. A statewide survey of farm employers in 1989 revealed that the most common planned personnel change in response to IRCA was to turn more seasonal employment over to FLCs (Martin and Taylor 1990). Forty percent of all respondents in the survey who planned changes in personnel practices expected to delegate more worker recruitment to FLCs. Between December 1988, when employer sanctions became effective for all farm employers, and December 1989, the share of FLC workers in the total California seasonal farm work force rose from 23 percent to 40 percent, and FLCs accounted for nine-tenths of the increase in seasonal farm employment in California between 1984 and 1989.

Preliminary results from a 1992 farm employer survey reveal that the trend toward FLC use on farms continued after IRCA's employer sanctions took effect. Almost half of the farmers responding to the survey used FLCs in 1992 to meet all or part of their seasonal labor needs, especially for harvesting. Forty percent of these farmers did not use FLCs five years earlier. The importance and growth of FLC use is particularly striking in the Central Valley, the largest agricultural region of the state, where 70 percent of the farmers reported using FLCs to meet their seasonal labor demands in 1992, up from 39 percent five years earlier. Payments to FLCs accounted for an average of 41 percent of the total labor cost on these farms.

Historically, FLCs have demonstrated what appears to be a willingness to bear the risks associated with hiring illegal immigrants. Under the Migrant and Seasonal Workers Protection Act of 1982 (MSPA), FLCs were the only employers for whom it was illegal under federal law to hire unauthorized immigrant workers prior to IRCA. Despite these sanctions, FLC activity expanded rapidly in the 1980s, and FLC reliance on new and illegal immigrant workers has been high. FLCs have also demonstrated an ability to evade immigration and labor laws. Indeed, efforts to survey FLCs reveal that one of the most salient characteristics of FLCs is in their elusiveness.

Farm labor contracting today represents perhaps the most fiercely competitive component of California agriculture. The contract prices FLCs receive from farmers are low: On a per unit labor basis, they may not cover the formal cost of labor to FLCs (the minimum wage plus payroll taxes, etc.). In between the contract price per unit of labor (hour) delivered to farmers by FLCs and the per unit cost of labor to FLCs, which can be imputed from the minimum wage, lies the mystery of the invisible profit margin in FLC operations. Raul Cavazos (1992) notes:

> If it takes 27 percent in the contract to pay the costs of unemployment insurance and workers' compensation insurance, and another contractor says he will charge a 20 percent commission, he will get the job. There is so much competition in the San Joaquin Valley that there must be at least 600 contractors there, but I believe that maybe 40 percent are crew leaders with certificates of registration. We are charging, on some jobs, five dollars an hour plus a 35 percent commission. I get angry because many of the growers, farmers, and packinghouses do not even ask the contractor for certificates of registration, state licenses, or evidence of workers' compensation insurance. It makes me wonder how these guys can run 6,000–7,000 acre operations, and they don't even ask the contractor for proof that he carries insurance. Some of these contractors have no offices or sanitation facilities, and they don't even ask workers for their I-9s.

Conclusions

California agriculture, with its labor intensity, seasonality, and reliance on new immigrants, has one of the most marginalized work forces of all sectors in terms of low earnings, unstable employment, and poor working conditions and worker benefits. Labor contractors, with their roots in field work, are the middle layer between well-organized and informed farmer-clients and vulnerable immigrant workers with few employment options, but often with a need for specialized immigrant services such as job matching, credit, transportation, and housing that are not available except from FLCs.

Farmers' risks and costs of recruiting and hiring immigrant workers increased after IRCA. Farmers complain that paperwork requirements to comply with IRCA's employer sanctions make it uneconomical to hire large numbers of seasonal workers directly. Many farmers also believe that FLCs are an effective "risk buffer" between themselves and immigration and labor law enforcers. For example, it is nearly impossible for government authorities to prove that farmers knowingly hire illegal immigrants through labor contractors. Under IRCA, FLCs are usually considered to be employers in their own right, with responsibility to comply with employer sanctions with regard to their own workers. Farmers, therefore, are not jointly liable with FLCs for knowingly hiring illegal immigrants.

IRCA may have increased the supply of FLCs to more than meet the demand. Growth in FLC activity, measured by numbers of workers, payroll, or the share of FLC workers in the work force, increased in the 1980s, but the number of FLCs increased more. The largest single spurt in FLC expansion came after IRCA's full implementation in 1988. Confronted by well-organized farmer-clients who regularly exchange information on the fees they pay to FLCs and by competition from the large number of new FLCs, contractors today are often forced to work with profit margins so tight that it has been suggested that they are engaged in "destructive competition." If the supply of workers to FLCs were not highly elastic, contractors would find themselves in a double squeeze: on the one side, farmer-clients willing to pay small mark ups above payroll costs, and on the other side, farm workers demanding high wages. Marginal FLCs would be driven out of the market and farmers would be forced to pay a higher mark up for FLC services. Fewer farmers, therefore, would use FLCs, but FLC worker wages and profit margins would rise.

The availability of an abundant supply of vulnerable immigrant workers at low wages, however, shifts the balance of market power in favor of farmers and FLCs. With access to a continued flow of new immigrant workers accustomed to wages less than one-eighth the minimum U.S. wage, contractors are able to pass on their low mark up to workers. In order to appear to comply with minimum wage and employment laws, they may do this by paying workers the minimum wage and then charging them for required services such as transportation. They also may recover profits by providing workers with inadequate working conditions, including health and safety, which new immigrants may be willing to tolerate because they are not much different from conditions in their home country; or by not disclosing their entire work force to unemployment insurance and social security authorities, thereby saving payroll taxes.

Given the minimum wage, mandatory payroll taxes, and unemployment laws, evading such laws appears to be the principal means of extracting a profit from the high-supply elasticity of immigrant labor. In practice, one of the major "services" FLCs provide their farmer-clients is that of circumventing state and federal employment laws to extract a profit from the abundant supply of immigrant workers. It appears that IRCA's employer sanctions have failed to end this "conspiracy of silence" between farmers and FLCs.

NOTES

1. These are reporting units with SIC codes 01, 02, and 07 in California's unemployment insurance (UI) system.

2. It is very difficult to find consistent data on farm-worker employment by commodity. A common approach is to ask farm advisors to estimate the hours of regular and seasonal labor required per acre to produce a commodity, but there are often unexplained differences between reports for the same commodity that is similarly produced in two counties. Mamer and Wilke (1990) report that the 40,000 acres of processing tomatoes in Yolo County required 6 regular and 38 temporary hours of labor per acre in 1989, while the 63,000 acres in Fresno County required 22 regular and 31 seasonal hours. These numbers were combined to generate a statewide average of 16 regular and 34 seasonal hours per acre. The major reason for the difference in hours is that in Yolo County irrigation hours were reported to be 0, while in Fresno they were 7 hours per acre. The Fresno report also included five hours of regular supervisory labor per acre, while the Yolo report had none. Both county reports estimated that harvesting required 11 to 13 hours per acre, or that harvesting required fewer hours per acre than thinning and weeding (14 hours).

At 12 hours per acre, sorting the tomatoes from 330,000 acres required 4 million hours of labor. Sorters sometimes work 12 hours per day and 6 day weeks; if they average 72 hours for 10 weeks, they average 720 hours. These calculations suggest that a total of 5,500 sorters would be required; at the usual wage hourly wage of $4.50 in 1993, sorters average $3240 each.

3. In one case, a Napa vineyard, whose offices were adjacent to an award-winning restaurant, had trouble recruiting workers despite higher-than-average wages. A farm labor consultant interviewed potential workers and informed the vineyard that it needed a hiring office away from the restaurant; farm workers felt uncomfortable having to go by it en route to the hiring office.

4. A January 1989 FLC conference at UCD heard that the FLC Alliance had 60 FLC members who had a peak of 25,000 to 30,000 employees, and that the National FLC Association had 425 of its 750 FLC members in California.

5. A 1989 California law made all housing with five or more farm workers subject to farm worker housing laws, regardless of who the landlord is.

6. The FLC labor market assumes that immigrant workers have no other demands on their time. Summer harvesting jobs typically begin at 5:00 or 6:00 a.m. and continue until 2:00 or 3:00 p.m.; workers must often wait for work to begin, and then wait for rides to their housing at the end of the workday. Workers often rest and typically drink beer late in the afternoon at their homes. The work week is five or six days. Workers are often paid away from the work site on a nonwork day, so they must pay for a ride to the check-disbursing point, then cash their check, often at a store, so that even being paid can take half a day and cost $4 or $5 for transportation.

11

IRCA's Effects in Oregon Agriculture

TIM CROSS, ROBERT MASON, AND LUIS CARABALLO

Introduction

The Immigration Reform and Control Act of 1986 (IRCA) was an attempt to control illegal United States immigration through the legalization of worker and employer fines and sanctions. Many special provisions were included in the law pertaining to agriculture because of the industry's reliance on a large migrant and seasonal work force. Many studies have been conducted to evaluate the impacts of IRCA on agriculture (USCAW 1993). These studies examined both regional and national impacts of IRCA, and most concluded that there were few, if any, beneficial impacts directly attributable to IRCA in agriculture. However, they did point out differences that exist at state or regional levels in the agricultural labor market.

Oregon agricultural employers rely heavily on seasonal farm workers for the production and harvest of a diverse set of crops. This chapter discusses the impacts of IRCA on agricultural employers and workers in Oregon by examining changes in labor demand and supply. The discussion is entirely descriptive, and is based on surveys and interviews of agricultural employers, workers, and labor contractors conducted during the past five years. Discussions with key informants in the migrant worker system are also used to describe perceived changes or impacts. These key informants were all closely tied to seasonal workers, and often functioned as intermediaries between workers and employers, government agencies, and other institutions.

A number of factors have influenced Oregon agriculture over the last few years. Among the more important are international trade agreements, natural resource policies, and consumer demand for food products, as well as immigration reform policies. Separating the impacts of any single factor from all the others is impossible. This chapter focuses on those changes that were likely to be highly influenced by IRCA, recognizing that many other factors may have been just as important in producing the observed changes.

Demand for Labor

Overall demand for labor in agriculture has remained fairly constant over the last five years, but changes within segments of the market have occurred. This section begins by examining trends in fruit, vegetable, and specialty horticulture (FVH) crop production in Oregon. These crops are typically the most labor intensive in the state. Benefits and labor management practices are explored next, followed by a discussion of farm labor contractor (FLC) use. The section concludes with a summary of wages reported by employers and direct payroll overhead costs associated with IRCA.

FVH Production

Agriculture is considered to be the largest industry in Oregon, with farm sales of over 170 different crops. The production of this diverse set of crops, with total sales of $2.7 billion in 1992, is concentrated in the Willamette Valley, with small pockets of intensive crop production also located outside the valley in irrigated areas. FVH acreage includes tree fruits (apples, pears, and cherries), vegetables for processing (sweet corn, broccoli, cauliflower, carrots, and beans), small fruits and berries (strawberries, blackberries, raspberries, and wine grapes), and specialty crops (Christmas trees and nursery crops).

Figure 11.1 shows total FVH acreage in Oregon since 1982. This graph shows that total FVH acreage has increased slightly over the last 11 years, with vegetable crop acreage increases contributing most to overall acreage variability. Little change has occurred in small fruit and berry acreage or in tree fruit and nut acreage. Specialty crop acreage includes only Christmas trees and bulb crops. No acreage figures are available for nursery and greenhouse crops.

The growth of nursery crops is not adequately reflected in the acreage reported in Figure 11.1. Most nursery crops are grown intensively, but use relatively little land. However, an examination of the farm-gate value of Oregon FVH sales, shown in Figure 11.2, does provide an indication of the growth that has occurred in the production of specialty crops. This graph shows that the total value of FVH sales has almost tripled over the last 11 years, and most of this growth is attributable to the increased value of specialty crops. Nursery crops account for the largest share of specialty crop sales, with Christmas tree and bulb crop sales accounting for most of the remainder.

Figure 11.1. FVH-harvested Acres in Oregon, 1982–92

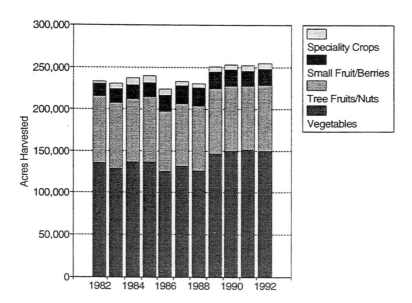

The growth in nursery crop production has led to increased demand for agricultural labor. Many tasks in nurseries are several-month to year-round activities. A survey of 21 nursery employers (Mason, Cross, and Nuckton 1993) found that they annually employed an average of 72 year-round workers and 32 seasonal workers per nursery. Peak nursery labor demand occurs from January to May, a time in which other FVH crops have low labor demands. In addition, increased acreage of Christmas tree production has led to greater labor needs in November and December. Sales of FVH crops accounted for 40 percent of all agricultural sales and 60 percent of all crop sales in 1992. The relative share of FVH sales has increased over the past 11 years.

The net effect of nursery crop growth has been to increase the length of employment for some seasonal workers and provide expanded year-round employment for other workers. IRCA has provided the legalization tool needed by migrant Hispanic workers to take advantage of this employment opportunity. Because nurseries tend to hire from the same labor pool as other FVH employers, this competition for workers has led to improvements in working conditions and benefits offered to nursery workers.

Figure 11.2. Farm-gate Sales of FVH Crops, 1982-92

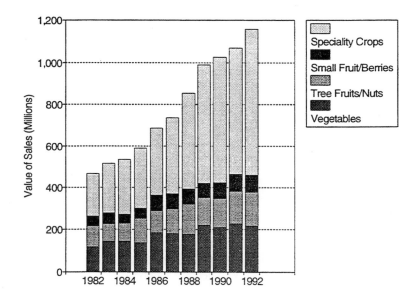

Benefits and Working Conditions

The lack of pre-IRCA baseline studies documenting benefits and working conditions in agriculture makes it difficult to quantify changes in these areas. Instead, recent surveys of employers regarding benefits offered and surveys of worker perceptions of changes in working conditions are used to examine changes due to IRCA.

Table 11.1 shows the results of a 1991 survey of employers in four crops (Mason, Cross, and Nuckton 1993): nursery, strawberry, Christmas tree, and pear. This table shows that, overall, a fairly low percentage of employers provide comprehensive benefit packages. However, considerable variability is evident among the crops examined.

Nurseries provide the broadest range of benefits (although many benefits are provided only to year-round workers), Christmas tree growers routinely pay bonuses, and strawberry and pear growers provide the highest percentage of housing. The average number of years benefits were provided suggests that benefit offerings have not changed significantly since IRCA was passed in the strawberry and pear industries, but benefits have increased in the Christmas tree industry.

An evaluation of workers' perceptions of working conditions is shown in Table 11.2. Workers were asked if they thought working conditions had improved, remained the same, or worsened over the last five years. Overall, the majority reported that working conditions have remained about the same. Pear and strawberry workers had the highest percentages of workers reporting improvements in working conditions. Worker perceptions are subjective measures, and questioning them about changes does not reveal the extent to which there are problems in working conditions. However, these findings appear to differ from national studies suggesting that working conditions in agriculture have deteriorated since IRCA's passage.

Labor Management Practices

Labor management in Oregon is largely performed by employer-owners, without the heavy dependence on farm labor contractors observed in California and some other states. This keeps Oregon growers in close contact with workers and provides direct communication links between the two groups. A drawback of this characteristic is that growers are not only labor managers, but must also devote time to other production, marketing, and financial management issues. This results in the use of fairly traditional labor management practices. Recruiting, training, hiring, and firing procedures have not changed significantly since IRCA was passed.

Most growers rely on workers returning year after year as their source of labor. Some growers have begun to implement training programs, but this is largely due to pressure from workers compensation insurance and is not the result of IRCA. Discipline policies generally consist of verbal warnings followed by termination.

Agricultural labor management practices in Oregon (as in most other states) are not, by most accounts, very modern. One sign that labor management practices in Oregon need improvement is the fact that during the past five years, several interviewed growers experienced labor shortages, in spite of the apparent ample supply of labor.

Table 11.1. Summary of Benefits Provided by 87 Agricultural Employers in Oregon, 1991.

Benefit	Nursery (N=21)			Christmas Trees (n=32)			Strawberries (n=23)			Pears (n=11)			Overall (n=87)	
	Number	%	Years	Number	%	Years	Number	%	Years	Number	%	Years	Number	%
Housing	5	24	NA	5	16	3.8	15	65	7.9	11	100	23	36	41
Bonuses	17	81	NA	21	66	6.8	7	30	5.4	9	82	14	54	62
Paid education	10	48	NA	3	9	2.3	3	13	NA	5	45	NA	21	24
Health insurance	16	76	NA	5	16	4.2	5	22	5.0	4	36	19	30	34
Paid vacation/sick leave	18	86	NA	6	19	3.5	4	17	11.0	5	45	NA	33	38
Transportation	1	5	NA	11	34	5.6	15	65	12.3	4	36	NA	31	36
Work equipment	18	86	NA	18	56	9.9	14	61	20.7	11	100	19	61	70
Profit sharing	11	52	NA	0	0	0	2	9	NA	1	9	12	14	16
Child care	0	0	NA	0	0	0	2	9	NA	0	0	0	2	2
Other	8	38	NA	3	9	6.5	4	17	NA	4	36	NA	19	22

*NA = Not Available

Table 11.2. Worker Perceptions of Working Conditions, 1991.

Working Conditions Have:	Nursery		Christmas Trees		Strawberries		Pears		Overall	
	Number	%	Number	%	Number	%	Number	%	Number	%
Improved	9	32	6	24	10	29	16	49	41	34
Remained constant	17	61	18	72	21	60	15	45	71	59
Worsened	2	7	1	4	4	11	2	6	9	7

Some of these shortages may have been due to economic circumstances, a locational disadvantage, or weather-related abnormalities (such as crops maturing earlier than usual). In other cases, it was more likely a case of needed improvement in management.

Farm Labor Contractors

Many areas of the United States have experienced marked increases in the amount of seasonal labor that is hired through the services of a farm labor contractor. In Oregon, use of FLC services varies by commodity and region. The apparent trend is toward increased use of FLCs only for production of those commodities which have used FLC services in the past.

Employment data regarding FLCs in Oregon are sparse. Surveys have shown that 80 to 90 percent of strawberry (Mason and Cross 1992a) and cranberry (Mason and Cross 1992b) growers use FLCs as one method of recruiting a portion of their workers. The same surveys show that growers have relied heavily on other recruitment methods as well, including employee referrals, workers who return year after year, the state employment office, and newspaper want ads. Unfortunately, the estimated percentage of employment obtained through each of these methods is unavailable. However, one frequently finds FLC and non-FLC workers side by side in Willamette Valley FVH fields.

FLCs are used by many Christmas tree and strawberry growers. Strawberry producers hire both FLC and non-FLC workers, and Christmas tree growers tend to employ crews that are entirely supplied by FLCs. No FLC activity was found in the nursery crop and pear industries. Nursery crop growers provide higher levels of year-round employment than most other agricultural commodity producers, so their low use of FLCs is to be expected. The pear growers surveyed were all in Hood River County, which is located outside of the Willamette Valley. These growers have a long history of providing housing and other benefits as a way of attracting and maintaining seasonal workers, instead of relying on FLCs.

According to a 1991 survey (Mason, Cross, and Nuckton 1993), on average, Oregon FLCs were 39 years of age; they employed three foremen and 340 workers; they worked for five growers per year; and their annual 1991 payroll was about $370,000. All 26 FLCs surveyed were licensed, and 8 of the 26 felt that unlicensed contractors hurt their business by paying lower wages, ignoring withholding tax requirements, and hiring illegal workers. Only 5 of the 26 still expected to be in business in 1995 because of reduced supplies of legal workers, increased paperwork, and increased levels of potential fines.

This survey also pointed out a sharp contrast between FLCs providing strawberry workers compared to FLCs providing Christmas tree workers. In general, Christmas tree worker FLCs provided more days of employment for workers and paid higher wages than strawberry worker FLCs. Both reported average work days of seven to eight hours. Most strawberry workers were hired through FLCs during harvest, while most labor provided by FLCs in Christmas trees was for nonharvest activities (planting and shearing).

Growers of commodities that require large numbers of short-term seasonal workers will probably continue to pursue employment of FLCs. However, the wide variety of crops grown in Oregon, and the specialized skills needed by workers to produce some of those crops, will encourage some degree of specialization by commodity among contractors. Many growers are diversifying crop production to level out their labor requirements during the growing season and thus reduce their need to recruit workers several times during the year. Growers also report increasing levels of cooperation among themselves to provide longer employment periods. Diversification and cooperative agreements will tend to diminish demand for FLC workers.

Wages

Wages of seasonal agricultural workers in Oregon are influenced by Oregon's state minimum wage. From 1980 to 1988, the state minimum wage in Oregon was below the federal minimum wage. Since 1988, Oregon's minimum wage has been above the federal minimum wage. In 1992, the minimum wage in Oregon was $4.75 per hour, $.50 an hour higher than the federal minimum.

Another factor that influences Oregon's seasonal agricultural wages is location. Migrant workers traditionally move from California to Oregon to Washington during the crop year. Oregon's wages were set to attract workers from California and retain an adequate number for production so that all workers did not continue onward to Washington. IRCA encouraged some workers to settle in Oregon, potentially reducing the number of migrant workers in the state. As workers settle, incentives for offering higher wages due to a locational disadvantage are reduced.

Table 11.3 summarizes wages reported by growers and FLCs during 1991 (Mason, Cross, and Nuckton 1993). Wages for year-round workers were generally higher than wages for seasonal workers, with pears being the only exception. Seasonal nursery workers were paid about minimum wage levels; growers reported paying all other workers wages of $1 to $2 an hour above minimum wage levels. Results from a 1980 survey (Cuthbert) showed that wages averaged $4.76 per hour for seasonal workers during harvest and $3.42 per hour during nonharvest employment, so wages have obviously increased since 1978.

Table 11.3. Average Wages Paid by Oregon Growers and Farm Labor Contractors, 1991

Employer	Industry	Seasonal Workers ($/hour)	Year-Round Workers ($/hour)
Growers	Nursery crop	4.85	6.36
	Christmas tree	5.67	7.14
	Strawberry	5.59	6.00
	Pear	7.76	6.24
FLCs	Christmas tree	6.35	NA*
	Strawberry	7.02	NA

*NA = Not Available

One direct impact of IRCA on the demand for labor is the increased paperwork that was required after its passage. In particular, all employers were required to file completed, approved I-9 forms for each worker they hired. A 1990 survey of 139 strawberry growers (Mason and Cross 1992a) showed that I-9 forms took about ten minutes per form to complete and verify. Total cost of I-9 form reporting was about $60,000, or about $429 per grower. This illustrates one direct increase in the cost of labor to growers as a result of IRCA.

The Supply of Farm Labor

Labor supplies generally were adequate the past five years, but differences in the makeup in the state's farm labor force were observed. This section begins by reporting and comparing the magnitude and productivity of aliens and nonaliens in the work force, then discusses Special Agricultural Workers (SAWs) and social services, SAW migration from agriculture and other IRCA impacts, and, finally, the future of SAWs in the state.

Productivity

Alien migrants were a large percentage of the seasonal labor force before IRCA, and their proportion has increased markedly during the IRCA implementation period. For instance, undocumented aliens made up about 44 percent of all farm workers who harvested strawberries in 1987. They were estimated to be at least two-thirds of the work force in 1990 (Mason and Cross 1992a). Their rapid increase is due not only to the available supply, but also to the fact that domestic workers no longer apply in large numbers for work in agriculture. Moreover, alien migrants are the most productive of all types of hand-harvest workers. According to our estimates, they picked about three times as much fruit per day in 1990 as local teenagers and twice as much as local adults during the harvest season.

About 2 percent of the SAW applicants who signed up nationally were living in Oregon. A total of 25,855 aliens were approved and declared residency in the state through September 30, 1992. Informed observers report that twice that many SAWs work in Oregon who are residents of other states. Local observers believe that SAWs represent about 60,000 to 70,000 workers in agriculture. This value is far short of an estimated 100,000-plus alien farm labor force in the state, and suggests that many are carrying false documents (Dale 1993). With the number of SAWs in agriculture dwindling, firm numbers of workers are difficult to estimate.

For instance, surveys estimate that 26 percent of nursery workers, 66 percent of Christmas tree workers, and 40 percent of strawberry workers first entered the United States in 1986 or later, and that many, if not most, are illegal. About one-third of the nursery, strawberry, and pear workers and two-thirds of the Christmas tree workers said their relatives were here illegally (Mason, Cross, and Nuckton 1993). The latter group of new, illegal workers is young, single, poorly educated, and inexperienced, compared to legal SAWs. They comprise an estimated 25 percent to 35 percent of the FVH harvest work force in the Willamette Valley.

IRCA and Farmworkers

IRCA replaced a two-tiered work force, one legal, the other illegal, with a three-tiered system. The new third stratum is legalized SAW workers, many of whom (62%) brought their (illegal) spouses, children, and other relatives to the state. SAWs average 28 years of age (range 16–68); and half brought their children (3.2 average) with them. SAW workers average four years of schooling and 3 percent are fluent in English.

Supplies of workers were adequate and wage rates have either remained level or increased, according to surveys of workers. For instance, workers reported that they earned, on average, $6.69 per hour. With a labor surplus from an influx of illegal pickers, one would have expected depressed wages to follow, but that did not happen due to Oregon's higher state minimum wage. About a third of the workers said they worked an average of 8.5 hours a day. They worked an average of 3.5 years with any one grower. Pear workers averaged the highest year-to-year employment with one grower: 6.5 years. Year-round earnings were difficult to estimate, since many respondents worked seasonally in Oregon, other states, and Mexico, and in nonfarm jobs. Many had difficulty in recalling the wages they received.

An estimated 5 to 10 percent of the SAW work force leaves agriculture each year for employment elsewhere, usually minimum wage service jobs in the state's restaurant and tourist industries. A working knowledge of English appears necessary for employment outside agriculture. The speed up of binational communities in the state is expected to hamper migration outside agriculture. Their development probably will hinder rather than enhance the incentive to learn English, and agriculture therefore will remain the most attractive source of work for SAW labor.

SAWs and Social Services

Social services — housing, medical care, schools — have not kept pace with the demand for increased supply of alien migrants. For example, an apartment that housed four to six single workers before IRCA now must house one worker and his family. Health care is spotty and has become available only when a group trying to form a clinic is successful in attracting outside funding. Schools generally face a growing demand for classrooms and teachers, and there are cutbacks in state support. IRCA has contributed to the increased demand for services, primarily from the shift to families now accompanying SAW workers to the state.

The influx of SAWs with families has speeded the establishment of a Hispanic presence in the state, such as Woodburn, a community of nearly 4,500 Hispanics — one-third of the population. The magnitude of these farm labor-supply communities has not reached the size or organization of binational communities found in California or Washington state. However, they are shifting from a transient to a year-round population and, if present trends continue, will become well integrated, year-round binational farm labor supply communities in a few years.

An effect of this shift is the increased demand for State Legalization Impact Assistance Grants (SLIAG). The flow of these monies provides an accurate picture of social service uses by IRCA-legalized aliens.[1] The trend for year-round settlement in the state points to a demand for services, particularly health services, that exceeds the supply. Off-season demand now is exceeding the demand at peak harvest. Service providers attribute this to decreased migration and increased population settlement. As a consequence, workers are seeking attention to health problems sooner instead of later, as well as significantly increasing demand for treatments of seasonal ailments, such as colds and flu.

Although SLIAG has cushioned some of the financial burdens of IRCA on state and local government, its long-term ability to provide adequate services is not secure. The eligibility for most SAW aliens will cease in December 1993 and a property tax limitation environment points to a sizable reduction from state sources.

The Future for SAWs

In the meantime, IRCA incentives to improve the status of alien farm labor have fallen far short of its objectives to improve wages, working conditions, and the general well-being of alien farm workers. Other forces, such as an increase in the state's minimum wage and the shift to year-round hiring in some commodities, have improved the economic status of many alien migrants. Because of their seniority, some SAWs have moved into supervisory roles that extend their employment period.

The surplus of farm labor, fostered primarily by workers carrying false documents, delays the balancing of labor supply with demand for legal workers, to the detriment of both employers and workers. Few FVH employers are motivated to work together to extend the seasonal work of SAWs as long as labor surpluses abound. A stable work force enables growers to escape turnover costs, for example, costs of recruitment, hiring, and training. The magnitude of this cost has not been fully researched; nor has the benefit been studied for SAWs achieving greater year-round earnings, but they should be.

No viable procedure has been established in Oregon agriculture that allows an efficient match between jobs and workers.[2] Until one is constructed, SAWs may have little opportunity to improve their year-round earnings. They will continue to be among the chronic rural poor in Oregon. Growers will continue to pay the costs of an inefficient farm labor market, and taxpayers will pay welfare costs for the state's expanding rural underclass.

Conclusions

Oregon's FVH agriculture was already experiencing a number of economic changes in the last half of the 1980s, when IRCA was passed and implemented. The passage of IRCA produced an additional uncertainty — this one over the supply of competent farm labor. The penalties for hiring illegal aliens were viewed as a real threat, particularly when growers were subject to discrimination complaints if they did not hire a particular worker. The ability of SAW workers to leave agriculture after 90 days of working in the industry fueled additional uncertainty over the supply of competent help. The industry had little objective information about the magnitude and productivity of their harvest work force. IRCA appeared on the scene with little solid information with which to predict its impact, and raised major concerns over the farm labor situation in the state.

Oregon agriculture has passed through its adjustment of IRCA. This chapter has shown the results of studies to evaluate IRCA's impact on the state's agriculture industry. Disentangling IRCA's effects from all the social and economic forces at work is difficult at best. Data about worker supply and demand, wages and benefits, the use of farm labor contractors, and demand for social services provide a picture that shows the following:

- A two-tiered work force, one legal, the other illegal, was replaced by a three-tiered system. The new third stratum was

IRCA-legalized SAW workers, many of whom brought their (illegal) spouses, children, and other relatives to the state in large numbers. This group represents a fourth to a third of the FVH harvest work force in the Willamette Valley.

- Supplies of workers were adequate and wage rates either remained level or increased, according to surveys of employers and workers. Improved wages stemmed more from the phasing in of a higher minimum wage than from IRCA.

- Benefits improved for nursery workers, but that stemmed more from the industry's shift to a year-round work force than to IRCA incentives. Benefits for seasonal workers generally remain poor.

- Farm labor contractors play a role in providing workers, but their use is not widespread. One finds strong evidence of mixed recruitment methods. Increasing cooperation among growers to provide longer employment periods and diversification and cooperative agreements will tend to diminish demand for FLC workers.

- Social services — housing, medical care, schools — have not kept pace with the demand for them. IRCA has contributed to the increased demand for services, primarily because of the shift to families now accompanying SAW workers.

- The influx of SAWs with families has speeded the year-round establishment of a Hispanic presence in the state.

- An estimated 5 to 10 percent of the SAW work force leaves agriculture each year for employment elsewhere.

- A procedure is not in place that matches jobs with SAW workers to extend the employment period. Until one is established, SAWs will have little opportunity to improve their year-round earnings.

IRCA appears to have had only a few short-term effects for the state's agricultural labor. It has created a work force of legal SAWs and also has created an illegal group of false-documented workers. The prevalence of illegal workers stigmatizes Hispanics generally as illegal residents, making their integration into the larger non-Hispanic community more difficult. The surplus of farm labor, fostered primarily by workers carrying false documents, delays the balancing of labor supply with demand for legal workers, to the disservice of employers and workers.

Long-term effects that go beyond agriculture are another matter. The possibility of a large Hispanic influx, changing the state's demography, would produce social and political forces that might have profound effects in the years to come. For instance, children of Hispanic workers, now entering and living in the state in large numbers, will be likely to follow the same pattern of earlier waves of immigrants. They will become better educated than their parents, achieve higher income and occupational status, become active politically, and leave their mark on the social fabric of the state. IRCA may well claim some responsibility for these long-term effects.

NOTES

1. To receive SLIAG funds, local service providers must distinguish IRCA-legalized aliens from other service recipients. Aggregated SLIAG costs for Oregon through federal FY 1992 totaled over $15 million (U.S. Dept. of Health and Human Services 1992).

2. One may argue that FLCs, who supply labor for a fee, could fill this role, even though their use is not widespread in Oregon agriculture. Martin (USCAW 1993, p. 163) estimates that FLCs today supply more than half the harvest labor in many U.S. labor markets. He notes that FLCs, under IRCA, are employers in their own right. Many recruit the "new-new" immigrants, arriving since IRCA, who need intermediaries to help them find work and provide social services. We doubt the ability of FLCs to organize large numbers of legal SAWs, who typically have worked directly with Oregon growers or who have networks established, to find jobs and social services.

12

IRCA Impacts in New Mexico

CLYDE EASTMAN[1]

Introduction

As the prospect of immigration reform loomed large in the early 1980s, New Mexico farmers and ranchers began to echo the concerns of their brethren around the West. Their major concern was the prospect of losing the cheap, tractable labor force that has been part of the socioeconomic landscape in New Mexico for a very long time. They were quite convinced that workers would move out of agriculture to more lucrative urban employment as soon as their status was legalized. After a cursory initial study in 1983, a comprehensive study was initiated in 1988 to systematically assess the impacts of IRCA on New Mexico agriculture and to assess the impacts of legalization on the lives of the aliens themselves. Several publications have already appeared (Eastman, 1984, 1991, and 1992).

Labor-intensive agriculture in New Mexico is highly concentrated in four southern counties – Dona Ana, Luna, Chaves, and Eddy. The first two are immediately adjacent to the border. Of the various agricultural commodities produced in these counties, two stand out from the rest in terms of labor requirements. Chile and onions are still harvested with hand labor and the acreage of both commodities has substantially expanded since 1980. These two commodities have consistently ranked among the top dozen commodities in value over the past decade in New Mexico. In 1991, chile ranked fourth, with $59,219,000 in sales, followed by onions, with $44,538,000 in sales (USDA and N.M. Agricultural Statistics Service, 1991b). Mechanization of the harvest of both crops appears to be at least partially feasible, but cheap, plentiful labor has discouraged adoption of this technology. A few other commodities are also labor intensive, for example, apples, but the amount produced is small (less than $1,000,000 in 1991). Milk production, which is also labor intensive, provides a different example. Since they provide steady, year-round employment, the dairies have little problem attracting and keeping a stable work force.

[223]

A number of observers have argued, before the passage of IRCA and since, that international migration is a social process driven by powerful forces in both the sending and receiving countries and cannot be controlled by laws (Cornelius and Bustamante 1989; Tienda 1989; Stoddard 1986; Donato, Durand, and Massey 1992). Perhaps the most succinct statement comes from Tienda (1989, p. 109):

"After forty years of steady development, Mexico–United States migration is now so institutionalized, so widespread, so much a part of the family strategies, individual expectations, and community structures, in short, so embedded in social and economic institutions, that the idea of controlling it is probably unrealistic."

Of the various theoretical perspectives that have guided research on agricultural labor, Varden Fuller provides insights that are especially relevant to the current New Mexico situation (1991, pp. 92–93):

"Migratory laborers do not exist because the farm economy needs them; they exist because our society has a large backlog of unsolved social and economic problems. . . . The people have not become poor from working in agriculture; they have become agricultural workers because they are already poor. . . . Temporary work in agriculture is taken mainly by persons who chronically or intermittently can get nothing better to do; when something better appears, they leave."

The backlog of economic problems in this particular case is located in rural Mexico. "El Paso del Norte," the Pass of the North, has long been a major stop along the migratory route from the impoverished villages of rural Mexico to the land of economic opportunity, the United States. Many migrants get their first orientation to the U.S. work place in El Paso and the surrounding communities in low-wage seasonal employment. Southern New Mexico's agriculture is one of those employers.

Chile and Onion Production

Approximately 300 New Mexico farmers produced almost 30,000 acres of chile in 1991. The distribution of production among growers appears in Table 12.1. Acreage increased in eight out of ten years since 1982, more than doubling in that interval (Figure 12.1). Chile is one of three or four enterprises on the typical farm. Since production needs to be rotated to avoid buildups of soil-born pathogens, it is virtually impossible for a grower to specialize in chile. The search for uncontaminated soil is the primary reason for the movement of production away from the area immediately adjacent to the abundant labor supply of the El Paso–Ciudad Juarez metropolitan area.

Table 12.1. The Structure of New Mexico Chile Production:
Proportion of Growers by Size of Operation, 1991

Size of Operation	Distribution	Mean Acres within Each Category
< 100 A	73%	35 A
100 – 249 A	20%	135 A
> 250 A	7%	409 A
Total	100%	

Note: There are approximately 300 commercial growers in New Mexico.

Source: National Agricultural Statistics Service, Personnel Communication from Charles Gore, October 26, 1992.

Figure 12.1. New Mexico Chile Acreage, Harvested Acres, 1982–92

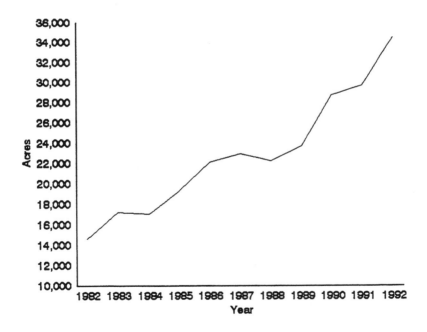

The New Mexico onion acreage is presented in Figure 12.2. Beyond these numbers, not many data are available on the structure of onion production. Knowledgeable observers agree that it is somewhat more concentrated than chile production. However, like chile, onions must be rotated to avoid a buildup of soil-born pathogens. Consequently, it is also virtually impossible for any grower to specialize in onions, and since onion prices are very volatile, there are economic reasons for producers to diversify their commodity mix as well. Whereas most chile is grown under contracts that specify acreage and price before the crop is planted, onions are not contracted. A volatile national supply-and-demand situation determines what the prices will be in New Mexico. Since essentially all onions produced in southern New Mexico are fresh market varieties, they have a limited market window and must be sold when they are ready to harvest. The New Mexico Department of Agriculture estimates that between June 10 and the end of July, from a third to one-half the onions sold in the United States come from southern New Mexico. In 1992, 4.9 million 50-pound bags were shipped from this area. The decline in onion acreage in 1991 was due to price fluctuations and other factors, not to the availability of harvest labor.

Figure 12.2. New Mexico Onion Acreage, Harvested Acres, 1982–92

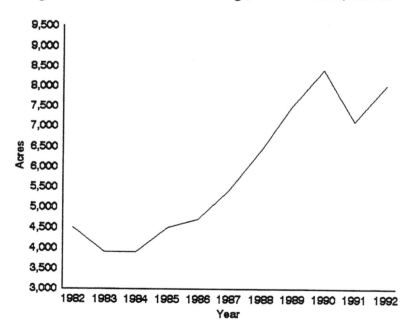

Employment and Mechanization

The most accurate estimates of employment trends in chile and onion harvest are extrapolations from acreage data. Since there has been essentially no change in harvest technology, trends in labor use must parallel acreage trends. Thus, it is fair to say that chile harvest labor use more than doubled between 1982 and 1992, following the acreage increases (Figure 12.1). In similar fashion, onion labor requirements also increased substantially during the same period. Table 12.2 shows the total estimated seasonal work-force requirements for both commodities in 1990. Figures were compiled by the New Mexico Chile Commission from a survey of growers in that year. The monthly variation in labor requirements is large, in spite of the fact that the onion harvest is essentially over before the chile harvest peaks. Also, there is still a fair amount of specialization; not all onion workers do chile, and vice versa. Nothing in these data indicates that IRCA has had any adverse impact on local agriculture. This finding is consistent with early findings in California and the findings of the USCAW 1993, Martin and Taylor 1989, Martin 1990).

Table 12.2. Total Estimated Seasonal Work Force Needed, by Month, 1990

Month	Chile	Onions	Chile Processing	Grand Total
		----------------Number of Workers----------------		
May	0	94	0	94
June	0	1,769	0	1,769
July	1,010	1,453	81	2,544
August	9,394	1,343	651	11,388
September	15,723	147	1,238	17,108
October	13,295	0	835	14,130
November	9,708	0	647	10,355
December	4,397	0	831	5,228

Source: Data compiled by New Mexico Chile Commission.

Prototype pepper-picking machines have been tested in New Mexico for at least two decades. A small number of machines were utilized during the 1992 season, but still accounted for only a small fraction of the total harvest. It must be emphasized that chile is not a homogeneous commodity, particularly for mechanical harvesting.

There are green chile, red chile, jalapeno, cayenne, and pimento, each with its own harvest characteristics. Even within one type, some varieties machine harvest more easily than others. Consequently, successful mechanical harvesting requires matching an effective machine with the proper variety, and even this may also require some additional processing steps. For example, machine-harvested chile may have to be cleaned of additional foreign matter. Therefore, the decision to switch from hand to machine harvest involves coordinating several disparate elements and implies some risks. Consequently, the incentives to mechanize must be substantial to overcome the inherent tendency to stick with proven production methods.

As long as labor is plentiful, there is little incentive to engage in a complex adoption process of switching to mechanical harvest. However, while wage rates have only inched upward, the chief incentive to mechanize appears to be the increased cost and trouble of meeting all of the more stringently enforced regulations mandated, among others, by the Migrant and Seasonal Agricultural Workers Protection Act (1983). Once the labor force was legalized, several agencies, including the Immigration and Naturalization Service (INS), the Department of Labor (DOL), and the Internal Revenue Service (IRS), have paid much closer attention to agricultural labor than was previously the case. Legal workers are much more visible than illegal ones. They can also more easily file lawsuits. Workers are filing, and winning, more and larger liability judgments against both producers and labor contractors.

IRCA resulted in increased paperwork. Employment Eligibility Verification Forms (I-9s) have to be processed for every worker, buses now have to be inspected and insured, social security and income taxes withheld and reported, working conditions and regulations posted, toilets and water provided, and the list gets longer every year. While labor contractors bear the first responsibility for complying with most of these regulations, growers can also be held responsible for violations. These various liabilities seem to loom larger every year in growers' perceptions.

Farm Workers[2]

Whether migrants or local residents, the majority of adult farm workers were born in Mexico. Unless they arrived in the United States young enough to enter the school system, their level of education is generally six years or less of primary school in Mexico. Few have more than a rudimentary ability in English. Those who have even a few years in the American school system can function in English and have gained other skills as well.

Harvest workers in southern New Mexico may be local residents, residents of the El Paso–Juárez metropolitan areas, or migrants from the outside. Migrants may come with their families or they may be single men whose families are in Mexico. For the onion harvest, migrant families typically come from the lower Rio Grande Valley in south Texas. These arrive in Dona Ana County in late May and remain until harvest ends in late July or early August. These migrants are typically accompanied by school-age children who help in the fields. The majority of these families return to their homes in time for the children to enroll in school.

Chile harvesters tend to reside in the El Paso–Juárez metropolitan area or small surrounding towns. Part of these workers will spend the several winter months, when little casual work is available, in other parts of Mexico. The population of several southern New Mexico villages is also swelled by individual migrants from the interior of Mexico. Nearly all of these men have legalized their status under the Special Agricultural Worker (SAW) program, but have not yet established the permanent residency envisioned by IRCA. Some have long histories of seasonal work in the United States and some are relative newcomers. They stay only as long as they can find work and then return to their families in Mexico. One reason they do not stay longer is the lack of affordable housing. Many rent sleeping space by the night, often from labor contractors. Others live under bridges, in abandoned automobiles, or in farm buildings (Gemoets 1990). The living conditions of these workers have received substantial coverage in the local media from time to time. While migrant families find apartments or houses to rent, single males have special problems. Some tend to drink heavily and engage in rowdy behavior and to be very destructive of housing. Consequently, many rural residents are not enthusiastic about having single farm workers in their communities. Efforts to provide more farm worker housing have been stymied by stringent federal requirements and costs that are very high when amortized over the short harvest season.

A National Agricultural Statistics Service survey during the 1991 harvest season found that chile pickers earned an average of $7.10 per hour for green chile and $5.70 per hour for red chile. There was substantial variation in earnings (from less than $2 to more than $15 per hour), according to the skill of the worker and field conditions (USDA and N.M. Agricultural Statistics Service, 1990, 1991a). Contractors are required by law to pay no less than the minimum wage, regardless of worker productivity. Every contractor who spoke on this point assured us that they do indeed pay inexperienced workers at least the minimum wage. They also said that any new worker not earning at least the minimum wage by the end of the second day would be dismissed.

Hourly wage rates are somewhat deceptive without taking into account annual earnings. The most commonly cited complaint about farm work is that it is so unstable. Workers literally do not know from one day to the next whether they will be employed. They may even endure a long commute to a distant work site only to find there is no work that day. It is also significant to note that during the 1992 onion harvest there were cut-rate labor contractors with workers willing to accept $.05 to $.10 per bucket less than the going rate of $.65 cents. Established contractors were put under pressure by some producers to lower piece rates.

The earning patterns differ considerably among the different groups of workers. Adult migrant workers reported *individual incomes* for 1991 in the range of $3000 to $7200. These workers typically worked six to eight months in two or three commodities and spent several months with their families, either in Ciudad Juárez, but more frequently in the interior of Mexico.

Local workers reported 1991 annual *individual incomes* in the range of $1200 to $7000. This segment of the labor force varies widely in the number of householders involved in agriculture. They reported one to six workers per household. The length of time worked also varied widely. Agricultural employment may represent the primary source of income for principal wage earners who work six to nine months per year, or it may represent supplemental income for wives or teenagers who work only several weeks during the summer.

Migrant families reported annual *household incomes* in the range of $5000 to $13,000. The families had two to seven workers ranging from nine- to ten-year-old children to middle-aged adults. All able-bodied members help out in the field with varying levels of effectiveness. A principal reason for incurring the expense and inconvenience of migration is the opportunity to find seasonal employment for all able-bodied family members.

In only a few cases were the annual incomes of farm worker households above the official poverty level, defined as $13,812 for a family of four, including two children less than 18 years of age (USCensus 1991). Many more than report receiving them could qualify for food stamps.

Legalized aliens report few or no immediate benefits from their new legal status. However, most think citizenship will eventually bring a better job and intangible benefits, usually phrased as a "better life." Most applicants made their way through the legalization process; however, many family members have not applied and new ones are still coming in. Some but not all of these could be accommodated under Family Fairness or subsequent programs. Often family members do not apply for legalization because even a $75 application fee, when multiplied by several family members, becomes a burden. Few migrants reported applying for or receiving any social services while in New Mexico. The most commonly applied for assistance is food stamps. Several migrants complained that the local offices often refused to accept their income reports, saying migrants were not able to accurately estimate their piece-rate earnings. Migrants feel that they were being discriminated against in favor of local recipients.

The Labor Market[3]

The labor markets for onion and chile workers overlap to a substantial degree and have many common features; consequently they are described as one market. The unique elements in the two work forces have already been described. Agricultural producers needing one or a few workers for occasional or longer-term tasks generally hire them directly. However, large-scale, short-term agricultural work forces in southern New Mexico are usually mobilized by labor contractors. Some contractors mobilize temporary local labor crews of as few as a dozen workers for harvest work on one or two relatively small operations. At the other extreme in size, some contractors mobilize several hundred workers for 20 to 30 different producers for harvest and to perform various smaller tasks throughout the year. Some of the workers are certified as legal by the New Mexico Employment Security Commission or Texas Employment Commission and are referred to as a contractor or employer. Some workers have a very casual, impersonal relationship with the contractor, but many are family members or acquaintances who have a long-term relationship with the contractor.

New Mexico labor contractors typically start their careers as farm workers, and in a few cases they may continue to do day work themselves, for example, as tractor drivers, during the off-season. While some earn essentially all their income from labor contracting, others have developed business interests unrelated to agriculture. Most contractors have only a limited command of English; however, one in our sample was a graduate of a United States university.

Contractors provide several vital functions in the agricultural labor market:

1. They recruit and deploy laborers. Since any individual producer requires a substantial amount of labor for a short time period, some way is needed to mobilize and allocate workers in an orderly manner. Contractors perform this essential service.

2. They supervise the workers. Contractors and their subordinates, if the crew is large, oversee the workers while they are in the field. They assign work areas and tasks, resolve conflicts, and ensure the quality of the effort.

3. They keep time of work records. They also check documents and fill out the I-9 forms required by the INS. They are supposed to calculate, withhold, and remit social security and FICA taxes.

4. They complete pay slips and pay workers, most commonly daily and in cash. This requires contractors to manage a substantial amount of capital resources. Some of the largest contractors are beginning to use bookkeepers, who do all the paperwork and prepare weekly pay checks.

5. They may provide transportation. Some contractors operate buses and pick up workers coming across the international bridges in El Paso or from other designated pick-up points. However, not all workers ride on the buses, as a substantial number provide their own transport or share rides with fellow workers.

6. They may provide other assistance, such as with housing or with personal problems. This is especially true with contractors who hire migrant workers. For example, first-time migrants may be helped to find suitable, affordable housing and to get settled into a strange community.

This list of responsibilities illustrates the substantial entrepreneurial ability required of a successful labor contractor. Contracting can provide a good living for large-scale operators or an attractive seasonal income for smaller operators. The large turnover among contractors indicates that a great many are not successful.

While farmers themselves could conceivably perform all the other necessary functions, some way would still be required to mobilize and allocate workers in an orderly manner. Since individual producers require a substantial amount of labor for a very short time period and processors require a steady flow of product through their facilities, close coordination of harvest labor provides the most effective mechanism to accomplish these ends. Labor contractors have emerged to perform these functions. The activities of labor contractors are regulated by Title I of the Migrant and Seasonal Agricultural Workers Protection Act. They are required to obtain certificates of registration for themselves and their supervisory employees. However, registration requirements are easily met and the lure of apparently easy money attracts unscrupulous operators who lack the necessary skills and who are willing to cut corners to get contracts. Penalties and legal judgments are often difficult to collect from unbonded operators who have few assets. In addition, operators who run afoul of the regulations can easily register a spouse or other relative and continue in business.

Workers with their own transportation will often move around from contractor to contractor, depending on field conditions from day to day. These workers look for the fields with the best working conditions. Since they are usually paid on a piece rate, they will try to go where they can make the most money. Several contractors complained privately that this movement made their job more difficult and more uncertain. Workers may also stretch the weekend by a day or two after a weekly payday. When asked how they minimize such inconveniences and maintain some semblance of worker loyalty, several contractors said it was important to treat workers well. This might consist of a small gesture like providing a cold soft drink during the hot afternoon. Having fair but firm work rules was also mentioned. One contractor reported buying turkey and fixings from a local cafeteria and then personally sharing the traditional holiday meal with her chile harvest crew on Thanksgiving. An experienced observer of the local farm worker scene corroborated the effects of differences in management styles. He reported that there is a noticeable difference in morale from one contractor's work site to the next. While workers on one site are joking and obviously enjoying their work, the workers on another site may be sullen, suspicious, and reluctant to talk to any outsider who appears in the field.

Most harvest work is paid by piece rate, but the law requires that workers earn no less than the minimum hourly wage rate, which was raised to $4.25 per hour before the 1991 harvest season. Each type of chile has a different piece rate, which is frequently adjusted to reflect field conditions. When moving into a new field with marginal conditions, the contractor will watch a few of the better workers to gauge whether the piece rate needs adjustment to compensate for poor conditions. Workers sometime refuse to enter a field until the rate is adjusted. Sometimes harvest conditions are so marginal that a flat hourly rate is paid. Ninety percent of green chile workers reported working between six and eight hours per day. Red chile harvesters reported a slightly shorter work day (USDA amd N.M. Agricultural Statistics Service, 1990, 1991a). The length of both the work day and the week may vary considerably depending on the amount of work available, working conditions, and energy of workers. When they provide their own transportation, workers exercise considerable individual freedom as to when they arrive at work and when they quit for the day.

Labor-intensive agriculture in New Mexico is concentrated along the southern portions of the Rio Grande and the Pecos River and increasingly in Luna County between Deming and Columbus. This geography is an important dimension of the labor market because daily commuting from the El Paso–Juárez metro area governs access to the substantial pool of labor residing there. Some workers are reluctant to commute daily as far as Deming or Hatch and cannot commute to the Pecos Valley. Consequently, these areas are more dependent on single male migrants who move in during the growing season. As chile production expands in the Pecos Valley, contractors with crews from the Las Cruces and El Paso areas are moving in to augment locally recruited harvest labor.

During harvest season, hundreds of workers cross the Santa Fe Street Bridge from Juárez to El Paso between 3:00 and 5:00 a.m. to board buses for the ride to the harvest fields. Others drive across the border in their own vehicles. The corner of El Paso Street and Sixth Street is the first intersection after the U.S. Customs complex at the end of the Santa Fe Bridge. Within 50 feet of the corner there is a parking lot where a dozen or more large former school buses park and load. Most of the workers board their buses in this lot. Most come across to work for a particular contractor. Small groups of workers and assorted other vehicles cluster along the first two blocks of each of the two streets forming the intersection. Smaller work crews are assembled at these locations. Buses with empty seats may make one or several passes down these streets to pick up more workers before departing for their work sites. There is considerable negotiation between bus drivers and a few of the workers as a last-minute attempt is made to fill the departing buses. Some workers appear to hold out to the last minute for a little better piece rate. The market is essentially over by about an hour before daylight. After that, workers continue crossing in vehicles and on foot to jobs in El Paso. A much smaller flow of workers cross the border from Palomas to Columbus to work in southern Luna County. These workers either have families living in Palomas or seasonal housing there. There are also a number of single male migrants in the work force who stay around Deming during the agricultural season.

Expressions of dissatisfaction with contractor performance are more frequently heard from regulating agencies than from farmers or workers. There is a high turnover among labor contractors, caused in part by citations for various violations of the myriad regulations governing their activities. Contractors are cited for not completing I-9s properly, not filing social security taxes on time or at all, and many other infractions. If they transport workers, the buses must be inspected and be insured. The U.S. Department of Labor created a stir in Dona Ana County when strike forces of 12 to 15 investigators conducted sweeps of contractors' vehicles and work sites in both 1990 and 1991. After word got out, many contractors quit working for several days rather than risk fines, causing a great deal of anxiety among producers whose chile was ready to harvest (Diven 1990).

Workers interviewed in the fields were rarely ever willing to express any criticism of either their contractor or contractors in general. This is not surprising, since they are beholden to contractors for their livelihood on a day-by-day basis. However, newspaper reports and the staff of Centro Legal Campesino indicate that there are unscrupulous labor contractors who shortchange workers in every way they can.[4] In theory, the open competition among many contractors should operate to correct the more serious abuses; workers should migrate toward those who provide fair treatment and away from those who do not. However, in a situation of surplus labor there may be such a scramble for available work that contractors and workers alike are forced to cut corners in order to survive. Also, the fact that cut-rate operators are willing to undercut the going rate of $.65 per bucket by $.05 or $.10 during onion harvest indicates a labor surplus, at least at that time.

It should be emphasized that since this area is immediately adjacent to the border, the border patrol has a substantial presence here. In any but the most isolated fields there is always the credible threat of a spot check of documents by the border patrol. While it is too much to say that no undocumented workers slip into the work crews, it is also probably true that there are few if any flagrant, wholesale violations of the law in southern New Mexico. In a situation of abundant or even surplus labor, there is little need to risk fines or the loss of a contractor's license by hiring undocumented labor.

Recent harvest seasons have seen workers' protests and union organizing activities. The Border Agricultural Workers Union staged several small demonstrations, for example, one at the 1990 Hatch Chile Festival. In the festival parade, they had an old bus loaded with workers shouting "*huelga*" (strike) and handing out leaflets. The union won one contract to provide harvest workers to one grower in 1991 and won a few lawsuits or settlements with contractors, but support among workers remains limited.

Conclusions

- The largest users of agricultural labor in New Mexico have continued to expand since IRCA without serious labor shortages. Spot shortages may have occasionally occurred, but all the evidence points to an adequate, even surplus labor force.

- Mechanization of the two major labor-intensive commodities, chile and onions, appears to be technically feasible. Some breeding work remains to be done to provide varieties better adapted to machine harvest than some current varieties. Harvesting machines also need some improvements. However, no one close to the scene suggests that it could not be accomplished for substantial segments of either commodity.

- At the present time, the biggest incentive for mechanization comes from the stricter enforcement of labor regulations. Before IRCA, enforcement of many provisions was rather lax, social security deductions were not always withheld or submitted, buses were not inspected or insured, housing regulations were not vigorously enforced, etc. Now the added paperwork of I-9s and the added burden of complying with the various regulations weighs heavier every year. More than the cost or availability of labor, the cost and uncertainty of meeting the regulations plus the fear of lawsuits have many growers seriously entertaining thoughts of mechanization.

- Labor contractors were involved with essentially all large-scale seasonal operations before IRCA, and that has not changed. Some contractors try very hard to be professional, follow all the regulations, and treat their workers fairly, while others operate on a shoestring, cut corners, and take advantage of their workers whenever they can. The latter group contributes greatly to the very high turnover among labor contractors and also get the news coverage. Labor contracting requires a substantial entrepreneurial ability and it can provide a good income for large-scale operators. In a production system where thousands of workers must be mobilized and deployed on a daily basis, labor contractors or some functional equivalent are required. The Border Agricultural Workers Union had a contract to provide workers to one grower in 1991, but that contract was not renewed. The union, despite several years of effort, has not been able to attract much labor support. Few workers, dependent as they are on daily work assignments, dare to "bite the hand that feeds them."

- Newly legalized farm workers report few immediate benefits from their new legal status and many family members remain without documents. Many have not been able to move beyond seasonal work. While few report receiving any public assistance, many would qualify.

- The evidence indicates that the flow of undocumented illegal aliens continues. Since documents are required to obtain employment, documents are being obtained illegally. Observers who check documents began to report seeing bogus documents in 1992.

There are also a number of aliens arriving without documents. Typically, they have a contact, often in the interior, but have to earn travel and subsistence money as soon as they cross the border. These people, usually single men, have to search longer and settle for the most casual jobs, but they have not quit coming. As was mentioned in the introduction, several observers argue that laws cannot stop illegal migration. To quote Tienda: "Expectations are too well ingrained, the networks of supportive friends and relatives too well established, and the economic incentives are too great to close down the flow with a law" (1989, p. 109). With each passing year's experience that argument gains credibility.

The El Paso–New Mexico sector of the border is a well-established crossing point for Mexican immigrants, who continue to arrive in substantial numbers. Until conditions in rural Mexico improve substantially, migration will continue to be attractive. As Fuller so aptly pointed out, farm workers have not "become poor from working in agriculture; they have become agricultural workers because they are already poor" (1991, p. 92). The same can be said for casual workers in general. These immigrants have found it difficult to move beyond casual employment in agriculture, construction, or the service sector because of their low proficiency in English and because they have few skills to offer. As long as they continue to flood into the region, agriculture will have a plentiful labor supply, but real progress in incomes and in living and working conditions is likely to be slow.

NOTES

1. Development Sociologist, New Mexico State University. The author acknowledges the support of New Mexico Agricultural Experiment Station Project 1–5–27412, also data and helpful comments from Charles Gore, State Statistician, NASS/USDA and many conversations with Larry Salazar, Agricultural Labor Specialist with NMDOL. Yolanda Delgado did many of the farm labor interviews and Lois Stanford provided helpful comments on an earlier draft. Finally thanks are also due to Gail James who typed numerous versions of this manuscript. Questions and/or comments may be addressed to the author at Box 30003/Dept. 3169, NMSU, Las Cruces, NM 88003.

2. The primary data cited throughout this chapter were gathered by personal interviews with farm workers and labor contractors in Dona Ana and Luna counties. Drawing anything approaching a random probability sample of farm workers is very difficult, expensive and was well beyond the means of this study. Consequently, primary data are reported as ranges or in qualitative terms. In addition, it must be recognized that income data tend to be less than precise for several reasons. Occasionally, a respondent will produce an income tax return to document the previous year's income but that is the exception rather than the rule. More commonly, all that can be obtained is a rough estimation of the total household income much of which was paid in cash on a daily basis to various household members at various times during the year. Since income tax is not withheld or paid on much of this, no precise total amount for either an individual or the household is ever calculated.

3. This section draws heavily on material previously published by Clyde Eastman "Impacts of the Irrigation Reform and Control Act of 1986 on New Mexico Agriculture," in the *Journal of Borderlands Studies*, Vol.VI, No. 2.

4. Personal communications from the staff of Centro Legal Campesino, Southern New Mexico Legal Services, Las Cruces, N.M.

13

Immigration and *Colonia* Formation in Rural California

REFUGIO I. ROCHIN AND MONICA D. CASTILLO

Background

California's population, which increased by some 8 million in the 1980s, is continuing to grow by a net amount of about 600,000 a year, or 1,650 every 24 hours. Most of this growth is in metropolitan areas, but a large "spillover" of population is moving to rural communities. Many of the rural bound are Mexican immigrants and Latinos/Latinas[1] from other parts of Latin America. A great many end up in the seasonal work force of California agriculture. Since the passage of the Immigration Reform and Control Act of 1986 (IRCA), more Mexicans and Latinos have settled permanently in California's rural communities.

In past decades, Mexican and other Latino immigrants settled temporarily within *barrios* of rural communities as numerical minorities, among populations that were mostly non-Latino or white. During the 1950s and 1960s rural Chicanos/Chicanas moved from rural to urban areas for other jobs and housing. However, during the 1970s and 1980s many Chicanos and Latino immigrants made rural communities their permanent homes. As their numbers increased in rural communities, the numbers of white people decreased in absolute and relative amounts.

No one knows for sure what the impact of IRCA has been on rural communities. However, a report of the University of California Task Force on Latinos (SCR 43 Task Force 1989) noted that at least a half million Latinos were immigrant settlers in rural areas during the 1980s. According to the the UC report, the vast majority of rural Latinos were clustered in some 100 communities where they could get jobs in agriculture. Moreover, the Special Agricultural Worker (SAW) provisions of IRCA assured California agriculture an adequate labor supply, at least to the point that there has not been a problem for growers to find workers when needed.

Colonias constitute the focus of this study. They connote a different reality from the past idea of *barrios*, which were pockets within white rural towns. Today *colonias* refer to rural communities that are almost exclusively Chicano or Latino, and where the white populations are numerical minorities. This demographic transformation raises a number of interesting questions: Are Latinos and Chicanos better off in rural communities where they constitute the majority? Are they being empowered economically and politically within *colonias*? Are the Latino youth of *colonias* better prepared for life and work outside of *colonias* than Latinos of other communities? Are they being prepared for nonfarm employment? Are Latino residents becoming self-employed entrepreneurs at a higher rate within *colonias*? What is happening to the non-Latino population of *colonias*? Finally, are *colonias* significantly different from other rural communities in the Southwest? If so, why?

The Study

In this study, *colonias* are juxtaposed to issues of immigration and agricultural employment. Although past studies have predicted a reduced demand for immigrant labor and a greater use of farm machinery in California agriculture, in agriculture both machinery and farm labor were added during the 1970s and 1980s. In particular, since the 1960s, immigrants from Mexico have entered California by the tens of thousands to harvest and process agriculture's labor-intensive crops, especially fruits and vegetables. In fact, the need for more specialized seasonal farm workers revived agriculture's dependence on labor to the point where California's farm lobbyists convinced the U.S. Congress to make special farm worker provisions within IRCA. Since the passage of IRCA, over 1 million immigrant workers from Mexico have registered under IRCA to work in agriculture; many have also settled down in *colonias* with their families, and in the general process of rural settlement, more *colonias* have been formed.

This study used time-series and cross-sectional data to examine the effects of IRCA on those rural communities where Mexicans and Chicanos have tended to settle. The specific objectives were: (1) to analyze the economic conditions and sociodemographic traits of California's rural communities, especially where Latinos have settled in large numbers; (2) to determine the extent to which Latino settlement changes socioeconomic conditions and well-being within rural communities; and (3) to assess the policy implications of these changes in terms of these conditions and needed reforms.

Recently the term *colonias* has been applied to rural communities in a variety of different ways (Brannon 1990; Texas Dept. of Human Services 1988). The term is Spanish and literally means a colony. It has various connotations, ranging from a nice community to an unstructured community without paved streets, running water, or adequate waste-water disposal. In this study, *colonia* refers to the fact that Mexican settlers usually remit earnings to family and friends in Mexico, using the rural community as a place to generate income and to extract wages and earnings. Such remittances are noted in Fletcher and Taylor 1990, Garcia 1992, and Palerm 1991.

In addition, we call communities *colonias* where Latinos are the majority, and we are interested in knowing if the *colonos/colonas* (the residents of *colonias*) are becoming better or worse off in terms of income, employment, and other indicators. We look at whether or not *colonos* (mostly Latinos) are developing advantageous ethnic enclave conditions or rural underclass conditions with more poverty and disadvantages.

Literature Review

Since the mid-1980s and into the 1990s, a few studies have shown that Latino settlement in rural communities has increased and is partially binational, that is, closely tied to seasonal agricultural employment in California and winter migration to Mexico (Street 1990; Gunter, Jarrett, and Duffield 1992; Taylor 1992; Taylor and Thilmany 1992; Garcia 1992; Palerm 1991). For example, Palerm (1991) found in four agricultural communities (Arvin, Fillmore, Guadalupe, and MacFarland) that about 13 percent of the Mexican workers maintain homes on both sides of the border. He also identified a heterogeneous settlement of Latinos, some arriving before 1940 (old immigrants), some settling between 1940 and 1960 (middle immigrants), and others settling since 1960 (new immigrants). According to Palerm, each wave of immigrants differs from the others in terms of motives for immigration, place of origin, and family networks both in the U.S. and Mexico. Moreover, Palerm describes three distinct types of agricultural employment for Mexican workers. At the first level are the more specialized workers, who can command high earnings throughout most of the year as migrants following a particular crop within the state or as sedentary workers having steady employment with a single multiple-crop producer. At another level is a less skilled migrant farm group, which is likewise employed nearly the year round by following the peak harvest seasons; this group returns to a home community in the U.S. or Mexico. Finally, a group of unskilled migrant farm workers arrives at the height of the harvest season, constituting "the lowest-paid, least secure, and most exploitable form of agricultural employment in California" (Palerm 1987/88, p. 5). The implication of this differentiation of Mexican farm laborers is that the more recent settlers, including uncounted numbers of SAWs, may be in the third level. Other rural Latinos may include large numbers of settlers who qualified for amnesty under IRCA, that is, they may be in the first- and second-level types of employment.

Procedures and Data

The procedures used for studying rural communities (and *colonias*, in particular) involved a number of steps: a review of the pertinent theories on community formation; the use of census data to identify rural communities and sociodemographic changes of the 1950s, 1960s, 1970s, and 1980s; and statistical analysis of communities.

Most of the data are from the 1980 Census of Population (USDC 1982). The census provided initial data on 148 rural communities (including small cities, towns, and "places"). These communities were selected from an initial list of 200 "places" with 20,000 or fewer persons in 1980. They fit a set of criteria that was used for identifying "rural" communities (Rochin and Castillo 1993). When the study began in 1989, there were no better, comprehensive data on communities. More recently, data from the 1990 Census of Population (USDC 1993) have been released, but not with the detail. It is interesting to note that the number of *colonias* studied has increased from 49 in 1980 to over 65 in 1990.

Demographic information was also obtained for each community from the population censuses for 1950, 1960, and 1970. This earlier data gave a clue as to the importance of immigration and Latino settlement over previous years. However, the Census Bureau did not provide a consistent set of data over time with regard to the factors considered in our study and the specific communities considered. For example, for community A there may have been information on self-employment, whereas for community B there were no such data. Thus, the findings are sometimes based on different sample sizes, as noted in the original study (Rochin and Castillo 1993).

Our primary attention focused on rural communities that had over 50 percent Latinos in 1980. They had different traits than the "control" group of other rural communities with fewer Latinos. The control communities of the study had fewer Latinos than *colonias*. Their Latino populations ranged from 15 to 50 percent of the population.

Since over 140 communities were studied with varying proportions of Latinos, the communities were analyzed also on a continuum from a low to high percentage of Latinos. Thus, our arbitrary definition of a *colonia* was relaxed to some degree in our regression analysis. With over 25 bits of general information on each community, we made comparisons of several indicators to see if *colonias* were relatively promising or disadvantaged communities compared to other small rural communities where Latinos are a relative minority.

Theories and Hypotheses

Four types of theories were considered with regard to the possible transformation of rural and ethnic communities: (1) the theory of underclass formation, (2) the *barrio* exploitation theory, (3) the farm worker exploitation theory, and (4) the theory of ethnic economy enclaves. Each theory provided a hypothesis for this study.

Figure 13.1 illustrates the range of theories proposed by others. From left to right they run the gamut from communities with extreme poverty to relatively prosperous ethnic enclaves. Using the writings of the authors indicated in the figure, we hypothesized that "underclass" and "exploitative" conditions would be evidenced in *colonias* by such socioeconomic indicators as low educational achievement, high levels of unemployment, segregated occupations with low earnings (primarily in agriculture), and a high incidence of poverty. With these conditions we expect a process of community "immiseration" to be underway.

We also hypothesized that ethnic "economic enclave" conditions would be established if *colonias* showed signs of being relatively prosperous in terms of Latino-owned local businesses (wherein *colonos* enjoyed the effective fruits of local stores), by the signs of growth in local jobs, and the provision of services (such as police protection, education, and health). In addition, in an economic enclave, or business-oriented *colonia*, we expected to find that local government public expenditures would be higher per capita than public expenditures in our control communities that had fewer Latinos. That is, we hypothesized that high concentrations of Latinos would foster favorable enclave conditions covering local economic enterprises and public goods and services. Although we knew that some communities are considered "rich and advantaged," we could find no theory that explained their formation. In a few comparisons we used other information from the state of California, such as data on the revenues and expenditures of rural communities. This study also employed simple linear regression analysis, where the variable "percent Latino" was regressed against data on each community's average household income, education, employment status, fiscal status, etc. The regressions proved useful in two ways. First, they suggested the degree to which communities might differ in high school completion rates, for example, with increasing proportions of Latinos in a community. Second, the regressions indicated whether underclass or exploitation traits or economic enclave conditions were relatively significant in a community as the proportion of Latinos increased. In short, this analysis provided a way to infer if Latino concentration, which increased in recent years, gave rise to unique patterns of change in California's rural communities.

Findings

Demographically, the census data revealed that none of the study's rural communities had more than 23 percent Latinos in 1950. The highest concentrations of Latinos 40 and 30 years ago were in places like Calexico, along the U.S.–Mexican border. Back then, most rural communities were largely populated by white persons. But beginning in 1970, and certainly during the 1980s, the white/Latino relationship changed in all rural communities. Whereas the Latino presence was significantly lower in rural communities in 1950 (ranging from 1 percent to 22.6 percent of the community's population), the proportion of Latinos ranged from 15.1 percent to 98.2 percent in 1980. Moreover, whereas the highest concentrations were first found along the border in the 1950s and 1960s, the highest concentrations of Latinos shifted to rural communities in the Central Valley of California, in particular in Kern, Fresno, and Tulare counties, among the richest agricultural counties in the United States. These counties also had the state's highest rates of urban and rural poverty during the 1980s (Gwynn et al. 1988; Kawamura et al. 1989).

From 1980 to 1990, all but 4 of the 148 rural communities of our study had significant and dramatic growth in Latino settlers. The four places that lost Latinos were small to begin with, and shrank marginally in population overall. For instance, Piru in Ventura County saw

Figure 13.1. Continuum of Possible Colonia Conditions

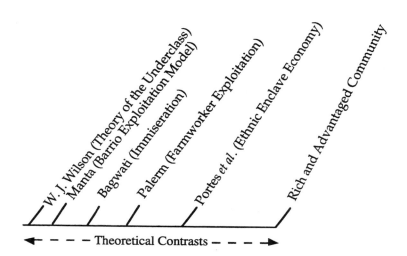

← – – – Theoretical Contrasts – – – →

Pathology of Poverty <u>Characteristics</u>

1. Labor is 'exported' primarily to the private wages and salary sector where exploitation and wealth accumulation within a neo-colonial capitalist system prevails.

2. Perfect substitutability of less educated workers within low wage, segregated employment areas.

3. No representation of minority owned/managed firms.

4. Physical isolation and segregation of the ethnic community highly prevalent.

5. Ethnic community is politically weak. Few public services.

6. Tend to be newer immigrant communities, i.e., have relatively recent concentration of Latinos[a]

Picture of Prosperity <u>Characteristics</u>

1. Labor is supplied to local private and public enterprises for the benefit of the minority enclave economy.

2. High differentiation of workers as unskilled, semi-skilled, and highly skilled managers and professionals.

3. High representation of minority owned/managed firms.

4. High degree of physical integration within the larger community.

5. Ethnic community is politically empowered and active.

6. Tend to be older immigrant communities, i.e., have fewer Latino settlers.[a]

[a] This condition is not necessarily characteristic of the models illustrated but rather reflects a condition specifically hypothesized among rural *colonias*.

its population decline from 1,284 in 1980 to 1,157 in 1990. Still, Piru's percentage of Latinos stayed at 75 percent. What was particularly noticeable during the 1980s was that most rural communities grew in population while the proportion and absolute number of non-Latinos (mostly white) declined significantly. Conversely, the Latino population increased by large amounts in most rural areas (Rochin and Castillo 1993).

These demographic changes not only indicate that inmigration and Latino settlement changed the ethnic composition of California's rural communities, but also that these communities have changed in several other ways. Indeed, this study found several disturbing socioeconomic conditions within *colonias*, especially with regard to the well-being of *colonos* (*colonia* residents).

In most cases, the mean values for indicators such as educational attainment, poverty, employment, business activity, and local revenues and expenditures on public services all pointed to worse conditions in *colonias*, compared to other rural places. For example, in terms of educational attainment, the rates of high school completion and college attendance were relatively lower for *colonos* who were 25 years of age and older, compared to residents of other rural communities. Rates of unemployment were significantly higher for *colonos*, as were the average poverty rates.

Regression analysis revealed important correlations between the percentage of Latinos in a community and several indicators for each community. These are shown in Table 13.1. In examining the table, it is important to look at the columns for the value of the R^2 (the coefficient of determination) and the size and significance of the regression coefficient. The coefficient of determination indicates whether or not the two variables move in the same or opposite directions and the degree of linear association. More informative is the regression coefficient, since it indicates how much the dependent variable changes as the independent variable (percent Latino) changes. Each regression coefficient with an asterisk is significant at the 5 percent level. A positive coefficient suggests that as the proportion of Latinos increases in a community, there is a proportionate increase in the value of the indicator. A negative coefficient (in parenthesis) shows an inverse relationship. As the regression coefficient approaches 0, so too does the correlation between the "percent Latino of a community" and the value of the community's indicator. The column showing the community's average conditions is for all the communities studied, in other words, N=148.

Table 13.1. Regression Results of Select Indicators against the "Percent Latino of a Community," 1980 and 1988

Indicators	Mean Value	Regression Coef.	Value of R²
1. Number of persons/household	3.2	1.779*	0.669
2. Rates of fertility/10,000	403.1	2.170*	0.349
3. Percent of population under 18	34.2	0.157*	0.442
4. Median age of family head	27.2	(0.125)*	0.356
5. Percent in poverty	15.3	0.176*	0.415
6. Percent under 18 in poverty	43.0	0.236*	0.368
7. Percent of adults with high school degree	50.0	(0.541)*	0.613
8. Percent of adults with some college	7.7	(9.687)*	0.182
9. Percent of labor force in agriculture			
Adult males	19.8	0.511*	0.513
Adult females	9.9	0.367*	0.551
10. Community revenues/capita			
In 1980	$199	(1.098)	0.020
In 1988	$488	(5.969)*	0.163
11. Community expenditures/capita			
In 1980	$190	(0.872)	0.013
In 1988	$484	(4.817)*	0.191

* Significant at 5 percent. Figures in parenthesis are negative, and asterisked figures are based on regression analysis of 148 communities.

Interpreting the regressions should be done line by line. For example, line 1 shows that communities with low percentages of Latinos (beginning at 15 percent) have fewer persons per household than communities with high percentages of Latinos. As the percentage of Latino increases in a community (up to 98 percent), the number of persons per household increases. The value of R^2 suggests a positive and significant correlation between average household size and percent Latino in a community. To read another line, the values for the indicator Community Revenues/Capita (#10), the coefficient value in 1980 was nearly 0. That is, in 1980, all communities had the same per capita revenues, regardless of percent Latinos. According to the average value of the variable, community revenues were about $199 per capita.

But in 1988, the coefficient was negative and the R^2 much higher than the R^2 for 1980. Although the average value of the indicator was $488 per capita in 1988, the regression coefficient indicates that revenue per capita was significantly lower in *colonias*, as the percentage of a community's Latinos increased.

Not shown in Table 13.1 are several other findings that were reported in Rochin and Castillo (1993). Namely, inhabitants of *colonias* were at highest risk of being unemployed. Regression analysis indicated that both male and female *colonos* were three times as likely to be unemployed as Californians statewide. Moreover, *colonos* were half as likely to be self-employed as their counterparts in communities with fewer Latinos. The unemployment problem appeared to be related to the occupations held by *colonos*. In 1980, six times as many *colonia* inhabitants were employed in farming, forestry, and fishing, as compared to the residents of the communities with smaller concentrations of Latinos.

Results from the employment and occupational analysis also indicated that *colonia* residents faced relatively lower earnings than other rural people. Moreover, exceptionally higher incidences of poverty were observed in *colonias*. Poverty appeared to be associated with the general employment in agriculture, but it might also be explained by the poorer educational attainment of *colonos* (Table 13.1). This result may be a function also of the absence of much self-employment in *colonias* and the high rates of unemployment of *colonos*. These conditions limit *colonia* access to investment capital. Also, the lack of entrepreneurial opportunities (owing to a lack of skills training in local businesses) deprives *colonias* from investment opportunities. In sum, the 49 *colonias* of this study did not have favorable ethnic enclave conditions. Both the availability of capital and business know-how appeared to be seriously lacking among the *colonias* studied.

The analysis included a closer look at the types of local business enterprises of rural areas. Although the data were difficult to find on each community, it was apparent that *colonias* did not have many local businesses covering such consumer needs as legal services, pharmaceuticals, medical services, recreational activities, and the like. However, our detailed analysis of the retail trade sector (again with limited data) revealed that clothing and general merchandise retailers were more likely to be found in *colonias* than any other types of local business. In general, *colonos* could shop locally for some groceries, clothing, and basic goods, but if they needed other services, such as medical attention, *colonos* would have to travel to another place, necessitating expensive transportation to other towns for these types of services.

Increasing Signs of Deprivation in *Colonias*

Given the poverty, unemployment, and limited business development outlined above, it is evident that *colonos*, by necessity, are selling their labor for primarily agricultural jobs. While their annual earnings are low, *colonos* still must travel to buy imported food and basic necessities to the extent that the terms of trade are against the *colonia* economies. The relative poverty of *colonos*, compounded by dollars drained by shoppers going to other towns, leaves a situation of low capital formation and limited local investment. The low business cycle is perpetuated in *colonias*, since few new jobs are created locally by the private sector that benefit the community. Furthermore, the continual inmigration of Latinos results in stiffer competition in the agricultural labor market for *colonos* to find and keep employment.

Juxtaposed to these findings, the regression analysis revealed that local government revenues and expenditures per capita are significantly lower today in *colonias* than in other rural places. *Colonias* became fiscally poorer during the 1980s. As indicated in Table 13.1, *colonias* and other rural communities did not have very different local government expenditures in fiscal year 1980. Evidently, between 1980 and 1988, a significant gap developed between the average government revenues and the expenditures of *colonias* compared to the fiscal situation of the other rural communities studied.

The findings support the conclusion that *colonias* have become increasingly disadvantaged in terms of public expenditures for public safety, transportation, community development, health, cultural events, leisure, and public utilities. Perhaps a combination of factors is responsible for declining public expenditures in *colonias*, including the passage of Proposition 13 in 1978, the eroding tax base of *colonias* (due to high unemployment and low personal incomes), and the economic recession of the early 1980s. But for the most part, *colonias* do not have the same living conditions today as rural communities with fewer Latinos. *Colonos* are relatively more vulnerable to economic hardship than they were in 1980.

In general, California's *colonias* are far from becoming ethnic enclave economies. There is little evidence to suggest that the inmigration and settlement of Latinos is being complemented by increasing local economic opportunities. There are very few signs of greater self-employment among *colonos* or indications that they are developing local enterprises and more nonfarm jobs. Thus, *colonias* appear more and more to be cases of classic "immiseration," a concept first coined by

Jagdish Bhagwati (1958) to explain how third-world economies become worse off over time with the growing demand for their most
abundant resource, which in the case of *colonias* is a redundant supply of fairly homogeneous labor. In such a situation Bhagwati suggests that the "terms of trade" could be shifted in favor of the poor by
harboring their most limited resource. In the case of *colonias*, labor
could be strengthened by organizing and collectively improving human capital formation.

Implications

As the population expands in a *colonia* and its human resources
become more concentrated in such occupations as farm work, the
terms of trade turn against them. Because local demand for goods
and services is not met by local businesses, *colonos* must shop in other
places, thereby importing their goods and services at higher cost. At
the same time, the increase in labor supply through more migrant
settlement in a *colonia* depresses local wages and earnings, resulting
in more poverty. Instead of becoming better off, population growth
in *colonias* makes everyone worse off. With the continuing influx of
Latinos and settlement into rural communities, there does not appear to be much relief in sight.

Labor is more difficult to organize into collective bargaining units
as unemployed labor is willing to work for less. Given the initial findings of several more *colonias* in 1990, the signs of immiseration should
not be ignored. Rural *colonos* are living under disadvantageous conditions due to a multiplicity of factors related to the demand for and
supply of their labor, as well as a declining revenue base.

Obviously, such conditions do not bode well for today's youth
and future generations. The indicated problems will be difficult to
reverse. In general, within California's rural communities, measures
are urgently needed to build the economic base of employment, education, and local business of *colonias*. State and federal measures must
address the critical conditions of poverty. The local governments of
colonias must also find ways to enhance the provision of goods and
services. Finally, it appears that local deprivation is not only the consequence of weak economic structure and exploitation, but also of
the limited power of *colonos* to set the terms and conditions under
which they sell their labor and develop their youth. *Colonos* will need
community empowerment to guide the course of their futures.

NOTES

1. The terms Latinos, Chicanos, and *colonos* used in this chapter also denote Latinas, Chicanas, and *colonas*.

14

IRCA and Washington Agriculture

RICHARD W. CARKNER AND JEFF JACKSICH

Migrant and Seasonal Farm Workers

A discussion of the impact of the Immigration Reform and Control Act of 1986 (IRCA) on Washington agriculture should begin with an overview of labor-intensive agriculture and how it has changed since IRCA. The dynamics of labor-intensive agriculture can best be understood by looking at the major labor-intensive crops and some of the variables that influence seasonal labor requirements. Washington's agriculture is very dependent on seasonal labor. The farm-gate value of agricultural production was $4.4 billion in 1991. Apples accounted for nearly $1 billion of the total. Other labor-intensive crops, when combined with apples, account for approximately 42 percent of the total, or $1.8 billion. "Other labor-intensive crops" include other tree fruits, such as peaches, cherries, and pears; and small fruits, such as raspberries, strawberries, blueberries, and cranberries. Vegetable crops are led by asparagus, but onions and other vegetables are also important. Ornamental nurseries, reforestation, and greenhouse plant production are also important labor-intensive activities in Washington.

Agricultural employment follows a reasonably predictable pattern from year to year (Figure 14.1). However, there are overlapping requirements between crops, which vary from year to year due to weather and other factors.

The high and low seasonal agricultural employment for 1991 ranged from 45,000 in midsummer to a low of 5,300 in winter. Further indication of the seasonal dynamics of Washington's labor-intensive agriculture is shown in Figure 14.2.

The farm labor season begins with fruit tree pruning, following harvest in the fall, and continues through the winter and spring. In March, planting, cultivating, and plant-care activities increase and nurseries are harvesting bedding plants for spring sales. Asparagus is the first major crop harvest activity that begins in April, with peak

[255]

Figure 14.1. Agricultural Employment in Washington State, Number of Workers by Month, 1990–91

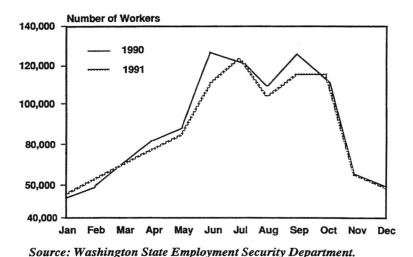

Source: Washington State Employment Security Department.

harvest employment of about 9,500 workers in May. Other activities during this period include flower bulb harvest, apple and pear thinning, hops training, and work in vegetable crops. The peak seasonal employment for May is about 24,000.

Labor requirements increase in June with cherry and strawberry harvest. The major seasonal employment peak for the year is in July, with about 45,000 workers. Major crop activities in July include apple thinning; stone fruit and onion harvest; raspberry harvest; and strawberry, blueberry, and summer vegetable harvest. August brings to a

Figure 14.2. Estimated Periods of Seasonal Agricultural Work in Washington State

	Apr	May	Jun	Jul	Aug	Sep	Oct	Nov	Dec	Jan	Feb	Mar
Asparagus												
Hop Twine/Trn												
Strawberry												
Cherry												
Apple Thinning												
Pear Thinning												
Apricot												
Misc Vegetables												
Raspberry												
Onion												
Nectarine												
Plum/Prune												
Peach												
Bartlet Pear												
Hop Harvest												
D'Anjou Pear												
Grape												
Apple Harvest												
Apple Pruning												
Pear Pruning												

Source: Washington State Employment Security Department.

close many of the July farm labor activities and brings some early apple harvest. In September there is more apple harvesting, as well as harvesting onions, potatoes, and summer vegetables. The peak apple harvest occurs in October, employing the majority of the 42,210 seasonal farm workers. Following the completion of apple harvest, the cycle begins again with pruning.

Washington's agriculture is dependent on migrant and seasonal farm workers (MSFWs) to plant, care for, and harvest its crops. Farmers growing perishable crops are concerned about having adequate farm labor, especially during peak harvest periods. Harvest activities occur throughout the state, many of them occurring simultaneously, and require thousands of farm workers. Many seasonal farm workers come from other states. The apple harvest in particular requires a large increase in farm workers, about 60 percent of which are from out of state.

The number of seasonal workers and where they come from is illustrated in Figure 14.3. Few agricultural areas within Washington have an adequate local labor supply to meet peak harvest requirements. In 9 out of 12 calendar months, interstate workers play an essential role in Washington agriculture.

Figure 14.3. Number and Origin of Seasonal Workers in Washington State, 1991

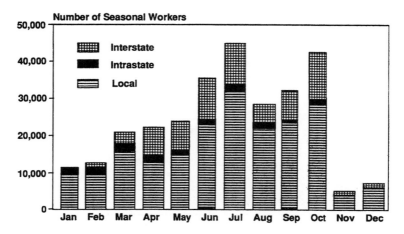

Source: Washington State Employment Security.

Anticipated and Actual IRCA Impacts

IRCA was designed to reduce illegal immigration. Concerns expressed as IRCA was being implemented were that the labor supply would be reduced since employer sanctions would limit the number of foreign workers (Langan and Melevin 1987). Another major concern for farmers was that newly legalized workers would migrate from agriculture and forestry to other occupations. This, however, has not happened in any significant way. The expressed concerns apparently had limited impact on plans implemented by Washington's farmers following IRCA.

Growth in Labor-Intensive Agriculture

Labor-intensive agriculture, with few exceptions, has experienced significant growth in Washington State. For the labor-intensive crops listed in Table 14.1, acreage has increased by approximately 72,000 in the 6-year period between 1986 and 1991. The only significant reduction in acreage and production was for strawberries. Asparagus acreage is about the same, but production per acre has increased over 20 percent. For most other commodities, significant production or acreage increases were noted. Apples, now the leading Washington crop, increased 25 percent in acreage and 39 percent in production. The apple crop employs the largest number of seasonal workers of any Washington crop. The expansion in acreage has been primarily in new, high-density production systems utilizing smaller trees. These new systems are designed to come into production early and to require less labor than older, low tree-density systems. While less labor may be required per acre, more total labor is required because of the net increase in acreage. New orchard systems also include new apple varieties that extend the harvest season and thereby lengthen the harvest employment window.

Table 14.1. Washington State Labor-Intensive Crops, 1986 and 1991

Commodity	1986 Acreage	1986 Production	1991 Acreage	1991 Production	% Chg Acreage	% Chg Production
		(lbs.)		(lbs.)		
Hops	17,400	35,496,000	28,200	53,551,000	+62%	+51%
Potatoes	118,000	61,950,000	141,000	75,435,000	+19%	+22%
		(tons)		(tons)		
Sweet cherries	11,900	62,500	14,000	50,000	+18%	−20%
Apples	126,000	1,550,000	157,000	2,150,000	+25%	+39%
Pears	21,700	266,000	25,200	336,000	+16%	+26%
Apricots	750	4,300	1,300	5,700	+73%	+33%
Grapes	27,900	156,000	29,200	196,000	+5%	+26%
Prunes	1,400	9,100	1,500	9,100	+7%	0%
Peaches	2,400	19,000	2,800	15,000	+17%	−21%
		(lbs.)		(lbs.)		
Asparagus	30,000	780,000	29,000	957,000	-3%	+23%
Raspberries	3,300	13,860,000	4,500	27,450,000	+36%	+98%
Cranberries	1,200	10,000,000	1,400	15,800,000	+17%	+58%
Strawberries	2,800	14,000,000	1,400	8,400,000	−50%	−40%
	364,750		436,500			

Source: Washington Agricultural Statistics Service.

While some of the increased agricultural production employs labor-saving production practices, as noted for apples, the demand for seasonal farm labor has substantially increased since IRCA. Farmers are using more seasonal farm workers for shorter periods of time. Washington State farmers have acted as if farm labor supplies will be available in the future, or they would have been less reluctant to invest nearly $18,000 per acre for establishing high-density apple orchards (Marshall et al. 1993).

Labor Supply

Agriculture is a major industry in Washington State. Labor shortages can cause losses in the millions of dollars to farms and the state's economy. Labor surpluses, on the other hand, can cause serious hardships to farm workers and place additional burdens on local services.

The supply of seasonal agricultural workers has been more than adequate in Washington State agriculture since IRCA. The supply and demand for labor in labor-intensive crop activities in Washington State varies a great deal throughout the year. There has been a surplus, except for a few spot shortages that were quickly filled with the assistance of the Employment Security Department. There is no data set that documents the surplus of farm workers. Reliable statistics describing the number of illegal workers are not available, nor is information on their location, wages earned, or working conditions. Some data sets, such as Migrant Seasonal Education Enrollment data or MSFW health clinic data, show significant increases in the number of MSFWs served.

The position of agricultural interests has been and continues to be to support a large supply of farm workers. There are many variables affecting farm labor needs for a given year, as well as the supply. Uncertainty is reduced for agriculture if a large pool of labor is available. The initial response to IRCA from Washington farmers was that labor shortages might occur because of the assumed effectiveness of employer-imposed sanctions. Registrations with Employment Security by MSFWs steadily increased from 1985 to 1989.

The initial increase may be explained by the reluctance of farmers to complete the IRCA employment eligibility verifications and the risks associated with noncompliance. Agricultural employers increasingly relied on the Washington State Employment Security Department. Farmers might also have relied on the Employment Security Department because of their expressed concern about labor shortages.

Over time, it became apparent that more MSFWs were available and workers increasingly bypassed Employment Security and went directly to farms. This is shown by the reduced number of registrations after 1989. This was also influenced by new federal regulations providing blanket protection for small farmers that did their own farm worker recruiting.

Farm worker responses to farm labor supplies were made to the Commission on Agricultural Workers, Hillsboro, Oregon, July 21, 1990. Six apparently documented Hispanic farm worker testified before the commission. Their concerns were similar. They testified about the glut of farm workers and resulting low earnings because available work was allocated over too many workers. Bad working and living conditions were consistently mentioned. This was related to worker over-supply and overcrowded living conditions. They mentioned the many illegals and the problems that caused for documented workers. It is very easy for employers to discharge documented workers because illegals are available to take their place. Promises not kept and exploitation by labor contractors were also mentioned. The resident Asian worker complained about being displaced by Hispanic migrant workers who were thought to be exploited by farmers.

Representatives of farm worker unions echoed the comments of farm workers at the Hillsboro commission meeting. Those testifying on behalf of farm worker unions indicated there were too many farm workers and not enough work. Oversupply conditions result in declining wages, poor working conditions, and inadequate housing. Cipriano Ferrel, President of Northwest Treeplanters and Farmworkers United, said, "IRCA hasn't changed anything fundamentally in agriculture" (USCAW 1993, p. 100).

Labor imbalances are the result of many interacting variables. Since IRCA, there has been a surplus of farm workers comprised of newly legalized farm workers and many illegal workers. Job opportunities in other states, influenced by weather and other factors, impact the supply of workers in Washington. This was especially true with the recent six-year drought in California that has increased the flow of farm workers to the Northwest over the last four years. Economic and political conditions in other countries, such as poor economics in Mexico and South America, have increased the supply of people willing to do farm work. Lack of local and state agricultural labor market information also contributes to labor imbalance. Market forces at the national and international level influence the demand for farm workers in the Northwest. Apples, in particular, are a major U.S. export commodity, requiring large numbers of out-of-state farm workers to assist in the harvest.

The Washington State Monitor Advocate report for 1992 estimated the number of MSFWs to be about 170,000, of which approximately 34,000 were legalized under IRCA. The report indicates that estimating the number of illegal workers is difficult, but that a labor surplus exists and has increased in recent years. The results have been that worker wages and living conditions have not changed since 1980. Growth in the number of MSFWs has also put a severe strain on local public service resources. This includes a large demand placed on social and public services by illegal aliens with fraudulent documents, according to Richard Smith, Regional Director for the Immigration and Naturalization Service (INS).

Farm worker surpluses have occurred during a time when agricultural production and the demand for labor have steadily and significantly increased. Surplus MSFWs have increased competition for jobs, increased MSFW unemployment, and reduced earnings for those who find jobs. Limited nonfarm jobs and modest educational and language skills on the part of most MSFWs reduce their participation in the nonfarm labor market.

Demographic Changes

As of October 1990, 9,267 MSFWs were legalized, based on their status as farm workers prior to 1982. Another 25,210 were legalized through the Special Agricultural Worker (SAW) process. The SAWs were about 70 percent male, whereas those in the group prior to 1982 were about equally split between males and females. A large number of those legalized under IRCA have settled in eastern Washington and Yakima County in particular.

An anticipated consequence of these newly legalized workers is that they will want to encourage other family members from Mexico or South America to join them in Washington. This will contribute to the supply of legal and illegal workers.

Employment Service

Washington's Employment Security Department took a more active role than some other states, using it as a marketing tool for offered services. All state offices did eligibility verification when requested, and some state offices promoted the service.

The response by employers was positive initially; they were nervous about liability issues and relied on Employment Security to relieve them of the responsibility. Table 14.2 indicates the fluctuations in registration, steadily increasing from 1985 to 1990, then beginning to taper off. As time went on, employers became more confident and did more eligibility verifications themselves. Table 14.3 shows a decline in agricultural employment eligibility verifications after 1989.

Employment Security has recently decided to phase out employment eligibility verification. The official reason is that it is an expensive task to perform in a period of declining resources. Perhaps another reason is that, with all the fraudulent documents floating around, the INS wants some help with enforcement. Employment Security views its role as a service provider, as opposed to being an enforcement agency. Eligibility verification activities by Employment Security puts the department in a dilemma. Where to draw the line becomes a difficult question, hence the decision to withdraw the service.

Table 14.2. Migrant and Seasonal Farm Workers Registering for Services with Washington State Employment Security, 1985–91

Year	Workers Registered
1991	19,345
1990	21,544
1989	25,700
1988	21,842
1987	16,331
1986	15,516
1985	13,339

Table 14.3. Agricultural and Non-Agricultural Certification Activity by Washington Employment Security Job Service Centers, 1988–92

	Year					
	1988	1989	1990	1991	1992	Total
Agricultural	11,299	12,252	9,500	7,147	6,018	46,216
Non-Agricultural	10,001	8,968	6,868	6,950	10,737	43,524
Total	21,300	21,220	16,368	14,097	16,755	89,740

Source: Washington State Employment Security Department.

Employment Security will continue to provide INS documents to employers and explain to employers what documents are required to prove employment eligibility. Actual document verification will be left to employers.

The agricultural clearance system is designed to respond to local shortages of temporary agricultural workers. If workers are unavailable in a local area, recruitment can be extended beyond the local area. This can be done within states (intrastate) or between states (interstate).

Prior to IRCA, the clearance system was used to recruit over 3,000 workers per year in Washington. This clearance system was used by asparagus growers for years. Since IRCA, the clearance system has been undermined by the available supply of MSFWs and other federal regulations, legal services, law suits, and attempts to organize the MSFWs by the Farmworkers' Union. These factors combined to discourage the use of the clearance system by Washington State farmers. Some farmers involved in the Interstate Clearance System felt they were being targeted by the Farmworkers' Union and the Evergreen Legal Services Corporation. An example is a H-2A job order in 1989 for 4,000 workers that was not certified because domestic workers were available through the strong efforts of the Farmworkers' Union and community-based organizations. Since these actions, agricultural employers have been able to fill their own job openings because of the ample supply of farm workers. Washington farmers have chosen not to continue their decades-long practice of using the Interstate Clearance System to recruit MSFWs from other states or to use the H-2A program to recruit foreign workers, as has also been the case in other states.

INS Enforcement

The INS enforcement practices in Washington State, like the nation, were changed by IRCA. They were constrained by inadequate funding and the increased complexity associated with enforcement under IRCA. The nature of the INS's border patrol enforcement practices shifted more to investigative and educational activities after IRCA. They have less staff for preventing national border crossings, catching illegal farm workers, and shipping illegal farm workers back to their home country. An example of the increased complexity for the border patrol is that IRCA contains a search warrant provision. This provision has made it very difficult for the border patrol to enter fields efficiently and effectively to check worker documentation. Often, by the time ownership of a field has been determined and a search warrant secured, farm workers will have moved elsewhere. Another factor has been the increased use of good fraudulent documents. Initially, these documents were very expensive, in excess of $400 in 1987, but now they are as cheap as $25. These are some of the factors that have forced the INS to do more investigative and educational work in support of enforcement. The legalization program under IRCA has led to a large increase in the supply of farm workers in Washington State, along with hard economic times in many of the countries of origin.

According to staff at the two local offices of the border patrol in Washington State, the initial enforcement activities after IRCA were largely educational activities aimed at farm employers and investigative activities leading to some employer warnings. In a few blatant cases, sanctions against farm employers for employing illegal farm workers were issued by the border patrol. These practices did little to stem the tide of illegal farm workers who were able to obtain the phony documents needed to secure employment in Washington State agriculture. There are no valid data on the number of illegal farm workers working with the aid of fraudulent documents. Given the lack of resources available to the INS, the number of illegals caught and shipped back to their home country has not been a good measure of the number working in Washington State agriculture. IRCA absolutely caused no shortage of workers in agriculture, according to INS Regional Director Smith. According to border patrol staff, even farmers have reported that more than enough farm workers are available since IRCA.

Farm Labor Contracting

The number of farm labor contractors registered with the state has changed little since 1986. Currently, the number of licensed farm labor contractors ranges from 150 to 180 per year. Many of the currently registered labor contractors are involved in reforestation activities and not in production agriculture.

In the opinion of an enforcement officer and an administrator for the Washington State Department of Labor and Industries, there are many more illegal or unlicensed contractors than those who are registered. Circumstantial evidence to support this is based on the sheer volume of workers who must find available work in a short period of time. They disguise themselves as company foremen, but they are responsible for workers moving from crop to crop. Because of the number of unlicensed contractors, it is difficult to compare labor contracting before IRCA and today.

Major changes since IRCA have been as much a function of state law changes as IRCA, but they are interrelated. Washington's employment standards covering minimum wages and child labor laws have had a large impact. Any improvement in farm worker absolute earnings is more attributable to rising minimum wages than the control of supply under IRCA. The net income of farm workers has actually declined due to reduced hours of work. There have been very few minimum wage complaints, in part because of the regulations outlining pay-stub reporting requirements that show earnings and hours worked.

There have been major changes in the employment of minors in the harvesting of major Washington crops such as asparagus and apples. These crops are harvested during the normal school year of September to June and, because of new child labor regulations, many employers have made it a policy to hire no one under 18 years of age. This has made it difficult for migrant workers, who travel as families, since their children can no longer find work. Increasingly, MSFWs are single males, in part due to child labor laws, which affect the ability of children to work in agriculture, but also due to inadequate housing and living conditions.

Conclusions

Washington, with nearly $2 billion worth of labor-intensive agricultural production, is highly dependent on access to migrant and seasonal farm workers. The anticipated response to IRCA by Washington's farmers was that employer sanctions would reduce labor supplies. In fact, employer sanctions have had little impact and the supply of farm workers has increased since IRCA. Many of these workers are illegal; however, no information base is available to identify the number, location, or earnings of these workers.

Since IRCA, there has been major growth in Washington's labor-dependent crop production. Apples in particular have seen significant growth. Even with increased demand for MSFWs because of expanded production, farm workers have remained in surplus.

To ease the imbalance between the supply and demand for farm workers, better farm labor market information is needed. Further study of MSFW populations is also necessary to improve our understanding of factors influencing the supply of farm workers. Unfortunately, federal funds have been cut, further weakening the data bases necessary to complete such studies.

15

IRCA and Labor Contracting in California

HOWARD R. ROSENBERG

Introduction

The explicit goal of IRCA was to control unauthorized immigration to the United States, but diverse interest groups had other, dubiously compatible aims in supporting this law. One was to reduce the relative isolation of the farm labor market, tighten labor supply, and thereby improve conditions of employment in agriculture. IRCA held promise for both new and old kinds of alternatives to widespread employment of workers who were here illegally (Rosenberg 1988). The logical goal of the new direction was a legal resident work force and a stabilized labor market operating more like the rest of our economy. The old direction pointed to institutionalized reliance on guest workers employed under more heavily regulated conditions.

Underlying the special treatment of agriculture by IRCA were assumptions about buyers and sellers in the farm labor market. The impact of IRCA on agriculture would be shaped through individual employer and worker responses to the inducements and penalties it created. Employers were required to conform to certain hiring standards and pushed to rethink their nonregulated management practices. They would face decisions, not only about new legal obligations but also their labor relations in general.

Some responses were rather immediate and far reaching, but the major impacts would take form gradually. The very context of these decisions would be fluid. Provisions did not all kick in at once, and some key implementing regulations and administrative policies took months or even years to establish.

Long-term effects of IRCA might be reflected as changes in: (1) who performs farm work, (2) mobility and occupational choice of legalized former farm workers, (3) worker exercise of employee protections under the law, (4) union organizing activity, (5) pay and other terms of employment in farm work, (6) use of farm labor contractors, (7) technological change substituting machinery for labor, and ultimately, (8) viability and structure of labor-intensive agriculture in the United States (Rosenberg and Mamer 1987).

Nearly six years later, overall supply of labor available to farms in California remains larger than demand, the real earnings of hired farm workers have eroded, and employment by farm labor contractors has markedly increased. Is the expansion of FLC activity an effect of IRCA? The Deputy Labor Commissioner observed in his 1947–48 study of FLCs (Bruce 1948) that there was simply more call for their services. Growers believed that contractors relieved them of difficulties, uncertainties, and costs associated with the direct employment of workers. A key to explaining the more recent growth in FLC activity is likewise to understand those who purchase the services they sell.

Initial Employer Responses

As it has for more than a century, California agriculture relies to a large extent on recent immigrants performing field jobs that do not appeal to most settled U.S. residents (Fuller 1991). Farm employers were understandably concerned about what IRCA would bring. In spring 1987, general confusion about the new law, regulations that restricted farm workers in Mexico from entering the United States to file SAW applications (and to work), and spot shortages of farm labor fed fears of summer harvest disruptions. Agriculture took a regular place on the nightly news, and government agencies readied to cope with crisis. The INS convened a public meeting of grower, labor, and agency representatives. The Employment Development Department in California initiated a weekly farm labor report (CalEDD 1987–92).

The most pessimistic scenarios were hardly realized. Transitional rules and facilities were set up to help with the entrance of pending SAW applicants from Mexico. Temporary relaxation of documentation standards for proving work eligibility eased the employment of SAW applicants from either side of the border. Harvests progressed through the summer and fall with little abnormality.

In November 1987, a year after IRCA was signed, a survey of agricultural employers in California explored their initial adjustments to the new law (Rosenberg and Perloff 1988).[1] Findings summarized below are based on 456 California-based firms that provided location, work-force size, and commodity identification data. Employers are counted in the CDFA reporting region where they produce output of greatest value. Respondents noted up to three types of commodities from which they derived most revenue, and their answers are aggregated with others in each respective crop mentioned (the sum of response shares by crop thus exceeds 100%). A large majority (71%) of respondents produced only "SAW crops," in which 90 days of work between May 1985 and May 1986 was the key qualification for obtaining legal status as a Special Agricultural Worker. Commodity groups not fitting the "SAW crops" definition are dairy, poultry, other livestock, and other (mostly silage).

Information and Understanding

Whether employers comply with any law depends first of all on whether they understand it. Many farmers were justifiably uncertain about the new law and what it required. Information on IRCA was unevenly available, and official guidelines were slow to reach many. Only 62 percent of respondents had yet received the official "Handbook for Employers" from the INS.

Employer associations, educational and social service organizations, and news media were advising employers and aliens long before the INS launched any substantial effort to inform. Survey respondents said that their most useful information sources were periodicals, seminars, and newsletters. About 40 percent specified questions or topics on which they needed clarification. Most frequently mentioned were: documentation required to establish employment eligibility, new sources of farm labor supply, the deferral of sanctions for employers of workers in SAW commodities, and unity of families in which not all qualified for legal resident status.

Only 8.1 percent felt very certain about what IRCA required. Employers in this group did better on a straightforward 12-item "exam" portion of the survey than groups who were less sure they understood; but they still missed, on average, more than one-third of the questions.

Aliens in Agriculture

Survey responses confirmed that California agriculture did depend heavily on labor provided by aliens, in contrast to findings of the USDA Hired Farm Working Force Survey (Oliveira and Cox 1988, 1989). In both 1986 and 1987, 85 percent of agricultural employers hired one or more aliens (Table 15.1). Virtually as many farms hired illegal as legal alien workers in 1986. In 1987, fewer firms (55%) reported hiring illegal aliens and more (77%) legal aliens. About as many legal aliens (37%) as illegal aliens (38%) were hired in 1986. In 1987, while the percentage of jobs going to illegal aliens fell to 31 percent, the share to legal aliens rose to 41 percent. All aliens thus constituted nearly as many of the hires in 1987 (72%) as in 1986 (75%).

Self-reported (probably understated) employment of undocumented workers was most common among producers of SAW commodities and least common in non-SAW crops, to which the enforcement grace period did not apply. Illegal entrants were a larger share of the farm labor force in southern California and among the larger employers. Producers of grapes, fruits, and ornamentals were more likely to use illegal labor than were producers of livestock, cotton, dairy, and grains or field crops, which utilize more capital-intensive production technologies.

Legalization Assistance

The survey found a high level of employer involvement in legalizing alien workers. Farmers provided information about the new opportunities to obtain legal status through individual discussions (55.2% of employers), group meetings (20.3%), short written notices (15%), and detailed written explanations (9.3%). To facilitate the application process, 35 percent of employers supplied INS forms and 32 percent referred workers to Qualified Designated Entities. Respondents specified several other types of help, including money to pay fees, transportation, and personal completion of forms. One said he gave workers "whatever they need."

Letters or documents to verify past employment qualifying workers for the SAW program were the type of assistance by far most commonly provided (78% overall). Farmers who had hired illegals in 1986 or 1987 were understandably more likely to provide documents (90%) than those who did not (55%). Not all employers were pleased with the fruits of their help to employees. Two lengthy letters from respondents expressed frustration that newly legalized SAWs were leaving for other employment. Both farmers considered themselves good employers and felt rather betrayed by workers moving on, as well as by the federal government cutting off their supply of affordable labor.

Table 15.1. Hiring of Aliens in California Agriculture, 1986–87

	Respondents N	Respondents %	Percent Employers Hiring: Any Alien 1986	Any Alien 1987	Legal Alien 1986	Legal Alien 1987	Illegal Aliens 1986	Illegal Aliens 1987	Percent Illegal of All Hires** 1986	Percent Illegal of All Hires** 1987
			···Percent Hiring One or More···							
All Employers	456									
Complete hiring data	392	100	85	85	71	77	71	55	38	31
Incomplete hiring data	64									
Only SAW crops	283	72	93	93	77	84	79	64	40	34
Mixed crops	61	16	76	79	67	72	69	39	27	17
No SAW crops	48	12	51	48	36	40	34	19	12	3
Commodity*										
Poultry & dairy	36	9	68	69	45	61	48	27	23	19
Livestock	52	13	56	50	46	44	43	23	21	7
Ornamental & nurs.	35	9	97	94	84	91	70	57	38	32
Grapes	107	27	97	99	73	87	93	81	63	49
Nuts	90	23	91	90	79	84	72	51	34	26
Tree fruit	126	32	95	92	82	85	87	68	52	38
Other fruit	29	7	95	94	83	86	86	70	25	34
Vegetables	48	12	98	98	84	83	79	76	21	14
Grains	59	15	71	75	62	69	50	37	21	8
Cotton	12	3	91	100	82	100	82	58	13	4
Edible field crops	48	12	84	87	77	83	58	47	25	11
Other	54	14	68	70	58	61	47	39	22	15
Hires (peak minus year-round employees)										
0 - 6	106	27	64	64	49	58	43	27	38	15
7 - 19	76	19	84	81	71	77	64	46	35	17
20 - 49	99	25	95	94	82	87	82	64	48	32
50+	111	28	100	99	82	86	94	78	37	32

* Total exceeds 100 percent because most respondents have multiple crops.
**Weighted by farm size within classification.

Compliance with the New Rules

Rates of agricultural employer compliance with the new law in 1987 were tempered by both confusion about its requirements and knowledge about the deferral of sanctions. Fewer than one-quarter of survey respondents reported having already fired or refused to hire any worker for not having proof of employment eligibility (Table 15.2). Farms that had been visited within the past two years by the INS border patrol were 62 percent more likely than the others to have screened out illegals, and southern California employers were also much more likely to have reported starting this early.

Table 15.2. Employers Complying with IRCA, 1987

	Percent		
	Completing I-9 (after May 1987)	Fired or Refused to Hire Illegal	Intend to Hire Only Legal
All Sectors	55	24	55
Only SAW crops	58	25	50
Mixed crops	44	21	62
No SAW crops	44	19	21
Commodity			
Poultry & dairy	46	24	62
Livestock	37	21	70
Ornamental & nurs.	71	48	57
Grapes	61	18	34
Nuts	54	25	51
Tree fruit	59	26	48
Other fruit	67	19	55
Vegetables	64	25	60
Grains	39	20	65
Cotton	77	46	83
Edible field crops	50	25	62
Other	48	17	65
Size (employees at peak)			
0-6	41	23	69
7 - 19	51	24	55
20 - 49	57	23	52
50+	69	26	40
Border Patrol Visit			
Yes	67	34	52
No	52	21	55

A higher proportion (55%) of firms intended to hire only legally eligible workers in 1988 than reported doing so in 1986 (29%) or 1987 (45%). Smaller farms (0–6 employees at seasonal peak) were planning more than the sample average to hire only eligible employees, and farms with 20 or more employees, less. Intent to hire legally was generally highest among employers that hired no illegals in 1986 or 1987 and those who regularly attended association meetings.

Most respondents willing to continue hiring ineligible workers in 1988 (34%; another 11% said they were uncertain) indicated that they would if they could not find enough legals to perform the work. Several comments and supplementary letters indicated that the desire to operate legally was outweighed by the resolve to get the work done by whatever means practical.

Rules and deadlines for verifying employee eligibility were different for hires made before November 7, 1986, from then till May 31, 1987, and after June 1, 1987. Only 14 percent of respondents had completed the I-9 form (optional) for hires made before November 7, and 28.1 percent for hires from November to June (mandatory to have done by September 1, 1987). A majority had begun to complete the I-9 (mandatory within three days of beginning work) for hires beginning June 1987, and they had started to do so, on average, in mid-July.

Adjustments in Management

Little crop loss due to labor shortage was reported, but circumstances in 1987 did not present a fair test of labor market adjustment to the new law. Respondents wishing to send a message to policy makers may have overstated their losses. Only 10.6 percent, however, reported any 1987 crop loss due to labor shortage. They estimated losing an average 17.9 percent of potential crop value. Few respondents cited management changes in 1987 to avert potential disruptions, 30 of them specifying major business adjustments to IRCA already made or contemplated, most commonly (1) mechanizing or changing crop mix to reduce the need for labor and (2) scaling down or leaving agriculture.

Though only 37 percent used farm labor contractors to recruit workers in either 1986 or 1987, 13 percent relied on them more and 2 percent relied on them less in the latter year (Table 15.3). Referrals by supervisors, other employees, and grower acquaintances, used by a large majority of the sample, were other means to which employers resorted more, but to a much lesser extent, in 1987. Walk-in recruitment, a source for the largest share of respondents (71%), was used less in 1987 by 13 percent and more by only 4 percent. Written advertisements, visits to worker homes, and EDD referrals were each used in either year by less than one-fifth of employers.

An overall reading of the initial response to IRCA was that farmers were inclined to comply with the new law, but not if it meant injury to their business. A majority had begun to document employee eligibility, even though penalties for not doing so were not yet applicable to most. A large minority acknowledged readiness to hire illegals if faced with labor shortages, or to make other changes that were needed and feasible. Two important factors likely to bear on the need for and feasibility of near-term adjustments were: (1) the vigor and ingenuity of INS enforcement efforts and (2) the relative availability and attractiveness of alternative jobs for SAW-legalized aliens, as well as other potential farm workers. So far, the rules had changed, but the players looked much the same.

IRCA Implementation

In December 1988, (1) the SAW application period ended, (2) full applicability of employer sanctions in agriculture began, and (3) the question burning for two years — "Will there be an agricultural labor shortage in 1989?" — was about to be answered. Size of the *real* agricultural labor supply would depend on the effectiveness of the new hiring and verification requirements. Enforcement was the great wild card in the 1989 labor outlook.

While attention was being called to the use of fraudulent papers supporting SAW applications (Ramos 1989), as of December 1988 the critical arena of fraud shifted from the INS legalization office to the work place. Wittingly or not, employers could end up hiring ineligible people presenting imitation green cards, social security cards, birth certificates, and other seemingly legal documents. A farmer's response to fraudulent papers might depend on how they looked, the market for legal labor, and the odds of getting hurt by sanctions.

Employers meet their primary obligation under IRCA by certifying on the I-9 form that documents presented by workers "appear to be genuine." INS officials described a standard, "the 100-foot rule": If a document is not obviously fake from 100 feet away, one cannot be expected to know it from the real thing. Even growers intent on more than meeting their obligation could not be sure to avoid hiring illegals.

Table 15.3. Changes in the Use of Four Recruitment Methods, 1986–87

	Survey Sample		FLC			EDD			Referal by Manager			Walk-in		
	N	%	Not Used	Less	More	Not Used	Less	More	Not Used	Less	More	Not Used	Less	More
All Employers	456	100	63	2	13	82	2	3	37	3	9	29	13	4
Only SAW crops	322	71	60	2	15	83	2	3	30	3	10	24	13	5
Mixed crops	72	16	52	3	13	72	2	7	40	3	10	31	16	5
No SAW crops	62	14	92	0	0	88	2	0	73	2	0	54	7	0
Commodity														
Poultry & dairy	45	10	87	0	0	84	0	3	81	0	0	55	8	0
Livestock	61	13	80	2	4	79	2	4	57	2	4	50	7	2
Ornamental & nurs.	48	10	84	0	8	82	0	0	26	0	13	15	18	2
Grapes	121	27	45	2	26	88	1	2	23	5	13	19	11	8
Nuts	100	22	64	1	11	85	1	2	29	2	13	28	13	3
Tree fruit	140	31	54	3	18	87	2	4	26	4	10	26	13	3
Other fruit	32	7	69	0	10	68	4	4	18	4	10	13	17	0
Vegetables	56	12	46	2	16	76	2	2	29	4	4	16	14	0
Grains	71	16	59	8	8	73	4	4	35	0	9	34	18	0
Cotton	13	3	15	0	31	92	0	0	38	0	8	15	31	8
Edible field crops	55	12	46	0	10	73	2	0	38	0	6	16	31	8
Other	66	15	63	2	10	74	2	5	43	5	7	27	16	4
Hires (peak minus year-round emp.)														
0 - 6	137	30	82	0	3	81	0	3	58	2	4	38	10	0
7 - 19	85	19	72	2	12	88	3	1	45	2	9	34	16	6
20 - 49	116	25	55	4	14	85	1	2	29	6	10	18	12	10
50+	118	26	44	1	20	70	2	7	13	1	9	16	15	1

Employer Use in 1987 Compared to 1986 — Percent

Entrepreneurs were turning out phony forms that conscientious, law-abiding employers would not suspect to be counterfeit from even 100 millimeters. The task of ascertaining genuineness was and is complicated by the variety of documents that workers may use to prove their legal right to work. The paper chase had come to the farm (Rosenberg and Billikopf 1989).

Farm Labor Contractors

As growers assessed their options, labor contracting was expanding. The California Employment Development Department (EDD) payroll tax files show that the average employment level of some 50,000 under the industrial classification code for farm labor contractors (SIC-0761) in 1982 rose by more than half to 77,300 in 1990 before dipping a bit to 74,300 in 1991. Meanwhile, employment in other agricultural production declined. The real value of FLC payrolls statewide similarly rose nearly 50 percent, while others declined 10 percent from 1982 to 1991.

The EDD file is only one source of estimates of labor contracting activity. Differences in how government agencies define FLCs are reflected in 1990 lists of those (1) registered with the U.S. Department of Labor — 2,896, (2) licensed by the state Department of Industrial Relations — 1,136, and (3) filing unemployment insurance as farm labor contractors (the EDD file) — 1,080. Only 506 of a total 3,580 entities on any of the three lists are on all of them (Figure 15.1), and an unknown number of persons who act as labor contractors are on none of the lists.

A special study reported to EDD (Rosenberg et al. 1992) was based mainly on interviews with 180 labor contractors in the Fresno, Imperial, Monterey, San Joaquin–Stanislaus, and Ventura–Santa Barbara County areas during 1991. The project included smaller complementary surveys of growers and workers. Broad questions addressed were: Who becomes an FLC and why? How are their businesses organized? How and where do they market their services? How do they manage employees and deal with government regulation? What is their outlook on the farm labor market? A wide range of business practices in relation to customers, employees, and government agencies was found, including evidence of contractors who are professional, businesslike providers of service to participants in the farm labor market. There was also some justification for new or revised regulatory measures.

Figure 15.1. California FLCs on Government Lists, 1990

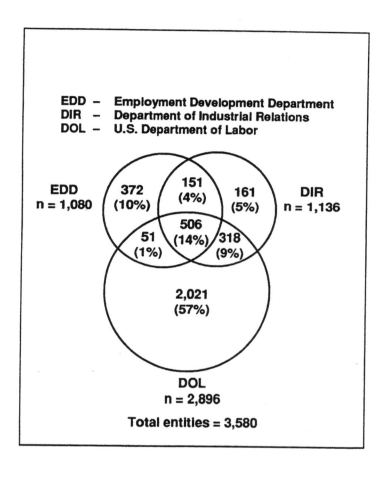

Project interviewers were much more likely to reach "visible" FLCs, selected from government agency files, than unlisted operators. More than half (55%, Figure 15.2) of interviewees were officially known as FLCs in records of all three agencies from which the sample was drawn, compared to only 14 percent of the total population known.

Most in the sample have enduring relationships with customers and direct ownership interest in at least one related business (63%). One-third of the contractors are currently also involved in the operation of a farm. They operate as FLCs an average of nearly nine months a year (a third operate all 12 months), and almost 70 percent have been contractors for a minimum of five years. More than a fourth had been growers or farm managers before entering the FLC business.

Labor contractors have various business sizes and "product lines," just as their customers, with different crops, organizational structures, and tastes for direct involvement in farm operations, have different sets of needs that they hire FLCs to serve. Contractors in the survey provide services to an average of 15 growers or packing houses, but some have only one customer, and even some large FLCs do all their work for a few. Less than 20 percent put their arrangements with any customers into the form of a written contract. The structure and amount of FLC charges to customers vary more between than within crop and regional groupings. A sizable minority of the contractors say that they accept commission rates determined by customers, but hardly any of the 30 growers interviewed in the study share this view.

Contractors mediate between cultures as well as factors of production. Labor contracting in California today most typically engages native Spanish speakers of Mexican descent for work on farms run by English speakers. FLCs are mostly of Hispanic background, about half are born in Mexico, and more than one-quarter are born in California. More than half speak Spanish at home. Most are in their 40s and 50s and average less than ten years of total schooling — though, notably, about one-fourth have some college education and have plenty of agricultural experience.

Most contracting firms are small, but the one-seventh with payrolls exceeding $1 million account for more than half of all FLC employment and three-fifths of wages paid. Peak employment among contractors in the study is 280 average, with a median of 150. Family members of the contractor are involved in two-thirds of FLC businesses, most commonly in office tasks. Nearly all contractors employ first-line supervisors — foremen, crew bosses, or *mayordomos* — to deal directly with workers. The number of foremen year around aver-

Figure 15.2. Survey Interviewees Identified as FLCs on Government Lists

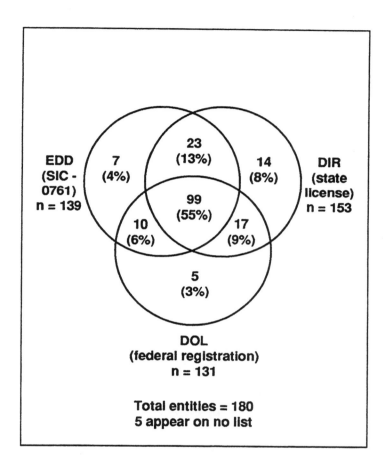

ages 4 and at peak season nearly 8, ranging up to 62. Half of the FLC firms have six or more at peak, and only 6 percent have none.

Core functions of the FLCs include recruiting and hiring workers, directing them to the work site, supervising their work, and paying wages. Contractors generally provide drinking water and field sanitation facilities and often furnish tools. Less typically, some contractors provide worker transportation, housing, food, and check-cashing services. Foremen are involved in most aspects of personnel management — recruiting, hiring, instructing, assigning tasks, and enforcing work rules.

Although four in five make efforts to stay in touch with employees during the off-season, all contractors must find at least some new workers each year. About half the work force, on average, is composed of returnees. FLCs recruit new employees through the same methods, to similar extents, as growers. The most heavily used means are accepting walk-ins, asking current employees for referrals, and delegating the responsibility to foremen or supervisors. FLCs in the survey do not exercise a competitive advantage by hiring workers abroad and bringing them across the border illegally. A mere 8 percent recruit outside the counties where they work, including only two in other states and two others in Mexico.

An intrepid 38 percent had placed orders with the EDD job service, though none reported finding their best workers and nearly none finding most employees this way. Satisfaction with the EDD service varied only slightly by FLC size but dramatically across survey regions, from 29 percent to 86 percent "usually" or "sometimes" satisfied.

To enter the contracting business requires little or no capital investment, and most in the study did not find the license examination and bonding requirement difficult. Operating as a contractor in full compliance with the law, however, is much more challenging. Four out of five FLCs in the survey had been inspected by at least one government agency during the 1987–90 period. Contractors expressed concern about the irony of greater regulatory attention being given to those who are more stable and observant of laws.

IRCA's Influence on FLCs

Can we discern differences between those contractors who went into business respectively before and after January 1987, which might suggest an impact from immigration reform? Nearly one-third of the survey sample entered the business in 1987 or later. Their 1990 payrolls tended to be smaller than those who began during 1977–86, as would be expected of younger firms, but, unexpectedly, were about the same as operators in business prior to 1977 (Table 15.4). Post-IRCA entrants are well represented throughout the range of size groups.

Table 15.4. FLCs Entering Business in Pre- & Post-IRCA Periods by Payroll & Previous Occupation

| | Year in Which Started FLC Business | | | |
	All	Through 1976	1977-86	After 1986
	%	%	%	%
Payroll				
<$250K	26.7	33.3	20.8	30.6
$250-$499K	20.6	25.6	14.3	26.5
$500-$999K	17.0	12.8	18.2	18.4
$1,000-$2,999K	25.5	20.5	32.5	18.4
$3,000K+	10.3	7.7	14.3	6.1
Total (N=100%)	165	39	77	49
Prior Occupation				
Ag. worker	9.6	15.2	9.1	5.6
Farm foreman/ supervisor	62.7	52.2	64.9	68.5
Farm owner	6.8	8.7	6.5	5.6
Nonagricultural	20.9	23.9	19.5	20.4
Total (N=100%)	177	46	77	54

Many FLCs said that they got into contracting because of a company restructuring where they had been employed or because of other organizational changes, and numerous comments referred to having entered business with the help of growers who had previously employed them as foremen or supervisors. A greater share of post-1986 FLC entrants than of other interviewees had been supervisors or foremen immediately prior to contracting.

The case of one FLC, with elements common to many others, illustrates the difficulty of making a confident attribution for the impetus to enter labor contracting. This contracting business was formed in 1985 by the then field supervisor of a grape production firm, after managers of that firm decided to concentrate on marketing, reduce the operations staff, and procure labor-intensive vineyard services from outside providers. Presumably these managers knew of proposals for immigration reform, but they also knew that sales management and labor management were requiring ever more sophistication in separate realms.

The grape firm helped the supervisor prepare for his FLC licensing examination, obtain a start-up loan, and arrange for payroll services, and it became his first customer. He initially hired from among field workers who had also been employed by the firm, and for the first year most of the same people did the same tasks on the same vines as before. By 1991, however, his business had expanded to include contracts with nine different growers, total revenues increased more than threefold from 1986 (the first full year), and employment at peak rose from 22 to 45.

Even if the grape firm had made its move later and the ex-supervisor had become licensed as an FLC in 1987, it would have only arguably been in response to IRCA. One way that firms have adapted to a plethora of burdensome regulations, by no means limited to IRCA, throughout the economy has been to downsize and contract with outside providers — temporary and other "contingent" labor — for services that were once integrated within the organization (Belous 1989). Potential advantages of contracting out for agricultural services, as in other industries, are gains in organizational flexibility to pursue new ventures or markets, ability to access specialized skills and equipment needed for limited purposes, and concentration of core personnel on the functions that they perform best (Miles 1989).

The increased use of farm labor contractors and other production service companies is thus an expression of a trend by no means confined to agriculture and immigrants, but by now rather widely noted and disturbing to many (Morrow 1993; Castro 1993). As stated in the lead article of a recent *Wall Street Journal*, "This is the 1990s workplace. After spending years at one company, more American office and factory employees are getting transplanted overnight to a temporary or subcontracting nether world. They do the same work at the same desk for less pay and with no health insurance or pension benefits. Others are farmed out to an employment agency, which puts them on its payroll to save the mother company paperwork and cost" (Ansberry 1993). According to U.S. Department of Labor figures, temporary, part-time, and contract workers are now a quarter of the work force, and their numbers grew ten times faster than overall employment between 1982 and 1990.

Reasons given by growers (in the small complementary survey of FLC customers) for engaging FLCs instead of hiring more workers directly reveal a set of practical business considerations that include but do not focus on IRCA-related issues (Table 15.5). Growers are moved to patronize contractors to get work done when needed by people who can do so without unduly complicating their lives. FLCs can reduce personnel transactions, legal liabilities, and considerable technical difficulties of personnel management (for example, in recruitment, selection, pay administration, supervision, and employee relations). An economist might see it all as cost reduction in various disguises.

Table 15.5. Grower Reasons for Contracting with FLCs

	#	%
Short-term availability	24	80.0
Reduced paperwork	12	40.0
Reduced need for supervision	12	40.0
Easier recruitment	7	23.3
Reduced costs	6	20.0
Quality of work	6	20.0
Worker reliability	2	6.7
Labor dispute	1	3.3
Liability under laws	1	3.3
Language advantage	1	3.3
Specialized equipment	1	3.3
Other reason	2	6.7
Total Growers*	30	100.0

* Because growers cited multiple reasons, totals do not equal column sums.

Even a most disciplined farm manager cannot always make employment plans far in advance. Vagaries of weather and the marketplace unpredictably affect both how much and when labor is needed. For many short-term tasks, the farmer's ideal labor supply would be flexible, skilled, and abundant. But the people who supply labor have their own personal needs and schedules, different sets of abilities, and limited information about job openings. FLCs provide for the efficient mobilization of people and equipment to meet short-term needs when and where they are.

Hardly any farmers turn biological material and processes into marketable products by themselves anymore, and neither do they provide all the supervision to guide the hired workers who make it happen. Most do it through an intermediate level of employees that are crucial in farming, labor contracting, and all other types of businesses. Cultural and language differences between California's farmers and the persons hired to perform farm work compound the challenges of direct recruitment, selection, supervision, instruction, and other job-related communication. As intermediaries, FLCs and their hired foremen carry labor market information that bridges gaps between farmers and workers.

Farmers are exposed to a host of legal liabilities and constraints attached to the institution of employment (Rosenberg, Egan, and Horwitz 1993). National and state legislation over the past two decades has narrowed gaps between employee protections in the agricultural and nonfarm sectors. Developments applying to all industries have given employees more legal rights in their jobs, providing some of the benefits for which unions have traditionally bargained. For employers these protections and accompanying paperwork have raised the costs and risks of maintaining a directly hired work force. Increased regulatory complexity and paperwork associated particularly with agricultural employment have added to reasons for contracting out seasonal labor tasks. IRCA is only one base of vulnerability for farmers. Although growers and contractors may be deemed jointly liable for violations of many employee protections, farm operators have reduced or eliminated exposure to some charges of wrongdoing by using FLCs.

More easily recognizable as cost reductions are not having to carry underutilized staff and sharing in any savings realized by the FLC, either through efficiencies or the lowering of labor standards (Mines and Anzaldua 1982). Growers feel pressured to hold the line, not only on direct wages, but also on such mandatory benefits as unemployment and workers compensation insurance and such optional fringe benefits as health insurance and vacation pay. When FLCs are able to economize, they can pass on savings to customers. Growers may find flexibility paving the way to additional cost savings, as contractual arrangements for labor impinge little on decisions to alter production, technology, staffing, and terms of employment in the future. Direct employment, in contrast, resembles to some farmers more of a fixed overhead than a variable operating cost.

Two ways of economizing illegally are to underpay wages or payroll taxes, and farm labor contractors have been suspected of both. The predominant method by which California contractors calculate charges to their customers is to add to direct wages an "inclusive" percentage commission that covers workers compensation insurance, unemployment insurance, all other payroll taxes, business expenses, and office overhead. Workers compensation accounts for roughly half the total commission, depending on crop and experience modification. Since typical totals of required insurance and payroll taxes are, for example, 29 percent in tree fruit, 26 percent in vegetable crops, and 21 percent in strawberries, an inclusive commission of less than 30 percent in most crops suggests that the FLC charging it may be either cutting corners illegally or not long for the business. Survey interviewers picked up several comments to the effect that unlicensed contractors (none of whom could be identified) had been underbidding legitimate operators with untenable commissions. Only one respondent out of the 124 in our sample who reported using an inclusive commission, however, had a rate below 30 percent (Figure 15.3).

Do contractors come up short on payroll taxes? Survey responses about total payroll and employment can be compared to data collected in EDD unemployment insurance files. Total payroll declared in interviews exceeds the total recorded for the same FLCs in unemployment insurance (UI) files by 61 percent for those classified by EDD as farm labor contractors, 22 percent for contractors reporting under other industry classifications, and 52 percent overall (magnitude and direction of individual differences is shown in Figure 15.4). In contrast, the net difference between peak employment reported in interviews and maximum monthly employment in the UI records is very small, though many individual differences are offsetting. The payroll discrepancies may indeed be revealing seriously underreported wages and UI taxes, but there are other explanations for the data. Accuracy of the UI files warrants further assessment.

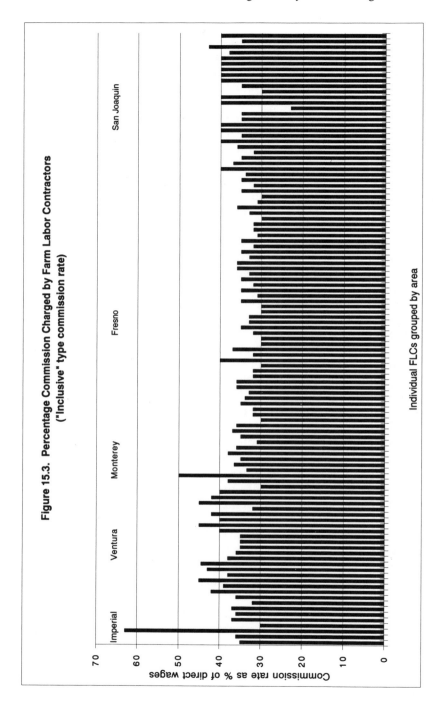

Figure 15.3. Percentage Commission Charged by Farm Labor Contractors ("Inclusive" type commission rate)

Figure 15.4. Differences (%) between Payroll Reported in Interview and Unemployment Insurance Files

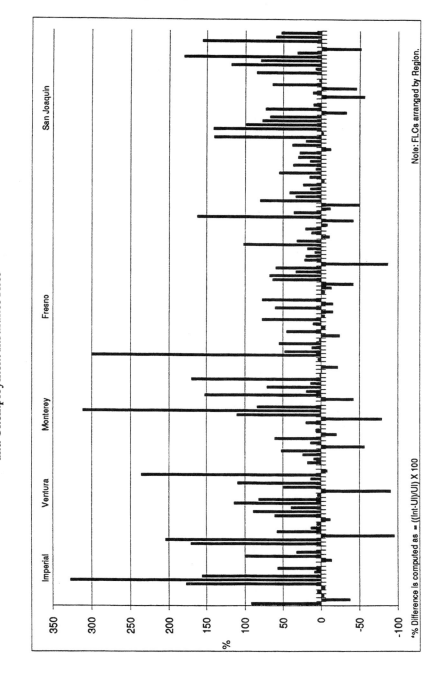

Regulating FLCs

Comprising the largest single employer category in an industry in which employees are obviously hurting, FLCs have been given special attention in new proposals to help workers. Legislators and agency administrators have moved to expand and more actively enforce farm labor laws in California. And they have placed particular emphasis on tightening their control over contractors (Rosenberg 1993).

A new law signed by the governor in September 1992, and effective January 1993, explicitly obligates labor contractors to provide all supervisory employees with training on laws that regulate terms and conditions of agricultural employment, including worker safety. This bill, AB 3146, also expands the scope of the FLC licensing examination, directs the labor commissioner to suspend contractor licenses for two failures within five years to pay wages due, and establishes a phone information line to advise workers or growers about individual FLC compliance with applicable laws.

To obtain a farm labor contracting license in California had already required filing a written application, paying a $350 annual license fee, depositing a $10,000 surety bond, undergoing an investigation of character and responsibility, and passing a test of knowledge essential to the occupation. These requirements have not, however, always translated into effective influence on the behavior of first-line supervisors who are responsible for the bulk of day-to-day employee relations in all but the smallest FLC businesses. Although contractors may be held accountable for the conduct of these supervisors acting as their agents, crew bosses often operate with a large degree of autonomy.

Another bill, with more mandates for FLCs, was vetoed in 1992. By administrative initiative, rather than legislation, DIR had already adopted a few of the ideas it contained. Since April the department had in place a toll-free phone line with bilingual personnel for farm workers to inquire about their rights and report labor law violations. It scrutinized farm labor contractors through a special mail audit of payroll records and customer lists, and through an effort to identify unlicensed operators. Results of this program, with which some 80 percent of all state-licensed FLCs readily complied, according to the labor commissioner's office, were then used in choosing contractors for enforcement visits. Additional measures to control contractors and penalize growers for the misdeeds of those they do business with have been introduced in both the state legislature and U.S. Congress.

Coordination of worker protection activity by the DIR and other public agencies was formalized in a new Targeted Industries Partnership Program (TIPP), which began November 1, 1992. Linking enforcement arms of the DIR with those of the EDD and the U.S. Department of Labor (DOL), TIPP has stepped up labor law enforcement in agriculture. The DOL Wage and Hour Division and the DIR Division of Labor Standards Enforcement has committed bilingual investigators and attorneys to the effort in nine areas of the state. TIPP initially targeted both the garment manufacturing and agricultural industries, because of their history of low-paying jobs and labor law violations.

Sweeps and violations found to date under this program have been heavily publicized to encourage more voluntary compliance (Bradshaw 1993). Teams of state and federal investigators are looking especially for violations involving minimum wages and mandatory overtime premiums; child labor; licensing and registration; workers compensation coverage; field sanitation, injury, and illness prevention programs; migrant housing; and unemployment tax contributions. While working mainly in areas assigned on the basis of crop production and garment manufacturing cycles, teams also move to locations where special needs are identified.

An employee outreach component of TIPP includes the distribution of cards in several languages outlining worker rights under state and federal laws. A toll-free phone number for farm workers offers information on both state and federal services. It is staffed by bilingual employees during periods more convenient than normal business hours to agricultural employees. The program is designed to also provide employers with information on their legal responsibilities through public seminars and a printed summary.

The new statute requiring FLCs to train their foremen raises a number of basic questions. Are FLCs to educate their supervisors in *all* of the myriad laws that pertain to hiring, wage determination, payroll deductions, meal and rest periods, workers compensation, disability and unemployment insurance, discipline and discharge, transportation, housing, child labor, collective bargaining, and more? How much information on any topic that is included will be enough? In what form is it to be given — will a one-sheet summary of everything be sufficient or should contractors start booking classroom space? Clarifying regulations from the DIR are still awaited.

While the labor commissioner considers how to define standards of adequacy for the required training, the more critical question is how well this law or any other can effectively prevail over economic forces in regulating managerial behavior. Whatever the rules, they are sure to meet some unexpected adjustments, creative compliance, and blatant disregard, especially in the current labor market.

Conclusions

Challenges in understanding and regulating farm labor contractors are not at all new. Undertaking his study of FLCs in 1947, Labor Commissioner Bruce noted the difficulty of determining whether a particular operator should come under the statutory definition of labor contractor (and therefore should be licensed) and the apparent inadequacy of the required (then $1000) bond as a guarantee of wage payment. His stated purpose was to figure out how to implement the Labor Code as it applied to FLCs, given that there were serious problems in (1) assuring that they were licensed as employment agencies, (2) collecting unpaid wages due to farm workers, and (3) dealing with other matters affecting the welfare of agricultural workers, growers, merchants, and the general public.

Only partly because of definition problems, we do not know for sure how many FLCs there are in California, but we do know that they are an established part of the agricultural industry. Perhaps the clearest finding of the 1992 research is that contractors are heterogeneous, both functionally and organizationally. They come in various shapes and sizes and serve a constellation of functions not necessarily centered on either illegal immigration or verification of employment eligibility. Certainly not all, and probably not most, fit the image that some have of them as ship captains of the nineteenth century who find workers abroad, provide them with counterfeit documents, become their first U.S. employers, and discourage them from complaining of labor law violations (USCAW 1993, pp. 155–166).

Although contractors who flaunt even the basic filing requirements were underrepresented in our survey, the report adds to evidence of FLCs trying to run their businesses effectively and fairly within the guidelines of public policy. New regulatory policies, even if well intended and coupled with vigorous enforcement, are unlikely to achieve their aims for farm workers if they disregard either the realities surrounding growers and contractors today or the lessons of experience under IRCA so far.

When the Immigration Reform and Control Act of 1986 was enacted, nearly all observers thought that the farm labor scene would change. With provisions of the law kicking in on schedule, lively debate over administrative rules, and employers and workers going about their business, changes have been everywhere. One of them is that labor contracting has expanded in California. The changes are partly because of IRCA, no doubt, but are also because of other laws and such assorted factors as the Mexican economy, California land values, expansion of agricultural production abroad, the drought, public concern about environmental toxins and food safety, prospects for a North American free trade, the state of public education, health care costs, the Persian Gulf war, and the dismantling of the Berlin Wall.

While complex immigration reform in 1986 may have added to reasons for procuring labor through contractors, other market factors had already and have since provided many more. IRCA is but part of the ecology of forces affecting decisions that determine agricultural labor supply, demand, and terms of employment.

NOTES

1. The California Agricultural Statistics Service, Department of Food and Agriculture, drew a random sample of 2,000 employers for this study. Of 1,938 employers who received the survey instrument, 487 (25%) responded. They were representative of all California agricultural employers, as characterized by the 1982 Census of Agriculture, in geographic and commodity distribution. Returns from medium-sized organizations exceeded and from small organizations fell short of their proportionate shares of the state population.

Part III.
IRCA and Agriculture
in Selected Commodities

Differences in the impacts of IRCA across commodities reflect both differences in the commodities themselves and among the regions in which the commodities are grown. Regional variations in recruitment and hiring practices, remuneration, and working conditions reflect the tightness of local labor markets, which in turn are influenced by their proximity to labor supplies for farm work and the labor-recruitment infrastructure available to match labor supplies with agricultural jobs. Many of these regional differences are evident in Parts I and II of this volume. For example, farmers in the eastern United States rely, sometimes heavily, on H-2A workers to harvest their crops. By contrast, California farmers do not participate in this formal labor recruitment program, instead hiring workers from pools of labor fed by migrants who follow networks of family contacts from villages in Mexico to farmers or labor contractors in California. Because of this, the California farm work force traditionally has been less legal than, for example, the Virginia tobacco work force, and both employer sanctions and farm worker legalization potentially are of greater consequence to the former.

Nevertheless, even within small regions, studies have documented significant differences among farm commodities with regard to such variables as worker characteristics (such as legal status), earnings, and recruitment and hiring (such as the use of farm labor contractors versus direct hiring. These commodity-specific differences may be due to several factors. Commodities differ widely with regard to the availability of labor-saving technologies and labor intensity. Seasonality and perishability can create demands for large numbers of workers over short periods of time to harvest some commodities (for example, peaches) and more stable labor demands with wider timing windows in others (for example, citrus). Western growers have argued that the H-2A program is not for them because the timing of labor demands is difficult to anticipate for the crops they grow. The combination of highly seasonal, labor-intensive labor demands and a well-developed infrastructure of informal labor recruitment from abroad traditionally make the western United States fertile ground for the expansion of unauthorized, intensive immigrant agriculture, potentially vulnerable to changes in U.S. immigration laws.

In light of this, the most striking findings that emerge from the commodity-specific papers in Part III are the minimal effect that IRCA has had on employment, remuneration, and working conditions in a wide variety of commodities, and what may be a convergence in hiring and recruitment practices and perhaps in working conditions across commodities.

Three common threads emerge from these chapters:

- First, IRCA has had a limited impact on the flow of workers into agricultural jobs. The authors find no evidence of labor shortages, increases in wages or piece rates, rises in worker productivity, or shifts to less labor-intensive technologies as a result of employer sanctions in IRCA. This finding is robust across commodities: from Florida nurseries (Chapter 20) to New York apples (Chapter 16), to citrus in Florida (Chapter 17, Part I) and California (Chapter 17, Part II), to table grapes (Chapter 19), to tomatoes (Chapter 18).

- Second, IRCA has accelerated structural changes in the farm labor market that were already underway by creating new incentives for farmers across a wide range of commodities to shift from direct hiring of workers to labor-market intermediaries. This structural change is best elucidated by Emerson and Polopolus (Chapter 17, Part I) for Florida citrus, but it is also evident in all of the major California crops, as well as in the data on apples in New York and Pennsylvania contained in North and Holt's research (Chapter 16) and on Florida tomatoes in Runsten and Griffith's study (Chapter 18). A shift away from direct employment is stimulated by economies of scale in recruitment and hiring, which are enhanced by the new paperwork requirements and hiring risks created by IRCA's employer sanctions. It has profound implications for the enforcement of immigration and labor laws, the expansion of immigrant-intensive agriculture in the United States, and farm worker welfare.

- Third, the chapters on IRCA and commodities reveal variations in wages and piece rates, working conditions, and wages and recruitment practices, but they are probably due more to differences in labor markets than in the commodities themselves. Moreover, there is some suggestion that these differences may be narrowing, with the waning role of formal recruitment (such as the H-2A program), the growth and extension of informal migration networks (especially from Mexico), and the expansion of labor market intermediaries, who draw from these migration networks as an alternative to direct hiring. Various labor market indicators presented in these chapters are consistent with these trends: the absence of capitalization and labor productivity growth (Chapter 20); flat or decreasing real wages or piece rates for agricultural workers (all chapters); deteriorating housing conditions in areas that have traditionally offered on-farm lodging for seasonal workers (Chapter 16), and the elimination of free services that contractors and farmers once provided tomato workers (Chapter 18).

16

IRCA and Apples in New York
and Pennsylvania

DAVID S. NORTH AND JAMES S. HOLT

Introduction

The apple industry nationally is strong: There is a good and growing domestic market for the product; the introduction of a controlled
atmosphere (CA) a generation ago made it possible for consumers to
buy fresh apples all year long; and the export market has been favorable in recent years, with more fresh apples being exported than
imported. Imports of apple juice concentrate, however, have grown
recently.

The New York and Pennsylvania apple industry is not quite as
strong as the industry nationally. This is the case because growers in
these two states have a higher percentage of processing apples, which
bring lower prices, than the national average. Furthermore, the region suffers from Washington state's extensive promotion of its red
delicious apple. The industry in the two states continues under family management, and is concentrated in four regions, the first three of
which are covered by this study: south-central Pennsylvania, the
Hudson Valley, western New York, and New York's Champlain
Valley.

Growing and selling apples is a complex business. There are a
series of interactions among the end-users of apples, cultural practices, and patterns of labor utilization. Careful handling of the orchard during the preharvest and of the fruit during the harvest can
bring good prices to the grower, but the reverse is equally true. In
recent years the two states' growers have moved toward more fresh
market production and toward smaller, easier-to-pick trees.

The labor market implications of these complexities are many:

- Picking for the fresh market requires more care on the part of
  workers, and more supervision on the part of growers, than
  picking for the processing plant.

- Changes in consumer tastes, which has reduced interest in late-ripening varieties, has shortened the length of the harvest season.

- The environmentally based decision to eliminate Alar has also tended to reduce the length of the harvest season.

- Growers' decisions to plant other-than-apple fruit trees (such as cherries, peaches, and pears) would tend to lengthen the season, but this is not the trend, and many growers do not have this option because of soil or weather conditions.

- Season-shortening factors, generally, make the New York–Pennsylvania fruit season less attractive to workers and tend to create more of a crisis atmosphere for growers during the harvest season.

- More intense cultural practices (for example, the use of trellises) increases the need for preharvest labor while decreasing the need for harvest labor.

- The use of smaller trees, which is a trend, makes it possible to draw from a larger, less elite labor force, and increases harvester productivity.

- Mechanical harvesting appears only to be a possibility, at the current level of development, with processing apples, and only for some of them.

- Variables such as spot picking versus strip picking, big trees versus small ones, and picking fresh fruit versus processing fruit tend to complicate compensation systems and labor-management relations.

- All else being equal, growers with integrated operations can afford higher labor costs than those without; similarly, fresh fruit growers near major urban markets can afford higher labor costs than those more distant from such markets.

The Apple Industry Labor Market

There appear to be four different patterns of recruitment in the area studied. There is the use of the H-2A program, the Employment

Service (ES), crew leaders, and the direct hire system (in which there are no labor intermediaries). There are advantages and disadvantages, for both growers and workers, in each of the systems. The Employment Service, other than servicing H-2A orders, plays a small role in the industry.

The range among the formal systems is from the U.S.–Jamaican system at one extreme, for bringing H-2A workers to the orchards — which operates smoothly despite some major bureaucratic complications — to the total lack of union-organizing efforts at the other end of the spectrum.

The direct-hire system, more frequently seen in smaller operations, involves the growers and workers in face-to-face contact with each other, without either side making any use of the H-2A program, crew leaders, or the ES. Some of these systems are potentially in conflict with each other, such as the crew leader system, which uses only domestic (nonimmigrant) workers, and the H-2A system, which is based on the employment of immigrants.

The growers, some of whom use two or three systems at the same time, reported this pattern of use:

Use H-2A system	32%
Use Employment Service	52%
Use crew leaders	40%
Use none of the above (direct hires only)	32%

It should be noted that *all* H-2A growers *must* use the Employment Service to secure their H-2As, so that apparent dominance of the ES is quite misleading.

When we asked the workers how they got their current jobs, we got these replies:

Through the H-2A system	19%
Through crew leaders	26%
Through the Employment Service	4%
Other ways (direct hires)	51%

There were 21 respondents who worked for crew leaders. Most were blacks (ten U.S. citizens [USCs], three Haitians, and one man from St. Lucia); the other seven were Mexicans. They estimated that the crews they belonged to consisted of about 28 workers.

Crew Leaders

Crew leaders are rural entrepreneurs who manage groups of domestic farm workers; the term farm labor contractor is rarely heard in the East. Usually the workers are either Hispanics or blacks, and the crew leader is usually, but not always, from the same ethnic group. Although this is impressionistic, typically the better crew leaders have pulled together a group of relatives and acquaintances from a rural area who have ties to one another. A crew strung together from unemployed urban workers, or, worse, recruited along the way, is less likely to be cohesive and productive, and is more likely to be exploited. An ethnically mixed crew is not necessarily a happy omen of rural democracy and mutual toleration.

The principal role the crew leaders play in the Pennsylvania and New York apple harvest is recruiting workers. They may also do some scheduling and supervise some of the work, supervise housing, and do recordkeeping, but the primary role is to bring the workers to the grower.

The principal advantage of the crew leader system to the grower is that it answers the omnipresent question: Who will harvest the crop?

The principal advantage to the worker — and this attracts some of the least advantaged of the workers — is that the crew leader will find work for him and transport him from job to job. There are hierarchies among migrant farm workers, and those on the bottom rung are those dependent on crew leaders. The next step up is to be a migrant who gets around and gets work without a crew leader, and the next step after that may be to become a year-round worker in one location or to become a crew leader himself.

Our sense was that there were relatively few crew leaders working apples in either Pennsylvania, where we talked to two of them, or in the H-2A-dominated Hudson Valley, where we talked with three crew leaders. There are clearly more of them in western New York, where we also talked to three of them. The ES counted 78 crews in all of New York agriculture, while 20 were counted in the Pennsylvania tree fruit industry in 1991.

The eight crew leaders we interviewed included three U.S.-born blacks, one Guatemalan (fairly new to the business), three people born in Mexico, and an Anglo woman who was a co-crewleader married to a Mexico-born co-crewleader. We also had several conversations with the area's one "super" crew leaders, Cliff DeMay, who is one of the reasons why there are a number of crew leaders working apples in western New York.

What DeMay does is to play the role of intermediary between crews and the crew leaders he recruits, on the one hand, and growers on the other. He also helps growers apply for the USDA's Farmers Home Administration (FmHA) worker housing funds. During the season, he moves his crews around to maximize the amount of work that they get, while simultaneously seeking to harvest the crops of his grower-clients. He screens workers' documents for IRCA compliance, but does not handle payrolls; that is done by the growers. "We are happy to pay Cliff his 3 percent because we regard him as insurance," one grower told us. "If our regular crew did not show up we would count on Cliff to find another one."

DeMay owns an orchard, and formerly managed one of western New York's largest (then corporate-owned) orchards before going into the super crew leader business about four years ago. He is a major factor in western New York, particularly in Wayne County, where "as much as one-third of the harvest labor force is affiliated with one major contractor" (NYES 1991).

This role is not without controversy. Some crew leaders not affiliated with DeMay regard him as a competitive threat. Some people in migrant-serving organizations worry that he has expanded the use of the crew leader system that they find inherently flawed. They also worry that his recruiting of predominantly Hispanic crews will further narrow harvesting opportunities for the remaining black workers. Further, DeMay is usually involved in some scrap with the Department of Labor.

Clearly DeMay's activities represent an interesting development, bringing some modern management systems to bear on a rough-and-ready business. To our knowledge, his operation, which involves close to 1,000 workers at the peak of the season, is the only one of its kind in the Northeast; it certainly is the only one in these two states.

Quite aside from the DeMay operation, we were impressed with many of the crew leaders we interviewed. One of the leaders had a presence in the commanding-to-menacing range, another was hostile, and a third was probably a scoundrel, but the other five seemed to be decent to admirable people working in a tough business.

A Portrait of the Workers

The apple work force in the two states is both heterogenous and changing. Nonimmigrant (H-2A) workers (from Jamaica) dominate the harvest in the Hudson Valley, play a small role in western New

York, and have no role in Pennsylvania. In recent years, Mexican nationals, including some SAW workers, have played an increasing role in the harvest, while native-born blacks and Puerto Ricans have played a diminishing one. Virtually all of the workers are interstate migrants who leave the area at the end of the harvest; only a few travel with their families. Many of the workers also pick oranges and other tree fruit; a large majority prefer the comparatively well-paid, if brief (6–8 weeks) apple harvest to other farm work.

Looking at this population in more detail, we find three broad types of civil status: U.S. citizens (USCs), nonimmigrant aliens (the H-2As), and the other foreign born. The USCs include migrant southern blacks and a few Puerto Ricans and local workers. (Many of the local workers are "settled out" black and Puerto Rican migrants and a few Mexican migrants. Very few of them are white.) Some of the foreign born are Caribbean and Central American, but predominantly they are Mexican. The foreign born include substantial numbers of legal permanent residents (green cards), some people who have secured Special Agricultural Worker (SAW) status under IRCA, some who have applied for that status but whose cases had not been decided when they were interviewed, and others with no documents, or, more likely, fraudulent ones.

Overlapping at least some of the legal categories noted above are the variables of language and national origin. There are three language groups: English, spoken by southern blacks and Jamaicans; Spanish, spoken by just about everyone else; and Creole, used by the Haitians.

There is yet another subgroup, this one directly attributable to IRCA. When the regulations defining "perishable commodity" were published, sugarcane was not so defined. Farm worker advocates sued to have sugarcane defined as perishable. During the litigation the court ordered the Immigration and Naturalization Service to accept SAW legalization applications from workers whose qualifying employment included sugarcane. As a result, some H-2 workers who had come legally to New York to work in the apple harvest decided to apply for legalization under the SAW program. Typically they based their claim for 90 days of farm work on work in both apples and sugarcane. The courts ultimately ruled that sugarcane was not perishable.

This is relevant here because few of the H-2s who worked in apples and applied for SAW status had enough employment in apples alone to meet the requirements. After the court ruled that sugarcane was not perishable, some former H-2A SAW applicants, apparently through INS staff-level benevolence, were given SAW status anyway; others apparently had pending applications at the time of our interviews. However, some of the former H-2A SAW applicants were denied status, in effect becoming illegal aliens. IRCA thus created illegals out of what had been legal workers.

During the course of the project we interviewed a study group of 80 apple pickers. They were chosen in such a way as to secure appropriate numbers of members of the various groups known to be active in the harvest. The 80 apple workers included 74 men and 6 women. Many spoke English (68%), but were foreign born (69%), with the Jamaicans having both characteristics. These were their birthplaces:

Mexico	28%
Jamaica	28%
U.S. mainland	25%
Haiti	11%
Puerto Rico	6%
Honduras	1%
St. Lucia	1%

The distribution of the workers' home bases, as opposed to places of birth, was quite different:

Florida	41%
Jamaica	20%
N.Y.–Penn.	18%
No base	11%
Puerto Rico	4%
Other U.S.	4%
Mexico	4%

In terms of civil status, they reported:

U.S. citizen	31%
Green card	23%
H-2A	19%
SAW (approved)	14%
SAW (no decision)	6%
Other	6%
Undocumented	1%

We assume that those in the Other category, all of whom were foreign born, lacked legal status in the United States. The citizens, incidentally, were all native born. None of the foreign born had become naturalized citizens; some of those with green cards had been here for many years and could have naturalized.

Most of the workers we encountered were males whose families were elsewhere, generally in the American South. These were not, by and large, young men. The average ages were, by place of birth:

Jamaica	42
U.S. mainland	40
Haiti	35
Puerto Rico	33
Mexico	31

Most of the workers (85%) reported that they had children, a variable that, as we will show later, has some unexpected potential financial ramifications.

Many of the workers, except for the H-2As, had from some to a lot of nonfarm work experience in the U.S.:

Long-term nonag. work experience	26%
Short-term nonag. work experience	31%
Non-H2A, no nonag. experience	26%
H-2A	16%

We defined long-term work experience as six months or more. About half of the respondents with nonfarm experience reported having their last nonagricultural job in 1990 or 1991, the rest at more distant times.

The jobs they had held were primarily unskilled or semi-skilled ones. The median gross pay reported (sometimes many years before) was $200, a level about $50 lower than they reported for their most recent work in apples. In other words, most of the workers were telling us (although the interview format was such that this was not stressed) that they were grossing more in the apple harvest than they had in their last nonagricultural job.

We divided the workers into three migration categories, as follows:

Year-round N.Y.–Penn. residents	14%
Mainland-based migrants	41%
Other migrants (who return to other nations or Puerto Rico)	45%

Most of the workers (86%) were migrants, in the sense that they said that they had lived in more than one state, territory, or nation in the course of the 14-month period we covered. The dominant pattern for the mainland-based professional farm workers usually revolved around long periods of orange picking in Florida, shorter periods of other farm work, often some unemployment, and then the apple harvest. Much of the other farm employment was in tree fruit. A significant portion of the apple pickers did pruning and other orchard work before they started their work in the apple harvest.

Given the seasonal and migratory nature of harvest work, we expected to see a higher average level of unemployment in this group than in the nation generally, and we did. Over the year the average unemployment level was 13.4 percent, about twice the national average for the work force as a whole. As best we could, we counted weeks when the individual was potentially able to work but was not working. What was interesting were the different patterns of unemployment:

Period of Unemployment	Percent of Work Force
No weeks	39%
1–4 weeks	19%
5–20 weeks	33%
21+ weeks	10%

The largest single group of workers was always working, or at least did not lose five consecutive days of work all year long, despite the often migratory and often uncertain nature of the work. (Some of this group simply returned to self-employment in Jamaica.)

A minority of the workers, 25 percent, said that they had collected unemployment insurance (UI) during spells of unemployment. It turned out, as one might expect, that those who were able to draw UI were unemployed on an average of 9.8 weeks, while the average for the whole group was 6.5 weeks. We also found that workers with access to unemployment earned more during the season than workers who had never drawn UI checks.

Somewhat surprising was the seasonal pattern of unemployment for apple pickers. As Figure 16.1 shows, the highest level of unemployment was in the summer, just before the apple harvest; for many other farm workers, mid-winter is the time of highest unemployment.

Figure 16.1. Unemployment of New York–Pennsylvania Apple Pickers

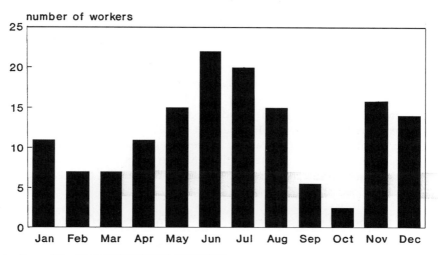

Each unit is a month in which a worker reported some unemployment
Source: UACAW 1993.

Earnings and Working Conditions

There is general agreement that farm workers do hard, useful work and that the income and economic status of many persons who participate in the hired farm work force are low. It is also probably true that picking apples, the focus of this case study, is more attractive than some other tasks in agriculture and better paid than many.

Apple harvest worker earnings appear to be stable and probably above the U.S. average for farm work, but well below those of the average factory worker. In most cases workers were living in free housing. There appeared to be no problems regarding the deductions of social security and similar taxes.

The Employment Service collects data on piece rates and determines a prevailing rate each year; H-2A users are obliged to pay at least these rates. Table 16.1 shows the prevailing rates, in cents per eastern apple box, for picking fresh fruit for the last seven years.

When the rates are paired with the presence or absence of H-2A workers, two comments can be made: (1) In places where non-immigrants dominate the labor market (the two valleys) the piece rates have been flat for six consecutive years, but there is more volatility in the piece rates of the other two regions and some tendency for the piece rates to rise; (2) After 1985 the average piece rates in the H-2A areas were *always* higher than in the non-H-2A areas.

A worker is paid a little over a penny a pound for picking fresh apples. A bushel holds about 42 pounds of apples, and the eastern apple box (which is the basis for the piece rates quoted) contains 47.25 pounds.

The surveyed growers estimated that their apple pickers earned $329 a week, gross, during the 1990 season; this is an average of estimates made by growers interviewed in the two states. The growers were not asked to show their payroll records.

The workers surveyed in the 1991 season (the 55 of the 80 who had all the needed data) reported, on average, gross weekly earnings of $260. The median gross was, as it usually is in earnings distributions, a little lower at $246. The median net was $186. (We did not work much with net earnings, on the grounds that the deductions for the H-2A workers for savings programs, for insurance, and in a few cases for meals, were usually much greater than for the other workers.)

Worker interviews were begun at the beginning of the harvest season, and workers were asked about their gross and net pay for the pay period immediately preceding the occasion of the survey. In a

few cases workers were interviewed before they had received a pay-
check, and most workers were interviewed early to mid-season. Therefore, their weekly earnings do not necessarily reflect a full-season
average.

Table 16.1. Prevailing Rates, in Cents, per Eastern Apple Box, for
Picking Fresh Fruit, 1985–91

Region	Fresh Fruit, 1985-91 Years						
	1985	1986	1987	1988	1989	1990	1991
CV NY	60.0	64.0	64.0	64.0	64.0	64.0	64.0
HV NY	50.0	60.0	60.0	60.0	60.0	60.0	60.0
W NY	50.0*	52.5*	57.5	60.0	60.0	60.0	60.0
AC PA	N/A	N/A	N/A	49.5	58.5	54.0	56.3

*In addition, a $.05-per-bushel, end-of-the-season bonus prevailed during these
years. The Pennsylvania reader should remember that the familiar per-bushel
piece rates have been multiplied by 1.125.
Key: CV=Champlain Valley, HV=Hudson Valley, W NY=western New
York, and AC PA=Adams County, Pennsylvania.

As a check on our data, we examined the subset of interviewed workers who had paystubs in hand for their last pay period. We found that they had an average of $272 in gross earnings, which is pretty close to the broader average of $260.

Several observations can be made about these numbers:

1.   New York and Pennsylvania apple pickers, in the fall of 1991, were making considerably less than the state's factory workers, who were paid $466.62 a week in New York and $503.78 a week in Pennsylvania (USDOL 1993a). Work in apples, however, appeared to be better paid than the kind of nonagricultural jobs that the surveyed workers had secured in the past. As noted earlier, the median earnings for such work were about $200.

2.   The surveyed apple workers, with an average gross of $260 per week, were in the same earnings range as several groups of California farm workers. For example, in October 1991, California agricultural workers generally had average weekly earnings of $269.86, fruit and tree nut workers had an average of $246.14, while deciduous tree fruit workers were paid $221.82 (CalEDD 1991).

3.   The New York–Pennslyvania apple pickers appeared to be earning more than some of the workers covered by other CAW researchers. For example, Kissam and Garcia (1993) found that farm workers in southwestern Michigan were earning $136 a week on average, and Mason found average gross weekly earnings in the lemon harvest of $137 per week in Tulare County, California, following the freeze in the winter of 1990, and $153 a week in the Yuma County, Arizona, lemon harvest (Mason, Alvarado, and Riley 1993).

Although our study group was too small to secure statistically significant data on this point, the variations among hourly earnings for the various subgroups of workers were interesting. As Table 16.2 shows, those who spoke Spanish made more than a dollar an hour more than those who spoke English. A little exposure to social service agencies and nonfarm employment led to higher wages than no exposure to these systems or to much exposure. Men earned more than women, and hourly earnings declined as workers grew older.

**Table 16.2. Median Gross Hourly Earnings
of Several Groups of Apple Pickers**

Group of workers	Number	Earnings
Language:		
Spanish-speaking	11	$7.53
English-speaking	25	$6.22
Place of birth:		
Born in Mexico	10	$7.53
Caribbean (black)	12	$6.71
U.S.-born (black)	10	$6.31
Recent social service usage:		
One-time user	10	$7.18
Non-user	15	$6.52
User more than once	11	$6.00
Civil status:		
Green card	13	$7.53
U.S. citizen	12	$6.31
Age:		
30-39 years	12	$7.18
≥40 years	14	$6.71
<10 years	10	$6.22
Exposure to non-farm work:		
Short-term non-farm work	11	$8.05
Long-term non-farm work	11	$6.31
No non-farm work	14	$6.00

Source: North-Holt survey for Commission on Agricultural Workers

Housing

The housing situation in the New York–Pennsylvania apple industry appears to be different than it is in some other agricultural industries. Most of the surveyed workers (89%) lived in grower-provided housing at the time of the interview; an almost equally high percentage of the workers (84%) were traveling alone; and workers with families present were the exception, not the rule. Typically the workers stayed in small, on-farm clusters of houses, trailers, and dormitories, known universally as camps.

Free housing for apple workers is mandated under the H-2A program, and has become the norm in the rest of the New York apple industry as well. The pattern in Pennsylvania is for growers to provide on-farm housing, with many of them making minor charges. In our survey of growers, we found that only 2 of the 18 New York growers charged their workers for housing, while 5 out of 7 in Pennsylvania did so. Rates were low, ranging from $5 to $12 a week. Workers have come to expect these housing programs, and it is one of the attractions to migrants.

The housing varies, from grim to quite acceptable. It tends to have no other use than to provide shelter for a temporary work force, although sometimes it is used for storage during the off-season. We were told by a Pennsylvania camp inspector that in recent years new housing is more likely to be dormitories for single persons than to be family housing. New farm labor housing must meet both local zoning and building code standards as well as state regulations for such camps. Local zoning can be a problem — we kept hearing about efforts by growers or Rural Opportunities, Inc. to build better housing for workers being vetoed by local zoning boards, with the suggestion that racism played a part in these zoning decisions. Rural Opportunities, active in both states, is a government-funded, migrant-serving agency.

A continuing friction about housing is set off by farm workers arriving, by definition homeless, and wanting to move into a grower's housing units before the grower needs the workers. Since growers tend to worry that their housing will deteriorate more rapidly with greater use, they are often reluctant to open camps before the harvest. The early arrival of the crews is another indication that (1) the apple harvest is a desired kind of work opportunity and (2) there is nothing better for the workers to do at the time, as reflected in the high levels of summertime unemployment in Figure 16.1.

Various FmHA housing programs are being used in the apple area to build labor housing, although this is often for year-round workers. Both Rural Opportunities and the super crew leader, Cliff DeMay, have played the role of broker in getting specific farmers to apply for these grant and loan programs. We understand that a disproportionate share of the nation's FmHA farm labor housing projects are located in these two states.

One of Rural Opportunity's initiatives has sought to reduce the sense of isolation that often accompanies farm labor housing. Though we do not have all the details, that agency has played a broker role among the growers, the crew leaders, the residents of the housing, and the telephone company to cause the installation of pay phones in several farm labor camps.

The government not only funds some farm labor housing, but also inspects and regulates it as well, a process that is often both complex and controversial, with strong feelings on all sides. Growers sometimes object to the substance of the inspections (nitpicking), but they are more likely to complain about their multiplicity and inconsistencies in the interpretations of the various inspectors. Worker advocates argue that standards are not tough enough, and that in some areas the enforcement is lax. However, they complain that inspection is far more lax in non-H-2A, nonapple camps than in the apple area.

A given labor camp is routinely subject to two or three rounds of housing inspection, depending on how the grower recruits his work force. The first round of inspections, which takes place in late winter or spring when the camps are empty, is done for the state. In Pennsylvania the agency is the state Department of Environmental Resources; in New York it is the county public health departments.

The second round, usually coming after the camps are occupied, is conducted by federal agencies; the feds usually only look at housing that has already passed the state inspection. Two units of the U.S. Labor Department divide the work; they are the Occupational Safety and Health Administration (OSHA) and the Employment Standards Administration. The latter also looks at wage records and uses OSHA's housing standards; OSHA inspectors apparently do not look at the growers' wage records. Any of these agencies can come back again if there is a complaint, and sometimes this happens. The Pennsylvania state agency also tries to do a complete set of postoccupancy inspections as well as the preoccupancy ones, but usually does not have enough staff to cover all the camps again.

An additional round of housing inspections is set in motion when the grower files a job order with the Employment Service to hire workers from outside the state's borders. This is an interstate clearance order, which must be filed to obtain H-2A workers. So all H-2A growers get at least three inspections, sometimes more.

Camps are rarely closed by inspectors, but they often require that work be done on them before issuing a certificate of occupancy. Sometimes the need for repairs is noted by the inspectors just as the season begins, to the irritation of the grower.

Meanwhile, on the other side of the issue, there are criticisms of the standards used. For example, the state of New York has not (some would say not yet) decided that the flush toilet is required in farm labor housing; many growers have installed them without waiting for a mandatory ruling.

The combination of the decentralization of inspections in New York and the multiple inspections required by the H-2A users has created an anomalous situation in Hudson Valley agriculture. It is generally agreed by worker advocates that the housing provided by the apple industry in this part of New York (mostly in Ulster County) is much better than that supplied by the onion industry (mostly in Orange County). Because of the H-2A program, much of the apple industry housing is thrice inspected, with the DOL inspectors regarded as fairly rigorous. It is used principally by nonimmigrant workers.

On the other hand, the onion worker housing is used by a mix of U.S. citizens, those with green cards, and probably undocumented workers. But since the onion growers never use the H-2A program and rarely use the Employment Service, that level of inspection is missing. The principal inspection of their housing is done by the Orange County Public Health Department, who is not known for the vigor of its inspections. So the best housing is inspected more often and by harsher critics than the worst.

We asked the workers, who presumably have more first-hand knowledge of the question than anyone else, "How does your current housing compare with other farm-worker housing you have seen?" We heard these answers:

Much better	40%
A little better	16%
About the same	19%
A little worse	4%
A lot worse	5%
No basis to compare	16%

Effects of IRCA

IRCA appears to have had little direct impact on the industry, other than legalizing the previously illegal presence of a minority of the workers. Sanctions are regarded as a nuisance by the growers and not a threat. The use of the H-2A program has not increased as a result of IRCA's changes in that program, and there has never been any local pattern, as there was in California, of open-field enforcement of the immigration law. The authors disagree on the indirect impact of the legalization (SAW) program, with Holt seeing it as having no impact and North arguing that it has helped to increase the size of the work force, thus helping to hold wages steady and facilitating the ongoing displacement of native blacks by Mexican nationals. The authors also disagree on the H-2A program, with Holt finding it appropriate and necessary and North calling it a needless expansion of the labor force in an already loose labor market.

Other IRCA Effects

The principal issue raised by growers was too much governmental regulation; too many agencies wanted too much information, made too many demands of them, and sent too many inspectors to visit them. H-2A users were particularly outspoken along these lines. The principal issue raised by workers was a desire for more pay.

It may be useful, if only negatively, to mention some of the subjects that we know are controversial topics elsewhere, and that were either not mentioned or not stressed in the New York–Pennsylvania apple industry.

While worker advocates were unhappy with the way growers treated would-be domestic workers, particularly those applying for work when the season was underway and the dormitories were full of Jamaicans, we did not hear (as one does in Florida) about the H-2A growers dismissing (or not subsequently rehiring) Jamaicans for not being productive enough.

We also did not hear worker advocates complain that apple growers barred anyone from their camps, a practice encountered in the New York onion industry, which relies wholly on domestic labor.

Nor did we hear from growers, as we did in Oregon, about the high cost of workers' compensation. The minimum wage was another issue that was rarely mentioned, although wage-hour inspectors were often discussed. Apparently wages are high enough that the federal minimum of \$4.35 and the New York minimum (during 1991) of \$3.80 an hour were not regarded as very significant.

A problem raised sometimes, by both growers and worker advocates, was the resistance to better farm worker housing that showed up from time to time in negative zoning decisions.

Recommendations

We agreed on three joint recommendations and differed a bit on a proposal to raise wages for apple pickers.

There is a need for much more focus in the *enforcement* of housing and other regulations in these two states. We have noted what appears to be a lopsided situation, in which there are more inspections (and more migrant advocacy work) in the less threatening situations (such as H-2A users in the apple industry) and less in the more troublesome ones. We suggest that enforcement resources would be better utilized if they were focused on known problem areas in agriculture, and if employers with good records were simply visited once every two years, not annually (unless there are complaints).

More focus in the delivery of *social services* would be helpful, too. It is clear that these programs will be needed for many years to come, so some long-range planning is in order. All else being equal, it would make sense to consolidate the management of these programs into two or three networks per state, as opposed to the current multi-network system. Why can't migrant health clinics and migrant daycare programs be co-located in the same place, for example, and perhaps run by the same contractor? Why not increase a grant by a percentage point or two if it is physically located in the same building as another migrant program run by another agency? Better funding for these programs would be useful.

We suggest that *crew leaders*, like migrant social service programs, will be around for years to come, and that the appropriate public policy is to seek to improve the institution, not simply to regulate it. In addition to a rigorous licensing process, we make three sets of suggestions:

First, that a technical assistance program be launched, on an experimental level, to provide crew leaders with information on how better to practice their trade. These programs, provided in both supply and demand states, would help crew leaders understand state and federal regulations, and would give them concrete information on how to work with banks, insurance companies, and regulators. Perhaps there could be exchanges of good-practices data, on how to motivate crews, how to negotiate with growers, and how to preserve grower-owned housing. These technical assistance programs, perhaps overtly, perhaps subliminally, could be used to encourage the crew leaders to move toward more respectable, law-abiding styles of operations.

Second, efforts should be made by both agribusiness and the government to encourage better crew leaders to stay in the business, to stay law abiding, and to meet their obligations. Perhaps below-market loans for new buses and vans could be provided by the agricultural credit system (to licensed crew leaders with several years of good records); perhaps agribusiness could work out some group purchases of vehicular insurance needed by crew leaders with transportation-approved licenses.

Third, some consideration should be given to creating model crews, as examples to others and to encourage a new type of crew leader. Perhaps funds could be found, in the Departments of Labor and Agriculture or in the private sector, to underwrite the launching of five types of new migratory crews:

1. The advocate model, crews run by DOL-funded entities like Rural Opportunities.

2.   The religious model, crews sponsored by churches, probably southern black churches.

3.   The Peace Corps model, crews run by bilingual returned Peace Corps volunteers.

4.   The corporate model, crews run by major agribusiness firms with interests in crops in several states.

5.   The extended family model, crews run by veteran migrant farm workers who are also elders in extended families.

North's recommendation for *increasing wages* immediately for many but not all apple pickers will cost the growers nothing but some paperwork. He suggests that the growers use an obscure existing provision in the tax code, an aspect of Earned Income Credits. EIC is an IRS-operated program for low-income, working families; it provides as much as 15 to 17 percent additional income (not tax rebates) to people who qualify. Virtually all apple pickers have low enough incomes to qualify, and a majority have children, a requirement as EIC now functions. EIC payments are usually, but not necessarily, made when the worker files a tax return.

North points out the employers *already* have the power to augment the apple harvesters' wages by *advancing* EIC payments to eligible workers, using the payments made to the workers to offset the money growers owe to the IRS. North also suggests that, to ease both the growers' paperwork and the lot of migrant workers generally, the IRS code be amended so that growers making advance EIC payments to currently eligible workers be allowed to make similar payments to migrants without children.

Growers who use this technique, North points out, will be getting a cost-free advantage over potential rival employers who do not use the technique: They can, in effect, raise a $.55 piece rate to $.63, for instance, with the government paying the difference.

Holt feels that the subject of the economic well-being of farm workers as an occupational group, while a vitally important social issue, is not germane to this report and was not the objective of the case study. Holt does not agree that one of the purposes of IRCA was to increase farm worker wages or change agricultural labor management relations. While it may have been an expected outcome by some individuals or groups who supported the legislation, the legislative history does not support the conclusion that this was the intent of Congress in enacting the legislation.

17

IRCA's Effects in Citrus

The citrus industry in the United States represents approximately 23 percent of the total world production (Hanneman 1990). Leading in U.S. citrus production is the state of Florida (303 million cartons in 1989), followed by California (165 million cartons) and Arizona (20 million cartons). Citrus production in each of these three states differs in significant ways. While Florida produces the majority of domestic processed orange products (juice and other by-products), California produces the majority of citrus destined for the fresh market. Arizona citrus includes oranges and grapefruit, but is dominated by lemons.

This chapter presents an analysis of IRCA and citrus production in the three leading citrus states. It consists of two parts. Part I, written by Robert Emerson and Leo Polopolus, examines the implications of IRCA for Florida citrus. Part II, authored by Herbert Mason, Andrew Alvarado, and Gary Riley, presents a comparative study of the citrus industry in California and Arizona.

17

Part I. Florida[1]

ROBERT D. EMERSON AND LEO C. POLOPOLUS

Citrus is the dominant agricultural commodity in Florida, representing 26 percent of cash farm receipts in 1990 (USDA/ERS 1991). Moreover, citrus production is rebounding, with large plantings following the severe freezes in recent years. With 69 percent of the U.S. production in the 1991–92 season (Table 17.1), Florida is the dominant citrus-producing state in the United States. Oranges account for 74 percent of Florida citrus production, as compared to grapefruit, 21 percent, and the remaining 5 percent of specialty fruits, including limes. A further important characteristic of the Florida citrus industry is that most (92 percent in 1991–92) of the oranges are processed rather than sold as fresh fruit.

The *1987 Census of Agriculture* (USDC 1989) for Florida reports hired farm labor expenses of $149 million and contract labor expenses of $131 million for farms classified as fruit and tree nut farms (SIC 017), most of which in Florida are citrus. These represent 21 percent and 47 percent of the total for all farms in the respective categories. Combining direct-hire labor with the contract labor expenses, the fruit-and-nut category represents 28 percent of combined labor expenses for Florida farms. The combined labor expenses constitute 36 percent of production expenses for fruit and nut farms.

Although labor expenses are clearly a dominant component, representing 36 percent of all production expenses, the nature of the citrus harvest market is such that the *Census of Agriculture* data miss a major component of the citrus harvest labor market. Labor contractors are the dominant type of employer in the citrus harvest, but such costs are included in contract labor expenses only if the grower is in the business of employing the labor contractor. It is not at all uncommon, however, for the employer of the labor contractor to be an entity other than the grower of the citrus. It has become very common for an intermediary to be the firm arranging for the citrus harvest,

Table 17.1. Citrus Production: Florida and the United States, 1991–92

Item	Florida				United States			
	Oranges	Grapefruit	Other	Total	Oranges	Grapefruit	Other	Total
Bearing Acreage (1000 acres)	444.3	104.7	34.4	583.4	639.4	136.8	107.5	883.7
Production (1000 tons)	6291	1802	417	8510	8861	2224	1319	12404
Percent	74	21	5	100	71	18	11	100
Percent processed	92	46	43	80	74	43	40	65
Value of Production ($1000)	853,287	281,500	95,345	1,230,132	1,206,027	334,644	322,391	1,863,062

Source: FASS 1993.

contracting with a labor contractor to harvest the fruit, perhaps no longer owned by the grove owner; in other words, the fruit has been sold while on the tree. One consequence of this arrangement is that a large component of the harvest labor market is missed by the *Census of Agriculture* approach. The value of harvesting contracts by labor contractors, for example, who are harvesting fruit for a citrus processor, would not be included in the census data; the census data include only contract labor expenses of agricultural producers. The census data are suggestive of the extent to which this phenomenon prevails. There were 8,584 farms with production expenses categorized as fruit-and-nut producers in Florida, but only 3,165 of them reported any direct-hire labor and 3,819 reported contract labor expenses. Furthermore, many farms reporting direct-hire labor are likely to be the same ones reporting contract labor. While many of the producers without labor expenses are likely to be small producers, harvesting one's own fruit by the proprietor and his or her family, regardless of the volume of fruit, was not observed.

Labor Market Structure

The Florida citrus harvest labor market is an excellent illustration of entrepreneurship in the classical tradition of Schumpeter (1961) and Knight (1965). As we have argued elsewhere (Polopolus and Emerson 1991), the structure of the harvest labor market has been guided by the same forces as guide other entrepreneurial decisions, namely the profit motive. As Schultz has emphasized, entrepreneurs play a critical role in resolving economic disequilibria (1975, 1980). As entrepreneurs see opportunities for profit, they adapt their organizations to take advantage of those opportunities. Within the context of the citrus industry, the employment of harvest workers may be characterized as routine although seasonal activities for the producer. Given time constraints, the entrepreneur must allocate time among competing uses, including the choice of technology, marketing decisions, resource allocation, and so on. When faced with the chore of the necessary decisions required for hiring a large number of workers to do relatively routine jobs for a short period of time, it may be more profitable to contract for the harvesting services, letting another firm specialize in the harvest employment activity.

We suggest that while it does not appear that IRCA has had a great impact on the Florida citrus industry, the employer sanctions component may have enhanced the shift of employment for harvesting to third-party entities. A relevant issue is thus a consideration of the determination of the firm's boundary between activities performed internally and those which are done under contract. The entrepreneur must determine which activities are to be carried out within the firm and which are to be conducted via contracts or market transactions. Two extremes may be envisioned within the context of citrus harvesting. One extreme could be a single firm growing the citrus, harvesting the citrus, processing the citrus into orange juice, and hiring all necessary employees for the activities. This corresponds to the familiar vertically integrated firm. At the other extreme may be separate firms carrying out each step of the process, from growing the citrus to processing it for the final consumer product, and in particular a separate firm used for the harvesting activity.

The introduction of the significance of transaction costs by Coase (1952) and its reactivation by Alchian and Demsetz (1972) provides a framework for explaining how activities are selected to be performed within the firm as opposed to being performed by different firms. The transaction cost approach argues that the firm exists as an efficient means of conducting transactions; it is less costly to conduct its activities internally than through the marketplace. A key component of this approach is the recognition of the importance of information pertaining to the transaction. The more unique the information, the more likely it is for the transaction to be internalized in the firm. A good illustration is a transaction that involves any type of proprietary information.

Most harvest activity is highly routine work, having little specificity other than the location. Harvesting oranges for one firm requires the same skills as harvesting for another. Likewise, there are likely to be few proprietary secrets involved in the harvest activity relative to the producing firm or the receiving firm. Similarly, negotiating a contract for a labor contractor involves few additional features over necessary contracts in the absence of a labor contractor. Consequently, in many cases utilizing a labor contractor would be expected to result in significantly lower transaction costs pertaining to the harvest than would be the case with the direct employment of workers for harvest crews. An obvious example is the owner of smaller fruit acreage requiring only a few days of harvest services. It would be relatively expensive for this firm to maintain all of the information to conduct efficient transactions in the harvest labor market. Alternatively stated, there are likely to be economies of scale in the harvesting operation, with the economies of scale centered around the cost of information. The employer of harvest workers on a large scale can afford to specialize in information unique to the harvest labor market. Examples are knowledge of the sources of workers, language, and the techniques leading to optimal efficiency in harvesting, none of which have much bearing on other activities in agricultural production. Most importantly, a firm specializing in harvesting can distribute the cost of this knowledge over several different agricultural producers rather than each individual producer bearing the cost.

On the other hand, very large producers are able to spread these information costs over a large volume of product and time. As a result, large producers would be expected to be more likely than small producers to hire harvest workers directly. It is important to recognize that the large firm need not be the producer. Processing plants and packing houses may integrate the harvesting activity into their operations. Their scale is clearly sufficiently large enough to accommodate this activity, and in addition, they are able to maintain direct control of scheduling the commodity into the processing or packing plant.

IRCA and the Labor Market

IRCA proscribed a new set of rules and regulations that must be followed in the U.S. labor market. An important component of this was employer sanctions, applying not only to agricultural employment, but to all forms of employment. The essence of the regulations is that employers are prohibited from hiring undocumented workers, and at the same time are prohibited from discriminating against applicants who may bear a resemblance to some perception of "undocumented workers." There are substantial civil, and potentially criminal, charges associated with violation of the terms of IRCA. As with most such regulations, there is a significant record-keeping responsibility for the employer.

It is our contention that one significant effect of IRCA is the impact that employer sanctions have had on the organization of the industry, although employer sanctions were given little attention in the recent *Report of the Commission on Agricultural Workers* (1993). While labor contractors have been a significant component of the citrus harvest market for some time, employer sanctions have been one more factor shifting the structure of the industry further toward the use of labor contractors for the harvest. In the context of the earlier discussion, employer sanctions add one more feature to the process of hiring a large, temporary work force for harvesting as opposed to the more routine aspects of producing citrus. Given the large number of workers to be employed for a relatively short time period for most growers, there is an additional incentive for the producer to have a different firm handle the harvest function. More often than not, this firm is a labor contractor. The labor contractor has the specialized expertise associated with hiring large numbers of harvest workers. Foremost among these are the necessary language skills, recruitment techniques, and procedures for handling large numbers of employment applications with the associated record-keeping requirements of IRCA.

One way in which this has evolved in Florida is the establishment of a subsidiary of the firm to handle the harvesting activity, particularly if the firm is a processor or packing house. While the subsidiary is operated as a separate business, the parent corporation can still maintain assurance of a smooth flow of the commodity into the processing plant or packing house. In addition, the firm may have succeeded in shifting the liability for any potential violations of employer sanctions away from the parent corporation.

While the above type of organization suggests a relatively professional type of organization operating as a labor contractor, there are numerous other illustrations in which the labor-contracting entity is a totally separate entity. A continuing issue is that completely independent contractors tend to exhibit the full spectrum of possible business ethics in the employment arena. Vandeman, for example, suggests that sanctions for hiring illegal workers have not been effectively enforced against labor contractors (1988). Baumol's hypothesis regarding entrepreneurship applies directly to the expansion of labor contracting in the presence of undocumented workers and the existence of employer sanctions under IRCA: ". . . it is the set of rules and not the supply of entrepreneurs or the nature of their objectives that undergoes significant changes from one period to another and helps to dictate the ultimate effect on the economy via the allocation of resources" (1990, p. 894).

Employment Relationships

As noted above, much of Florida citrus is harvested by third-party employers rather than the citrus grower. The ultimate employer may only harvest citrus or may engage in other citrus nonproduction activities, as well. The third-party employers consist of labor contractors, processing firms, packing houses, and independent buyers of fruit (Emerson et al. 1991).

The potential employment arrangements are many, and can be most easily described by referring to Figure 17.1. Agricultural employment relationships are most commonly thought of as the direct hire of workers by the grower, as illustrated on the left side of the diagram. As is clear from the diagram, however, this is only one of many potential employment relationships observed in the industry.

This same grower, for example, might choose to have a labor contractor (Sub #2) handle the harvest, in which case the grower would have no harvest workers; all workers would be employees of the labor-contracting firm. Another possibility is that the grower chooses to hire some of the workers directly and use a labor contractor for another part of the harvest.

Figure 17.1. Citrus Harvesting Employment Arrangements

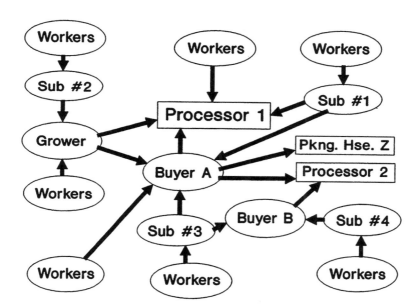

Another common practice by growers is selling the fruit to the processor while it is still on the tree. In this case, it is often the processor's responsibility to harvest the fruit or to arrange for the harvest of the fruit through a labor contractor (Sub #1). Still another common arrangement is the intermediary ("bird dog") who buys the citrus but is neither a grower, processor, nor packing house. He buys the citrus on the tree with the expectation of selling it at harvest time for a profit. Such intermediaries merely buy and sell the citrus, arranging for the harvest through another firm, typically a labor contractor. Alternatively, the intermediary may have its own harvest crews. Thus, as illustrated in the diagram, a grove owner may have fruit harvested on his trees by a number of different entities. In reality, while there are some pure forms of each type of firm illustrated, there are numerous instances of firms that combine different components of the functions in the diagram into a single firm. At the other extreme, we have found firms that merely *arrange* for labor contractors; they do not employ any workers, grow any fruit, buy or sell any fruit, or process any oranges. An important point to recognize is that the labor contractor is a viable alternative for harvesting by any type of economic organization, whether it is a grower, buyer, processor, or packing house.

As an entrepreneur, the labor contractor operates like any other entrepreneur. He delegates routine management activities to other employees. One such example is the organization of workers into crews. A crew leader is assigned to each crew to supervise the crew's activities in the grove, very much the same as occurs with harvest crews under direct employment by the grower.

The contractor's compensation is determined in a manner consistent with any other entrepreneur; it is the residual after all other expenses have been paid. The most important of these expenses is the harvest worker's payment, the piece rate. Any payment received through the contract above the piece rate represents the contractor's compensation for his entrepreneurial activities, crew supervision, administration of taxes and payroll, loading the fruit into trucks for hauling, machinery and equipment ownership, and hauling the fruit to the processing plant in some cases.

Detailed information on the various activities performed by labor contractors and the associated compensation is available from a series of prevailing wage and practice surveys designed and conducted under the supervision of the authors. The surveys were designed to identify employers of workers harvesting oranges for processing in Florida. Surveys are conducted twice each year and began with the 1989–90 season. The first survey of the season is at the peak of the early and mid-season variety harvest; the second survey coincides with the later Valencia harvest. The data are collected by personal interviews conducted by Florida Department of Labor and Employment Security personnel.

The 1992 Florida Valencia survey (FASS 1993) revealed that 72 (70 percent) of the 103 employers interviewed operated as labor contractors. This included two employers who harvested their own fruit in addition to operating in a contracting capacity. The corresponding 1990 Valencia survey revealed a similar 76 percent of the employers operating as labor contractors. The contractors offered a variety of services in addition to harvesting the fruit, although there are some interesting contrasts in the types of services reported in the two years. A summary is given in Table 17.2 for all crews operating under labor contractors in the two years. Nearly all reported roadsiding, maintaining the payroll, paying payroll taxes, and paying workers compensation in both years. However, 45 percent hauled the fruit to the processor in 1990, whereas only 18 percent did so in 1992. All provided the goat loader in 1990, whereas only 51 percent did so in 1992. There is a modest difference in the percentage reporting the payment of auto and liability insurance in the two years also, with more having reported doing so in 1990 than in 1992.

Table 17.3 itemizes the compensation received by labor contractors during the 1992 Valencia harvest. The average contract rate was $1.41 per 90-pound box of oranges, just under twice the average piece rate ($.74) for these same employers. This left $.67 from which the crew leader (supervisor) and other business expenses must be paid. The average crew leader compensation was $.12 per box, ultimately leaving $.55 per box for the remaining business expenses of the contractor. This is the return to cover equipment expenses (goat loader, trucks, etc.), taxes, insurance, other normal business expenses, and a profit to the proprietor.

There is considerable variation in the compensation received by labor contractors (the balance after paying the harvest workers). Figure 17.2 illustrates this for the 1990 Valencia season, suggesting a highly bimodal distribution of rates. A relevant question is whether or not there is a systematic explanation for the variation in contractor compensation. Labor contractor compensation is regressed against the various services provided by the contractors for the 1990 and 1991 Valencia seasons in Table 17.4. The explanatory variables are only binary variables indicating the provision of the service or not. Consequently, the variables in the regressions differ from year to year due to collinearities among the variables, since in some cases all contractors provided the service, or in others, all employers in the sample had exactly the same combination of selected services. The 1990 regression serves to explain the anomaly presented by the bimodal distribution of compensation for that year. Those who hauled the fruit as a part of the contract were compensated an average of $.35 more per box than those who did not haul the fruit. Recall that the 1990 harvest season was the year in which there was a severe freeze in the preceding December, resulting in frozen fruit. Due to the freeze damage, the trailers could only be half filled, thus substantially increasing the cost of hauling on a per box basis.

By contrast, the 1991 relationship suggests a much weaker role for hauling in the contractor compensation, as well as a substantially smaller numerical value for the coefficient. The main source of expla

Table 17.2.   Activities performed by labor contractors, Florida
                     Valencia harvest, 1990, 1992

Activity	1990 Crews		1992 Crews	
	Number	Percent	Number	Percent
Harvest the fruit	94	100	179	100
Roadsiding	92	98	168	94
Provide goat loader	94	100	91	51
Haul the fruit	42	45	33	18
Provide trucks for hauling	42	45	31	17
Insurance			119	67
Auto	82	87		
Liability	90	96		
Workers' compensation	94	100	172	96
Payroll taxes			172	96
Unemployment				
insurance	94	100		
Social Security	94	100		
Withholding tax	94	100		
Maintain payroll	94	100	176	98
Federal and State crew leader registration cards	94	100	na[a]	na

[a]na - Question not asked.

Table 17.3.   Average labor contractor compensation: 1992 Florida
                     Valencia harvest

Item	Cents per box
Contract rate	141
Worker piece rate	74
Balance after paying harvest workers	67
Crew leader rate	12
Balance for labor contractor expenses after paying harvest workers and crew leader	55

Table 17.4. Labor contractor compensation

Service performed	Valencia 1990	Valencia 1991
Intercept	58.0	49.10
	(2.21)	(5.66)
Haul oranges	35.0	9.41
	(7.03)	(1.81)
Roadside	4.0	
	(0.17)	
Provide goat loader		18.98
		(2.03)
Liability insurance	-3.0	
	(-0.17)	
Auto insurance	8.0	
	(0.86)	
R^2	0.57	0.12
Number of contracts	47	75

[a]Numbers in parentheses are estimated t-statistics.

**Figure 17.2 Contractor compensation: Florida Valencia
orange harvesting, March 28, 1990**

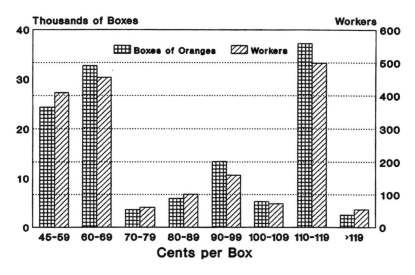

By contrast, the 1991 relationship suggests a much weaker role for hauling in the contractor compensation, as well as a substantially smaller numerical value for the coefficient. The main source of explanation was the provision of the goat loader, which added an average of $.19 per box (Table 17.4). The important point with these regressions is that there is at least a modest degree of systematic explanation for the contractor compensation; the contractor compensation is not merely random.

IRCA Effects

The results reported in the previous section were obtained from a survey designed to provide information on prevailing wages and practices rather than to establish the effects of IRCA. While much of the information is suggestive of IRCA's effects, it is necessary to rely on theoretical arguments to relate our observations to the likely implications of IRCA. Most importantly, there is no historical basis for comparison prior to IRCA.

It is argued that, to the extent that IRCA has had any impact on the Florida citrus labor market, it has been on the structure of the labor market, as described in the preceding sections. Employer sanctions were to be the cornerstone of the legislation following the legalization of SAW workers. With employer sanctions, it was argued, potential immigrants would recognize the futility of migrating with no hope for employment. However, in the apparent absence of enforcement of the employer sanctions, there is believed to have been little change in the flow of workers into agricultural employment.

Despite the apparent lack of effective enforcement of the sanctions and the ample supply of labor, it is suggested that the regulations associated with IRCA have induced a further shift in employment from direct employment by the producer to employment by firms specializing in harvest employment, most often, labor contractors.

Among these regulations is the additional record keeping necessary under IRCA, in addition to the requirements for determining an applicant's eligibility for employment without inadvertently discriminating on the basis of appearance. Both of these are activities more efficiently handled by entities specializing in the employment process.

Since full enforcement of IRCA became effective in December 1988, there has been an apparent abundance of labor for Florida citrus harvesting. It is unclear, however, that the apparent labor abundance can be entirely attributed to a lack of enforcement. The continuing effects of the crop destruction brought about by the freeze of 1989 significantly lowered the labor demand until the 1992–93 harvest season. This has also been a period of relatively high unemployment for the overall economy, further augmenting the supply of labor for agriculture and citrus in particular. Nevertheless, there is fairly general recognition of continuing flows of undocumented labor. One final piece of information that further confirms the ample supply of labor is the stability in the piece rate from season to season. The average rates were $.72, $.72, and $.73 per 90-pound box for the 1990, 1991, and 1992 Valencia harvest seasons, respectively. A tightening of the labor market would be expected to lead to markedly higher piece rates over these years rather than the stability observed.

To the extent that IRCA has enhanced contracting in the harvest labor market, some conjectures can be made regarding the implications of this change. The intermediary nature of the contracting agent is critically important and can serve to provide continuity in employment for the worker throughout the season. The extent to which this is the case is an empirical issue requiring further research. In contrasting direct employment for a few days each by several growers with a longer term of employment by an intermediary, one could argue that enforcement of labor laws and regulations would be facilitated. The extent to which this would occur, however, would be mitigated by the mobility of the intermediary and the resultant difficulty in monitoring labor laws and regulations in the case of intermediaries. Longer-term employment is also more likely to lead to the provision of worker benefits. But again, whether or not this is occurring is an empirical question to be resolved.

Summary

Our major premise has been that the primary effect of IRCA has been to influence the labor market structure. Incentives for direct employment of labor by the grower have diminished except for the very large grower, and the employment function has been further shifted to labor-market intermediaries such as labor contractors specializing in the employment of large numbers of harvest workers, thus spreading the fixed costs of hiring over a large number of growers. There have been few apparent indicators of a tightening of the labor market. Continued research needs to be focused on labor contractors to determine the implications on the labor market participants due to IRCA. The most urgent need is for better data.

NOTES

1. Part I draws from work done under an agreement between the Florida Department of Labor and Employment Security and the University of Florida.

17

Part II. California and Arizona[1]

HERBERT O. MASON, ANDREW J. ALVARADO, AND GARY L. RILEY

Introduction

Arizona's Yuma County and central California's Tulare County were selected as the citrus case-study sites. Economic, geographic, and agricultural differences between these two study sites make each unique, in spite of the desired singular focus of the citrus case study.

The major purpose of this case study was to determine the direction and magnitude of IRCA on the labor markets and citrus industries in the study sites. The bulk of the field research was conducted between November 1990 and May 1991. Beyond the collection and analysis of secondary data, primary data were collected via structured and unstructured personal interviews, as follows:

	Tulare County	Yuma County
Farm workers	52	20
Employers	6	6
Farm labor contractors	6	1
Crew bosses	—	6
"Seasoned" farm workers	4	3
Ex-farm workers	3	3
Industry, border patrol, and packing-house representatives	8	7

The farm worker sample in Tulare County was drawn by selecting six citrus growers from the 1989 unemployment insurance (UI) data provided by the Employment Development Department. The sample of growers was stratified by three size categories (large, medium, and small) based on the total payroll paid for 1989. In addition, farm labor contractors who were active in the citrus harvest were also included in the employer sample. All contacts with farm workers were made through these employers, and the employers helped schedule the interviews.

No database similar to the California UI listing of agricultural employment is available for Arizona. The citrus harvest in Yuma, however, is controlled by seven to eight harvesting companies. Six of the companies were contacted, and farm workers were drawn randomly from their employment lists.

Employers interviewed for the study were the same as those from whom the farm worker sample was selected. This allowed the interviews to be "reflective," in the sense that worker and employer perspectives on working conditions and related matters could be compared directly.

Crew bosses, farm workers with more than ten years' experience in the citrus harvest, ex-farm workers, and informed experts were also interviewed as part of this study. A structured questionnaire was used for all of these groups except the informed experts. This last group represented diverse views, including border patrol personnel, packing-house managers, employment service personnel, and representatives of industry associations.

Regional Overviews

Yuma County has about 165,000 acres of cropland. In 1989, the farm-gate value of agricultural products in Yuma County exceeded $443 million. Vegetables were the leading crop in terms of gross returns ($222 million in 1989). Field crops had a gross value of $101 million, tree and vine crops were valued at $50 million (with citrus at $45 million), and vegetable seed production was valued at $15 million (Univ. Ariz. Coop. Ex. 1990). The major citrus commodity in Yuma County is lemons. Rapid growth in labor-intensive crops, particularly vegetables, over the past decade suggests little concern over agricultural labor shortages. According to local observers and U.S. Census of Agriculture data (USCensus 1989), the majority of land in citrus production is owned by absentee landowners and operated by farm management firms.

Central California's Tulare County produces virtually every major agricultural commodity, having a combined farm-gate value of $2.2 billion in 1990. With favorable climate and soils, generally adequate water supplies, and a diversity of crops, agriculture is a year-round activity in Tulare County. The leading commodity in recent years has been milk ($412 million in 1990), with oranges in second place at $387 million and grapes in third at $278 million in 1990. Oranges, particularly seedless navel oranges destined for the fresh fruit market, are the dominant citrus crop in California. Agricultural acreages in Tulare County have remained fairly constant over the past decade at approximately 1.4 million acres.

Farms on the east side of the San Joaquin Valley are generally held by owner-operators who derive their main income from farming. In fact, many of the farmland buyers from outside the region have been citrus and dairy operators relocating from rapidly urbanizing southern California.

The Economics of Citrus Production

As with any agricultural product, the citrus industry experiences the economic vagaries associated with weather, crop sizes, and national and international markets. In general, prices for fresh market citrus are highly variable and are inversely related to the total level of production. The primary market for California and Arizona citrus is fresh; by-product returns are typically negative and reflect the salvage nature of this market.

One of the market advantages enjoyed by the citrus industry is that the crop generally ripens during the winter and spring months when summer fruit is not available. This market advantage also makes the fruit vulnerable to freezing winter temperatures, which can result in product loss and tree damage such as occurred in central California in December 1990.

Aside from this disastrous freeze, the decade of the 1980s was generally profitable for Tulare County citrus growers. The early 1980s were less profitable for the Yuma lemon industry, although returns improved substantially in the 1988–91 crop years.

Development of the Farm Labor System

As documented by several authors (Fuller 1991, Meister and Loftis 1977), labor-intensive agriculture in the western United States developed partly as a result of continuing supplies of immigrant workers. The citrus industries in Tulare and Yuma counties have followed this pattern and have relied heavily on immigrant labor since inception.

During World War II, the U.S. government became the guarantor of farm labor when the War Labor Board was permitted to contract directly with the Mexican government to import workers. This "guest worker" program evolved into the so-called *bracero* program, which was extended on an "emergency" basis until 1964.

It was during the period of the *bracero* program and the ensuing years that the citrus industry in central California and Arizona grew rapidly. Certainly other factors — such as the development of water projects, expanding markets, and shifts in production from southern California due to urbanization — affected the growth of the citrus industry. But the availability of *bracero*, "green card," and illegal immigrant labor from Mexico was critical in facilitating the expansion of citrus acreage, as well as in influencing employer attitudes toward the management of labor.

After the termination of the *bracero* program, there was concern that there would not be sufficient labor to harvest perishable crops. Although some spot shortages were noted, there were apparently enough former *braceros* and other workers with work authorization to provide labor to harvest the crops. In Tulare County, this legal work force was increasingly supplemented and replaced by illegal immigrants; most informed observers indicated that the citrus work force in Tulare County was approximately 50 percent illegal when IRCA was passed in 1986.

In Yuma County, undocumented workers have been less prominent than in Tulare County. Our field research indicates that this is largely due to Yuma County's proximity to the Mexican border and the apparent ease during the 1970s of obtaining work authorization for daily "border commuters." It should be noted that there is some confusion about the method of legalization for the Yuma citrus work force. Most employers interviewed indicated that the majority of their workers were legalized through the pre-IRCA green-card program, and therefore did not go through the SAW legalization program. Officials from the Yuma sector of the border patrol, however, claimed that the pre-IRCA citrus work force was highly illegal and that "80 to 85 percent of the current citrus work force are SAWs" (interviews with J. Lockwood and J. Elton, February 21, 1991).

Another important difference between the farm labor systems in the two study sites is that farm labor contractors have limited involvement in the Yuma County citrus industry. In the Yuma citrus industry, harvesting associations run by the packing houses and their crew bosses perform the labor recruitment and management functions that are often provided by farm labor contractors in the Tulare County citrus industry.

Organization of the Farm Labor System

According to Employment Development Department estimates, between 1,800 and 2,200 individuals were employed each month in the Tulare County citrus industry during peak harvest season (January through June) of 1990 (CalEDD 1992).[2] There were also as few as 7,000 and as many as 13,000 individuals employed by farm labor contractors during this same period, a large proportion of whom were undoubtedly working in the citrus harvest. Exact estimates of the total Tulare County citrus work force are not available, but EDD and others estimated that approximately 15,000 citrus workers were unemployed directly as a result of the December 1990 freeze. This total includes packing-house employees as well as those employed in harvest activities.

A variety of employment arrangements can be found in the Tulare County citrus industry. Some growers hire harvest crews directly, some packing houses hire crews and provide picking services for growers, and some growers and packing houses use farm labor contractors. There are packing houses that use both house crews and farm labor contractor crews for peak labor needs. In some instances, packing houses use farm labor contractors to recruit, transport, and manage the crews but retain the payroll function. Regardless of the structure of the employer-employee relationship, the first-line supervisor (crew boss, *mayordomo*) plays an important role in the recruitment and management of the work force. Although most operations have a centralized hiring office, the crew boss is generally responsible for deciding who works in the crew and keeping in touch with his or her crew during the off-season.

One clear trend in the Tulare County citrus industry is the growing importance of farm labor contractors (FLCs). Growers have long relied on farm labor contractors, but there has been a significant increase in the use of FLCs during the past decade. Informed observers estimate that at least 50 percent of the citrus harvest work force is employed by FLCs, and the actual proportion may be as high as 70 to 75 percent. The major reason for this shift toward FLCs, cited by industry observers, is that growers are overwhelmed with paperwork and other government regulations imposed on employers, particularly those in California. Examples given include IRCA forms (I-9s, ESA 92s), workers compensation, safety training, and payroll taxes. The general sense is that growers are increasingly willing to turn these compliance functions over to farm labor contractors.

Local experts in Yuma County estimate that there are about 5,000 workers employed in the area's citrus industry during peak harvest. Unlike Tulare County, farm labor contractors are not prominent in the Yuma citrus industry. Only one farm labor contractor active in the citrus industry was located in Yuma County. This FLC provides labor only for pruning, not in the citrus harvest.

Harvest activities are organized in Yuma County around the seven major packing houses. Each packing house runs a harvest cooperative, which provides harvest services for its growers. The harvesting companies employ crew bosses or supervisors, who are responsible for recruiting and managing the workers. Most workers live in the Mexican border town of San Luis de Rio Colorado, as do the crew bosses. The crew bosses assemble the crews and are paid partially on a contract (piece-rate) basis in relation to their crew's production. According to the crew bosses and employers interviewed, crew bosses in the Yuma lemon harvest are typically paid $5 to $7 per hour plus $1 per bin that their crew picks. The harvest companies and packing houses provide bus transportation from the U.S. side of San Luis to the picking sites, a distance that ranges from 30 to 70 miles. Crews are usually comprised of 25 to 30 workers. The larger harvesting associations employ up to 30 crews during peak season, which results in company employment levels that total as high as 800 to 1,000 workers.

Tasks Performed

Work efforts are organized in a similar fashion in the Yuma and Tulare citrus industries. By far the most important task is the harvesting of citrus. In mature groves, most of the picking is done on ladders, which requires the worker to carry a 50- to 60-pound sack up and down the ladder. Citrus for the fresh market must be clipped rather than pulled in order to maintain stem or "bud" integrity. Depending on the time of the season, fruit size, and market conditions, the picker may have to grade for size as well as quality and maturity.

The only significant change in the method of harvesting oranges in the past 50 years has been the shift to bins from field boxes (although pack-out rates are still reported on the basis of field boxes to cartons). This change reduces the amount of labor required, since bins are loaded with forklifts onto trailers, whereas the field boxes were loaded manually. But as one farm labor contractor interviewed in

Tulare County stated, "We still pick and prune oranges the same old way."

Once the fruit is delivered to the packing house, it is sorted and graded. The packing house "floor labor" has traditionally been supplied by women, primarily local and settled in the region. Changes in technology, particularly electronic and mechanical sorting and grading, have significantly reduced the amount of floor labor required in many packing houses.

Other tasks include irrigation, weed control, and other cultural activities. These tasks are usually performed by year-round employees. Pesticide application is increasingly done by professional application firms on a contract or custom basis.

Pruning is the final major task required for citrus production. Much of the pruning and suckering (removal of unwanted shoots from the tree base) is done by hand, although mechanical toppers and hedgers are used to keep the overall tree size manageable. Because of the desert climate and long growing season in Yuma, the groves require frequent pruning and topping to keep them at a height from which the fruit can be readily harvested.

Composition of the Work Force

The profile of the work force reported here is based largely on the interview data. The harvest work force in both Tulare and Yuma counties is virtually all Hispanic; in Tulare County, one of the 52 workers interviewers indicated that he was from the Philippines. In Tulare County 50 of the 52 workers interviewed and in Yuma County 19 of the 20 workers interviewed indicated that they were born in Mexico. This is an important statistic for long-run policy, since it is clear that the domestic work force has not "replenished" itself; rather, the replacement workers have been migrants from Mexico.

Of the workers interviewed in Tulare County, 90 percent were male, while 85 percent of those interviewed in Yuma County were males. These samples might underestimate the proportion of female farm workers in Tulare County; informed sources and previous research indicate that the Tulare County citrus harvest work force is comprised of approximately 20 percent females.

Forty-four percent of the Tulare County work force indicated they were married and currently living with their spouse. About 15 percent were married but not living with their spouse. Thirty-six percent were single or separated. In Yuma County, the married proportion was somewhat higher, at 75 percent of those interviewed.

In Tulare County, the mean average age of the workers was 36 years, with the median age at 32 years. The work force in Yuma County was significantly older, with the mean age at 43.3 years and the median age at 39 years. These averages are consistent with the observation that the Tulare County work force is dominated by recently legalized SAW workers and recent immigrants, whereas the Yuma work force is comprised mainly of pre-IRCA "green carders" and SAWs who had worked in the citrus industry for many years prior to IRCA.

The educational levels of the workers are quite low. In Tulare County only two of the workers had completed any formal education in the United States. The highest grade Tulare County workers completed in Mexico averaged 4 to 4.5 years. In the Yuma sample, none of the workers indicated that he had completed any formal education in the United States. The average mean grade completed in Mexico for the Yuma worker was 4.9 years, while the median grade completed was 6 years.

All of the workers interviewed in the two counties responded that their dominant language was Spanish. In Tulare County, 28 of the 52 interviewed indicated that they neither understood nor spoke English, with 21 responding that they had a "low" English understanding and speaking ability. In Yuma County, 8 out of 20 neither understood nor spoke any English. The remaining 12 responded that they understood and spoke "very little" English.

Farm workers in Tulare County indicated that, on average, they first entered the United States in 1977. Twenty-two percent, however, responded that they had migrated to the United States after 1986. There is a suspiciously large "bubble" of 16 percent who indicated they had entered the United States in 1985 (the last year in which qualifying farm work for the SAW program could be completed). For the Tulare County sample, the interviewees indicated that they had done farm work for a mean average of 11 years, with a median response of 8.5 years. Forty-two percent indicated that they had five or fewer years of U.S. farm work experience, which probably is a reasonable approximation of the work force proportion who were fraudulent SAW applicants or are currently working with fraudulent documents.

For the Yuma County sample, the mean year interviewees first came to the United States to work was 1969, with the median year as 1970. They had been crossing the border on a daily or weekly basis for a mean average of 9 years, with the median reported at 5 years. The mean average first year that the Yuma citrus workers came to the

United States to do farm work was 1970, with the median year as 1973. The average (both median and mean) number of years of U.S. agricultural employment reported was 15 years, with only three respondents indicating that they had worked for 6 or fewer years in farm work. Nineteen of the twenty farm workers interviewed claimed that they had started doing farm work in the United States by 1985 or earlier.

Farm workers in both Tulare and Yuma counties indicated that a substantial portion of their parents had migrated to do some farm work in the United States. In Tulare County, 21 of the 52 farm workers (40%) responded that their parents had worked in U.S. farm jobs, while 12 of the 20 Yuma farm workers (60%) said that their parents had done U.S. farm work. This suggests that well-established family and friendship networks, as discussed by Mines (1981) and others, remain important in determining migration patterns between sending regions in Mexico and crops and receiving regions in the United States.

A significant proportion, albeit a minority, of the farm workers in the sample stated that they had done some nonfarm work in the United States. Thirty-four percent of the Tulare County workers and forty-five percent of the Yuma County workers had done some work in the United States other than farm work. Most of those, however, had done so for fewer than two years and did not work at these jobs for long periods of time. Although the evidence is limited, this suggests that the citrus workers consider their primary jobs to be harvesting citrus, with some workers filling in during slack times doing short-term nonfarm jobs and farm jobs in crops other than citrus.

These worker profiles suggest similarities as well as differences between the Tulare and Yuma work forces. Both work forces are comprised entirely of immigrants from Mexico, the primary occupation is in harvesting citrus, and low education and English language skills mean that there is little nonfarm opportunity for the workers. Major differences in the two work forces are that Tulare County workers are generally younger, are more likely to have been legalized through IRCA or are not legalized, and are more likely to be in the United States without their families. Most of these differences can be attributed to the border location of Yuma County, which allows the workers to live in Mexico and commute on a daily basis to work in the citrus industry. Based on previous research and discussions with informed observers, it is likely that the Tulare County citrus work force profiled here is similar to the seasonal harvest work force found in all crops throughout the San Joaquin Valley. The Yuma County citrus work force is likewise representative of the seasonal work force in other crops in the area. Regional and geographic differences appear to be much stronger than work force differences that might be attributable to the crop type or other factors affecting the composition of the work force.

As part of the interview, workers were asked several questions relating to their legal work status. Approximately 10 percent of the Tulare County farm workers indicated that they had received work authorization prior to IRCA, and 54 percent had applied for legal status through IRCA (SAWs). About 31 percent of the Tulare County farm workers indicated that they were undocumented or fraudulently documented. This percentage of undocumented workers is probably low; most informed observers believe that the proportion of citrus workers in the San Joaquin Valley who are undocumented or fraudulently documented is closer to 40 or 50 percent of the work force.

The citrus work force in Yuma County is apparently largely legalized. All 20 of the Yuma workers interviewed claimed they had the legal right to work in the United States. Eight of the twenty workers said they were legalized prior to IRCA, with the remaining twelve indicating that they were SAW applicants. These survey results were confirmed by employers and other key informants. In January 1991, the border patrol checked papers for almost 4,000 farm workers in the Yuma area employed in the citrus and vegetable harvest; only 13 were found to have no documents or fraudulent documents (J. Lockwood and J. Elton interviews, February 21, 1991.)

The Effects of IRCA in the Citrus Industry

One of the assumptions underlying IRCA was that employer sanctions would reduce the economic incentive for workers to migrate illegally to the United States. This, it was believed, would substantially reduce the supply of new entrants into the seasonal agricultural labor market. Employers would then be forced to adjust to a legal — and presumably smaller — work force. As has been documented by several researchers (Heppel and Amendola 1991, Martin 1991), employer sanctions have not been enforced widely, and fraud in the SAW program and false work authorization documents have resulted in a labor market environment that is largely like pre-IRCA's "business as usual." Most of the major effects anticipated when IRCA was debated have not come to fruition in the citrus industries examined as part of this case study. Some of the impacts of IRCA that are identified and discussed here should be considered as minor or secondary; they do not represent major departures from pre-IRCA labor market conditions and employer-employee relations.

Effects on Workers

As part of the field research, citrus farm workers were asked several questions about their earnings, length of employment, and employment experience. Virtually all of the workers interviewed were paid on a piece rate for picking citrus. While the rate reported varied somewhat depending on the orchard (particularly tree size), fruit size and quality, and time of season, the Tulare County orange pickers reported an average piece rate of $10.50 per bin. Workers and employers reported that the average worker picks four to five bins per day and works five to seven hours each day. The average earnings per hour reported by workers in Tulare County was $6 per hour. Because of weather and marketing conditions, most workers are not employed for eight hours per day, five days per week. This results in relatively low average weekly earnings; workers reported that they had earned $137 in gross wages in their previous paycheck. This paycheck represented a self-reported work week of 25 to 27 hours. The weekly earnings data for Tulare County citrus workers should be viewed with caution, however, since the interviews were conducted immediately following the December 1990 freeze. Because the navel orange crop was late in maturing, the harvest was just gearing up when the freeze hit. The average earnings reported by workers is probably low relative to a normal year, when most workers average $175 to $200 per week, according to previous studies.

More recent data published by the Employment Development Department (CalEDD 1992) indicate that citrus harvest workers earned about $200 per week in January 1993. This is the result of a combination of relatively high hourly earnings ($9.44 an hour) and limited hours of work per week (21.1 hours per week).

Because lemons are smaller than oranges and it takes longer to fill a bin, the piece rates for lemons are higher than those for oranges. Yuma County workers and employers reported that the prevailing piece rate in 1991 for lemons was $26 per bin. Workers reported that they worked an average of seven hours per day, which is similar to the average reported by employers. Workers and employers estimated that the average picker could harvest 1.5 to 2 bins per day. Average hours of harvest work reported by the farm workers was 29 hours per week, with weekly gross pay averaging about $153.

Citrus harvest activities in the San Joaquin Valley occur almost the year round, with the summer months providing only limited employment opportunities. This does not mean, however, that the majority of harvest workers are able to find steady, year-round employment in the citrus harvest. Navel acreage in Tulare County is about twice as large as acreage devoted to Valencia oranges. Much of the Tulare County citrus work force is employed in the navel harvest only, with the more settled and senior workers continuing in the Valencia harvest. The stability and length of employment in the central California citrus industry has often been overstated because these production cycles have not been recognized; among the 52 workers interviewed for this study, the average number of weeks reported in the citrus industry was about 26 weeks in 1990.

Although the citrus industry provides a more stable employment base and a longer season than many agricultural crops, citrus farm workers still face significant periods of unemployment. Citrus workers in both Tulare and Yuma counties reported that they were unemployed for about three months in 1990. (The interviewees also stated that, on average, they were out of the country and not seeking employment for four to six weeks each year and that they were employed in crops other than citrus for eight weeks each year. A few workers in the sample also indicated that they did some nonfarm work during the year.) Virtually all of the workers stated that the reason for unemployment was lack of work due to the end of the harvest season. Fifty-six percent of the Tulare County farm workers said that they applied for unemployment insurance when laid off, while sixty-five percent of the Yuma County workers stated that they applied for unemployment insurance. Among those who did not apply for unemployment insurance, the most frequent reason given was because they believed they would not qualify for benefits.

Citrus harvesting appears to be a fairly specialized task. Only 19 percent of the Tulare County workers indicated that they did some citrus pruning work during 1990, and only one picker in Yuma County indicated that he did any pruning. About one-half of the Tulare and Yuma county citrus workers indicated that they worked in crops other than citrus, and they averaged about eight weeks of work in these other crops.

About 80 percent of the workers in Tulare and Yuma counties indicated that they did not receive any type of bonus payment from their employers. End-of-season and holiday bonuses were the typical type of bonus noted by workers who received some form of bonus payment. Only 35 percent of the Tulare County farm workers responded that they were covered by medical insurance (other than workers compensation); 50 percent did not have medical insurance, and 15 percent didn't know whether this benefit was provided to them. Eleven of the eighteen workers with medical insurance said that their employer paid for the entire cost, while five shared the cost with their employer and two workers paid for the entire cost of medical insurance.

In Yuma County, only 5 of the 20 workers interviewed stated that they had medical insurance; 13 had no medical insurance, and 2 did not know whether they had insurance. Three of the five with medical insurance had employer-paid medical insurance, while the other two shared the cost with their employers. These employee responses contradict information supplied by employers in Yuma County. Most of the harvesting companies indicated that they offered medical insurance for their harvest crews; premiums are comparatively inexpensive because the medical care is provided by clinics in Mexico. Three possible reasons for the discrepancies between employee and employer responses about medical insurance are that many workers do not work the minimum hours each month required to qualify, workers may not be aware of benefits provided, and the small sample size of 20 workers is biased in this regard.

Lack of knowledge about employer-provided benefits is also suggested in workers' responses to a question about workers compensation insurance. Although employers are required to provide workers compensation insurance, 20 percent of the farm workers in Tulare County indicated that they were not covered by workers compensation.

The same is true for unemployment insurance. Thirty-five percent of the Tulare County citrus workers stated that they were not covered by unemployment insurance, despite universal coverage in California. The perceived lack of coverage could be the result of the knowledge that a valid social security number is required to be eligible for unemployment insurance.

Very few of the workers receive paid vacations or holidays. In both Tulare and Yuma counties, 90 percent of the workers indicated that they did not receive paid vacations or holidays.

For many farm workers, transportation to the work site is a problem since they cannot afford their own vehicles. In Tulare County, 42 percent of the workers owned their own vehicles. Forty-four percent ride with others and share expenses, and twelve percent ride a labor bus. The average cost of rides for the two-thirds of the workers who pay for rides is $3.00 to $3.50 per day. In Yuma County, most of the workers ride in buses provided by the harvesting companies at no charge to the worker. In at least one instance, workers are paid for travel time to and from the border.

Workers were asked several questions designed to determine if working conditions — particularly pay rates — had changed since IRCA was passed in 1986. A key question was: "Have the wages you earn per hour gone up or down in the past five years?" In Tulare County, 27 percent of the farm workers indicated that wages had increased, 50 percent responded that there had been no change, 12 percent believed that they had decreased, and the remainder had not been picking oranges for five years. Among the 14 workers who indicated a wage increase, the average hourly increase was estimated to be about $.90. Those indicating a wage decrease estimated that the decline had averaged $.50 to $.60 cents per hour. Most of the workers who believed that wages had increased cited improved field conditions (smaller trees, better pruning, higher-yielding orchards) and the increase in the California minimum wage as the major reasons for wage increases. It is interesting to note that only 39 percent of the respondents correctly identified the prevailing minimum wage in California at $4.25 per hour.

In Yuma County, workers were split evenly between those who believed wages had increased and those who saw no change. Seven of the twenty workers interviewed indicated that hourly earnings had increased, another seven indicated that wages had remained constant during the past five years, and six workers had not worked in the citrus harvest long enough to respond. The average hourly wage increase was estimated to be $.34 by those workers who indicated that wages had increased. Reasons cited for the wage increases included increased piece rates, an increase in the minimum wage, and improved field conditions.

According to local experts and employers in Yuma County, the bin piece rate for lemons rose from $20 to $22 in 1985 to $25 to $26 in 1991. The major cause for the increased piece rate was an increase in the federal minimum wage. At the time of the interviews (February 1991), employers were also concerned about the impending change in the federal minimum wage, which increased to $4.25 per hour on April 1, 1991.

Workers were asked if they had belonged or currently belonged to a labor union. Twenty-one percent of the employees interviewed in Tulare County answered in the affirmative. It should be noted that our sample was biased in this regard because we purposefully sought crews who were working under one of only two union contracts in the Tulare County citrus industry. In Yuma County, none of the workers interviewed was or had been a union member.

A related question attempted to determine if workers had joined with other workers in the past five years in an attempt to improve working conditions. Less than 20 percent of the workers in Tulare County indicated that they had done so; this response rate is nearly identical to the proportion of the sample who had been involved in union activities. Only one worker in Yuma County indicated that he had been involved in this type of concerted activity with other workers.

Discussions with observers of the labor scene — and a review of California Agricultural Labor Relations Board and Arizona Agricultural Employment Relations Board data — indicate that there has been little or no unionizing activity since the early 1980s. A detailed discussion of the myriad reasons for this is beyond the scope of this study. What is important to point out is that the labor supplies in both the Tulare and Yuma county citrus industries have been abundant. If IRCA had managed to reduce illegal immigration and labor supplies, it might have created a more favorable environment for union-organizing efforts. But IRCA has not done this, and slack labor markets are antithetical to a successful organization of workers. In this sense, what IRCA has not accomplished relative to expectations is notable.

During congressional debate over IRCA and its precursors, agricultural employers were concerned that farm workers would leave agricultural employment as soon as they gained the legal right to work in the United States. Interview data, coupled with previous research, indicate that most farm workers do not expect to leave farm work in the near future. Almost 60 percent of the workers interviewed in Tulare County considered themselves permanent U.S. farm workers; only 8 percent planned to continue in U.S. farm work for one year or less, and 25 percent indicated that they would continue to work in U.S. agriculture for one to five more years. All 20 of the workers in Yuma County considered themselves permanent U.S. farm workers.

These results should not be construed to imply that the workers consider farm work to be their career of choice. Rather, they see little opportunity for other employment. They speak little or no English, have very little formal education, and do not believe that they have skills marketable outside of agriculture. Moreover, the information networks and support systems between rural Mexican villages and receiving regions in the United States revolve largely around agricultural jobs, and rural communities, and do not provide employment options other than the easy-access jobs in seasonal agriculture.

Effects on Producers

Although there was concern about the potential effects of IRCA on employers, the anticipated reductions in labor supply have not occurred. The effects of IRCA on agricultural producers have been largely procedural in nature and insignificant in magnitude.

The single most consistent effect of IRCA identified by citrus growers is the additional paperwork required to complete I-9s. Most employers do not view this as a large burden, although there was initial confusion about procedures and documentation requirements.

Several employers indicated that the labor situation had actually improved since IRCA. Many of their workers had become legalized through the SAW program, which has provided employers with a more stable and reliable work force. In some instances, legalization has allowed workers to settle in the local area with their families. SAW legalization has also permitted the workers to visit Mexico and return without fear of apprehension or through the use of *coyotes*.

Perhaps the most important effect of IRCA for some employers and workers was its requirement for search warrants and the elimination of the border patrol field sweeps. Several employers indicated that before IRCA border patrol sweeps of their fields often disrupted their harvest, albeit for short periods of time. Elimination of this enforcement tool is also one of the reasons that there are substantial numbers of illegal immigrants working in agriculture.

Little or no evidence of changes in recruitment methods or personnel management practices since 1986 was found. A few employers indicated that they had increased safety training and had developed programs to stabilize their work force through incentive programs (primarily medical insurance and end-of-year bonuses for workers who stay throughout the season). IRCA was not, however, identified as the cause of these changes.

There is no evidence that citrus growers have altered their planting decisions based on concerns about labor supplies. These decisions are based primarily on current and anticipated product market conditions. In fact, most growers get a puzzled look when asked if they consider labor availability in crop and expansion decisions. Perhaps this represents the growers' "Field of Dreams": "If I plant it, they will come."

Many of the projected effects of IRCA have not materialized because the law has not been implemented as anticipated. Only one of the employers we interviewed in the two counties had been audited by the Immigration and Naturalization Service or the Department of

Labor. All were aware of the extent of fraudulent documentation, and none had made any significant changes in his approach to labor management as a result of IRCA. As one grower in Yuma County stated: "Nothing has changed."

Conclusions

Case study research offers advantages in terms of depth of analysis and the ability to identify cause-and-effect relationships. The major weakness is that the results often cannot be generalized or extended beyond the immediate subject of study.

This case study of the citrus industry in some instances suffers from the limitations of case-study methodologies. The citrus industry is unique in that the employment opportunities are relatively stable and are offered over a longer season compared to other perishable crops. It is particularly difficult to generalize from the Yuma County situation, since its geographic proximity to large supplies of resident Mexican labor places it in an unusual labor supply position relative to most of U.S. agriculture. From the employer's perspective, the Yuma County situation is close to ideal: The Mexican border towns provide an abundant pool of legalized workers within daily commuting distance to work sites. Housing and social concerns about migrant workers are largely avoided because the work force lives across the border.

Despite these study limitations, it appears that the post-IRCA experience of the Tulare and Yuma county citrus industries has been quite similar to that observed in other labor-intensive crops and regions. The positive effects of IRCA have centered on the amnesty programs, which have allowed a large number of workers to legalize their right to work in the United States, localize their residence, and, in some instances, stabilize their employment. The negative effects are largely the result of what IRCA has not done; sanctions are largely unenforced, fraudulent documents are readily available and accepted by employers, and replacements and additions to the seasonal work force continue to come from Mexico. In Tulare County, these new entrants are illegal immigrants from Mexico. Because of its unique geographic situation, new entrants for Yuma County are Mexican residents with the apparent legal right to work in the United States. This supply of across-the-border commuters appears unlimited relative to current demands.

In both of the counties considered in this case study, the current solution to labor needs is quite satisfactory for employers. But the solution is uneasy and perhaps ephemeral. Tulare County employers are aware that if IRCA is enforced, they will face labor shortages; new supplies of immigrants will dry up. In Yuma County, the legalized work force is aging and long-term replacement options are not clear. IRCA has thus created a situation that, while temporarily workable, creates uncertainty for all involved.

Despite the uncertainties about labor supplies that have been created — or left unresolved — by IRCA, growers continue to make long-term investment decisions with no apparent concern about labor availability. Because of favorable market conditions, citrus plantings have increased since 1986. Growers and packers are generally optimistic about the future of their industry, and expansion decisions are based on product market conditions and the availability of water and suitable farmland.

Seasonal labor is not an immediate issue for the citrus industry. Supplies are abundant and there is little or no union activity. It is not likely that the industry will be concerned with labor-related issues until another "crisis" occurs. These crises occur periodically — for example, termination of the *bracero* program, union organizing, and the initial stages of IRCA. But barring this type of external stimulus, dealing with labor issues and changing personnel management practices will not be a high-priority item on the agricultural agenda.

Many assumptions have been made about the implementation of IRCA and its ensuing impact on agricultural labor markets. To date, the great social experiment continues to be untested. Illegal immigration continues almost unabated, and labor surpluses are admitted even by the most proemployer advocates. Because of these abundant labor supplies, wages and working conditions have largely stagnated or declined in a real sense. Since labor market conditions have not changed, employers have not altered their recruitment or personnel management practices.

The overall effects of IRCA on the agricultural labor environment are similar to the effects of other legislative and regulatory attempts to intervene in farm labor markets. During the policy debates and upon initial implementation of IRCA, promises were made by proponents about the salutary effects for domestic workers that would result from regaining control of the borders. Agricultural interests and their "fixer corps" (Fuller 1991) were nervous, despite generous concessions provided by Congress and the implementing agencies. But we find that once the dust settled, very little has changed in the five years since IRCA was signed into law. Congress can claim that it has done something about the immigration problem. But by allowing the implementation mechanisms to fail, Congress can also avoid the political wrath of employers who complain about increased government regulations and who are also worried about reduced labor supplies. This strategy can also mollify those who are concerned about any employment discrimination that might be caused by employer sanctions. The net result is the continuing conflict characteristic of farm labor policy: Although legislation to help farm workers has been passed, the political will to make it work is absent.

NOTES

1. For Part II, Martina Acevedo, research assistant, and Ricardo Ornelas, consultant, provided valuable assistance in interviewing farm workers and collecting other data. Research support was provided by the Commission on Agricultural Workers. This chapter is largely excerpted from our report to the Commission (Mason, Alvarado, and Riley 1993).

2. EDD no longer reports employment by county and crop, so more recent data are not available. Industry observers suggest that employment has been relatively stable since 1990, except for the post-December 1990 freeze.

18

Labor in the Tomato Industry: A Comparative Discussion of California and Florida[1]

DAVID RUNSTEN AND DAVID GRIFFITH

Introduction

This chapter examines fresh tomato production in the three major supply regions for the U.S. market: California, Florida, and northern Mexico. It also examines the impacts of IRCA in the context of the evolution of production systems in U.S. tomato agriculture. The research is based on extensive field interviews with workers, grower-shippers, and labor contractors in all three regions.[2]

Overview of the Fresh Tomato Industry

Florida and California are the dominant domestic suppliers to the U.S. fresh tomato market, while Sinaloa is the primary supply source from Mexico. From 1981 to 1986, both Florida and California market shares declined somewhat, while the market share for Sinaloa increased from 21 percent to about 26 percent. From 1986 to 1990, the market share for Florida exhibited no clear trend in varying between 39 percent and 46 percent, while California increased its market share from 23 percent to 25 percent. Sinaloa's market share declined after 1986 to about 22 percent in 1990.

Though there were no dramatic changes in shipments or market share for Florida, California, or Sinaloa, there was a substantial increase in shipments from Baja California. Before 1983, Baja shipments were relatively insignificant. By 1988, however, shipments totaled about 296 million pounds and represented 9 percent of all shipments in the U.S. market. Shipments captured by USDA from other regions in the United States have increased, but still account for less than 10 percent of total shipments.

All production areas "compete" in the U.S. market, but there is much complementarity among production regions that is determined primarily by climate. The "winter" market, defined as lasting from

[355]

November through May, is mainly supplied by shipments from Florida and Mexico. As shown in Figure 18.1, Florida is the dominant supplier during the November-December period and again during the April-May period. Sinaloa, in contrast, is the predominant supplier in January and February. During March, both Florida and Sinaloa supply significant quantities of tomatoes, which makes March the most intense period of competition with Mexico. During the summer and early fall, the primary supply areas shift to California and, secondarily, to Baja. As shown in Figure 18.1, shipments from California increase substantially in June and peak in October. Shipments from Baja peak during the June-July period, but Baja provides a relatively small but steady supply of tomatoes from August through December. Though there is overlap among the regions, nevertheless, the degree of complementarity is striking.

Figure 18.1. U.S. 1990 Monthly Fresh Tomato Shipments, by Origin

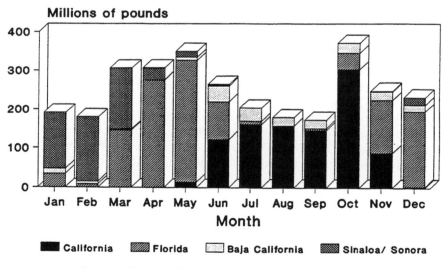

Source: Federal State Market News Service, June 1991

U.S. Fresh Tomato Acreage and Production

Over the 1980 to 1990 period, total U.S. acreage devoted to the production of fresh tomatoes increased from about 124,000 acres in 1980 to almost 141,000 acres by 1989, before declining to 134,000 acres in 1990. Total U.S. production over the decade increased by 32 percent.

Florida and California are the largest producing states; they accounted for almost 65 percent of total acreage during 1990. Since 1980, both California and Florida have exhibited increases in acreage, while the remaining states constituting the "other" category have registered a decline; nevertheless, 21 states produce tomatoes, mainly during the summer, and their combined acreage exceeds California's and competes primarily with California.

As noted, California is the second largest tomato-producing state, producing over 497,000 U.S. tons of fresh tomatoes in 1990. Production occurs from May to November with peak production in October, when California dominates the U.S. market. Primarily mature green tomatoes are produced at present, although vine-ripes were important historically and have been experiencing a resurgence in demand in the last two years.[3] The maturity of tomatoes at harvest (mature-green versus vine-ripe) is important because vine-ripes typically receive a price discount relative to mature-greens owing to shorter shelf life (although this price disadvantage may be reversing itself in certain markets as longer-lasting vine-ripe varieties are introduced). The level of maturity harvested varies regionally in California, along with yields and costs.

As with most industries, the fresh tomato industry in California has gradually become more concentrated and has involved fewer shippers. Whereas at one time there were many shippers who only produced locally for seasons that might last only a few weeks, this is no longer economically feasible. Two factors driving concentration have been the shift to mature-green tomatoes and the year-round demand for tomatoes by final markets.

Although mature-green tomatoes have been grown for decades, there has been continual improvement in the postharvest handling system. In particular, packing houses have become much more automated and gassing facilities have improved. This greater investment in fixed capital creates economies of scale if the production season can be extended. Thus, as the investment in packing houses grows, tomatoes are trucked from ever greater distances to fewer packers, thereby extending the number of weeks that the packing house can be run.

Simultaneously, the creation of ever larger buyers for fresh tomatoes, such as the supermarket or fast-food chains, who came to require year-round supplies of tomatoes, encouraged shippers to move toward multiple-area sourcing to be in the market for as many weeks of the year as possible. On the one hand, multiple-area sourcing allows one to sign a contract with, say, McDonald's, that needs tomatoes 365 days a year. On the other hand, being in the market the year round reduces market risks because returns can be averaged over the greatest possible length of time. Former tomato shippers interviewed in California for this study attributed their inability to compete to the relative shortness of their season.

These trends are very clear in California. For example, a packing house in the Stockton area will typically be supplied with tomatoes from as far north as Sacramento and as far south as Huron. A packing house in Merced will source tomatoes from around Merced, but also from Huron and Stockton. Packing houses in the Salinas valley truck tomatoes over Pacheco Pass from the San Joaquin Valley. Although this spreading out of packing-house supply continues to develop, tomatoes have been trucked from Huron to the Salinas Valley for over 20 years.

Similarly, there are any number of shippers in California who also grow or market tomatoes from other regions, such as Florida or Mexico. In fact, while the shippers do compete against one another, to speak of competition between regions, such as Mexico and California or Mexico and Florida, is not entirely accurate, because a great deal of production has been rationalized among the regions and is controlled by the same shippers.

One of the results of these trends has been a shift in location of fresh tomato production in California as the industry has rationalized its structure under the control of a smaller number of shippers. The total California area devoted to fresh market tomatoes was 38,000 acres in 1990 compared with 30,500 acres in 1980, or an increase of 25 percent. The growth in California acreage occurred in the San Joaquin Valley, while acreage was declining in southern California, principally because of urbanization pressures and the shift to Baja.

To summarize, on the one hand, there has been a decline in vine-ripe, pole tomato acreage in California, as mature-greens have taken more of a market share and as Mexico has become a bigger factor. Regions where pole tomatoes were grown, such as Cutler–Orosi in the San Joaquin Valley, Oxnard, or the south coast, have all become less important producing regions. On the other hand, the expansion of mature-green production has favored the San Joaquin Valley, particularly on the west side, where production complements the seasons of the long-established regions of Merced, Stockton, and Salinas. Whereas about one-half of the tomatoes produced in California in 1972 were mature-greens, now they are about two-thirds of production. The San Joaquin Valley regions together in 1990 accounted for 68 percent of total acreage and 60 percent of total production. This has risen from about 37 percent of production in 1972 (Jesse and Machado 1975).

Florida is the largest domestic producer of fresh tomatoes, with production occurring during the October through June period. In 1990, Florida produced 762,000 U.S. tons of fresh tomatoes, despite the aftereffects of the freeze in late December 1989. In 1989, Florida production was estimated to be 918,000 tons, a historical high.

As indicated earlier, although Florida is often and correctly described as a winter producer of fresh tomatoes, the vast majority of Florida shipments occurs around the late fall (November-December) and spring (April-May) peaks. Like California, Florida primarily produces mature green tomatoes, but, unlike California, it uses staked cultural practices with plastic mulch.

Since immigration reform, the few labor problems that exist in the south Florida tomato fields have been dwarfed by problems deriving from urban sprawl, environmental degradation, and the perceived (though unwarranted) problems of NAFTA. As part of one of the fastest-growing states in the country, south Florida's tomato-producing counties are among the most ethnically diverse in the nation, providing a broad foundation of network affiliation that provides immigrant labor to agriculture. Vegetable production is one of five major agricultural production regimes in the region, competing for land, labor, and other inputs with citrus, sugar, livestock, and ornamental horticulture. Within vegetable production, tomatoes are the clear leader, yet, as in California, the geographical heart of the industry has shifted in recent years.

Dominance in tomato production, for reasons unrelated to IRCA, shifted from the East Coast to southwest Florida in 1986. At the same time that southwest production emerged as the new south Florida leader, Dade County producers began displacing ground tomatoes with staked culture techniques, increasing yields. Staked acreages increased from 1,200 acres in 1987–88 to 5,750 in 1989–90, and at the same time ground acreage fell from 7,950 to 50 acres. We expect that southwest Florida will continue as the industry leader, with farm worker towns such as Immokalee and La Belle assuming more importance as centers of housing and services.

Total Mexican tomato exports to the United States were lower in 1990 (in physical volume) than in 1989. Acreages for the 1991–92 season in Sinaloa and Sonora were down relative to 1990–91. Excess tomato supply in the United States, low yields in Mexico, and rising dollar costs of Mexican production contributed to financial losses for Mexican tomato growers and their U.S. joint-venture partners in 1990–91. Consequently, U.S. investment to finance tomato production is currently decreasing in both Baja California and Sinaloa, which limits their ability to expand.

The Tomato Labor Market and the Impact of IRCA

California

There has been, at least since 1988, a pronounced farm labor surplus in all areas of California. This surplus results from continuing, and even increasing, migration from Mexico, combined with the low exit rate of SAWs from the farm labor force. In fact, the supply of labor is the one area where IRCA has had a significant impact. Far from limiting immigration, however, as the law supposedly intended, it has encouraged settlement of newly legalized workers in California, which has in turn provided new opportunities for other Mexicans to migrate to the United States.

Overall, IRCA legalized about one-half of the tomato harvest workers interviewed in California, although the proportion was higher in Fresno (73%) and San Diego (67%) than in Stockton (32%), because the Stockton labor force has more long-term, settled (or back-and-forth) migrants from traditional Mexican-sending regions. Over one-half of the Stockton workers were legally working in California before IRCA, while the labor force was heavily undocumented in San Diego and Fresno before IRCA. In fact, some Fresno tomato farm labor contractors reported legalizing 100 percent of their workers through IRCA (Vaupel 1991). Similarly, while only 8 percent of the workers interviewed in Stockton were not authorized to work in 1991, 27 percent were not authorized in Fresno and San Diego. No U.S.-born workers were found.

Continued immigration since IRCA has brought many indigenous workers from the southern Mexican highlands, particularly Mixtecs from Oaxaca, who compose the largest share of the tomato labor force in Fresno and San Diego. They are also the main source of labor in Baja and Sinaloa, and are actively recruited in Oaxaca by growers in those regions. Including the workers in Baja California, 41 percent of the tomato workers interviewed for the California CAW study were from Oaxaca (43% were Mixtec) and 36 percent were from Michoacan — the new- and the old-sending regions. The regional variation of origins was striking. In Stockton, 92 percent of the workers interviewed were from Michoacan, Guanajuato, or Jalisco, the traditional core-sending region, and the rest were Mixtec. In Fresno, by contrast, none were from the traditional region, two-thirds were from Oaxaca, and 80 percent were Mixtec (although this extreme difference may be due to sampling error). In San Diego, which has a very mixed labor force, 20 percent were from the core states and 40 percent (who were also Mixtec) were from Oaxaca. Finally, in Baja, 79 percent were from Oaxaca; the rest came from the core-sending states, especially Michoacan. Baja California and Sinaloa have acted as launching pads to the United States for many Oaxacan workers. In the CAW survey, 64 percent of the Mixtec tomato pickers who worked in California had previously worked in northwest Mexican agriculture. In an earlier survey of 130 Mixtec farm workers in California and Oregon, over two-thirds had worked in northwest Mexico before entering the United States (Zabin et al. 1993). This is in stark contrast to other groups of farm workers, because only 9 percent of the non-Mixtec tomato workers in California had previously worked in northwestern Mexican agriculture. All indications are that migration from southern Mexico is accelerating, which suggests that the share of unauthorized workers in the tomato labor force will continue to increase (Zabin et al. 1993).

The mean time for working in U.S. agriculture for all of the workers surveyed in California was 11 years; the median was 7 years. However, 21 percent had been working in U.S. agriculture three years or less, and 41 percent had been working in U.S. agriculture five years or less at the time of the interview — that is, they had entered since 1986. The mean time for all Oaxacan workers surveyed in California was 7.4 years; the comparable time in the United States from an earlier survey of Mixtec farm workers was 7.2 years (Zabin et al. 1993). By contrast, the mean time in U.S. agriculture for the non-Oaxacan workers in our study was 13 years.

All of the tomato workers interviewed in Baja, and virtually all (86%) of those interviewed in California, planned on remaining in farm work. Some of the possible alternative jobs mentioned included bricklaying, cannery work, landscaping, house cleaning, and various manufacturing jobs. Most workers spoke little or no English, had little education (average of 4.3 years), and did not know how to go about getting another type of job. Legalization did not have any significant effect on willingness to stay in farm work. The California economy is also in a serious recession, and a number of workers who had secured nonfarm jobs, particularly in southern California, had subsequently lost them and had to return to farm work.

Florida

Analysis of the CAW south Florida farm worker survey indicates that our sample conforms to images of U.S. farm workers portrayed in the most recent studies (Heppel and Amendola 1991; Mines, Gabbard, and Boccalandro 1991; Griffith et al. 1990; Polopolus 1989). Some of this has been due, of course, to the purposive nature of our sampling. In particular, we wanted to assure that nearly half of our sample consisted of Mexicans/Chicanos, since they predominate in Florida and in U.S. farm labor generally.

Besides the predominance of Mexicans/Chicanos in our sample, we found the farm workers in our sample to be highly mobile, relatively new immigrants, underemployed, young, working in a number of locations and for a number of employers or farm labor contractors during the course of a year, and relatively inexperienced in the U.S. farm labor market. These figures describe a young, male, foreign-born labor force, most of whom live with three or four other farm workers in rented housing. Less than 20 percent of the sample reside in Florida with any of their children, over 80 percent leave south Florida following the end of the vegetable harvests, nearly 13 percent were in Mexico immediately prior to moving to south Florida, and at least 75 percent of the workers interviewed reside in other states or countries during some portion of the year. Together, these statistics testify to a largely transient labor force, residing in the U.S. and south Florida for work as opposed to family or other noneconomic reasons. Among the foreign born, we see as well that they vary by their experience in the United States, a factor that becomes important when considering wages. Over one-third entered the United States since IRCA's passage and nearly half entered the farm labor market since IRCA; this supports the conclusion, in line with a number of other studies (Bach and Brill 1990, Griffith et al. 1990), that IRCA may have contributed to increased flows of migrants, since SAWs can now cross the border with more ease, and with each return trip spread more labor market information about El Norte (Massey et al. 1987).

Farm Labor Contractors, Crews, and Working Conditions

California

As noted above, the packing house is the central, organizing unit in fresh tomato production in California, as it is in fresh citrus or fresh stone fruit. In some other fruits and vegetables, such as lettuce, broccoli, cauliflower, or strawberries, methods have been developed to pack the product directly in the field, thus eliminating the need to handle the produce twice. This field packing gives rise to a more dispersed production structure, which is exactly the opposite of the tendency in fresh tomatoes. Field packing of fresh tomatoes is essentially blocked by the large number of grades and sizes that must be sorted and packed separately.

In mature-green tomato production, most shippers (but not all) have a large acreage of their own production, which is complemented by joint deals with contracted growers. We estimate from public records that about 90 percent of the fresh tomato acreage in California is leased.

In a typical arrangement, the shipper has the seed grown out in a nursery, and then transplants the tomatoes with his own crews. The grower prepares the ground and grows the tomatoes until harvest. The shipper then hires crews to harvest them and arranges for the tomatoes to be hauled to the packing house. The shipper packs and markets the tomatoes, deducts his costs from the sale price, and splits the returns in some previously determined shares with the grower.

The labor consequences of these arrangements are that the grower is usually responsible for more permanent labor, such as tractor drivers or irrigators, and the shipper is responsible for seasonal labor, such as transplant crews or harvest crews, as well as the packing. In California, shippers typically hire transplant crews directly and use farm labor contractors to provide harvesting crews. Some shippers hire one or more harvesting crews directly, and in San Diego all harvesting is done by direct-hire employees.

A relatively small number of shippers control most mature-green production in California, both through their own production and through contracts with growers. Whether the grower or shipper takes responsibility for the harvest, farm labor contractors are usually hired to supply labor for hand harvesting of mature-greens. It is more common for the shippers to hire the contractors, so that the same workers can be moved from field to field in a given region. In contrast, vine-ripe producers hire their own labor, because they need to harvest for much longer periods. This is equally true in San Diego as it is in Baja.

Working conditions were being impacted by the excess supply of labor. While this apparently had been occurring since the early 1980s, workers pointed to 1988 — the year people arrived from Mexico to apply for IRCA — as marking a turning point. Workers complained about the use of farm labor contractors and their practices, but the shift to contractors had occurred before IRCA.

Fresh tomato harvesting is highly seasonal in a number of regions of California, such as Imperial (six weeks), Bakersfield (two weeks), Salinas (two months), and Huron (two months and one month). Even the four-month seasons in Merced and Stockton are highly variable, with peaks and valleys of labor demand. Only the vine-ripe culture in San Diego and Baja provides long-term employment stability, which growers complement with other vegetables and strawberries.

While tomato harvest crews do move around within a given region, they do not usually travel from region to region like the lettuce crews. Growers and shippers are thus heavily dependent on farm labor contractors to supply them with seasonal labor. This has presented a considerable challenge on the west side of the San Joaquin Valley, where few people live and few other labor-intensive crops are grown. Workers are usually driven long distances from east valley towns to work in the Huron area. This situation is probably viable only with the large surplus of labor that currently exists.

Because tomato employers in California are operating in a surplus labor situation, relatively little consideration is given to improving conditions for the workers. Most of the attention is focused on having labor contractors compete against one another, bringing in new groups of workers — such as the Mixtec — to moderate the demands of old groups or preventing opportunistic behavior on the part of the workers. The large number of strikes and union elections in California tomatoes — in contrast to their virtual disappearance in many other crops — is evidence of a highly conflictive labor situation.

Efforts to mechanize the harvest in the early 1980s were prompted by labor organizing, but were abandoned when buyers rejected the poorer quality tomatoes. Even efforts to introduce harvesting belts have been abandoned. An ample supply of labor and declining real wages provide no incentive to change this situation.

Workers reported broad compliance with sanitation laws. Drinking water was universally present; toilets were almost as prevalent, although smaller farms in San Diego neglected them; only water with which to wash hands was irregularly provided. Interestingly, tomato farms in Baja California were almost as compliant as California employers with sanitary facilities, despite the absence of similar laws.

Workers complained that crew sizes had grown, which reduced the amount of work available on a given day. This was considered particularly onerous late in the season, when there were fewer tomatoes to pick. Numerous workers reported that their crews' size increased from 90 to 150 people as the season progressed. In addition, larger crews required two trucks simultaneously to absorb the quantity being harvested. Because they were not always available, there would often be considerable waiting in lines.

Picking tomatoes is not without skill requirements. In particular, workers are essentially asked to serve as sorters while they pick. They sort for both size and color, which might change even in the course of a day, depending on market conditions and orders. They also sort out a variety of misshapen or bad tomatoes. Workers also must remove the stem, if it does not come off as the tomato is picked, which takes time and often tears their gloves. Workers must wipe off the mud from the tomatoes, if it is stuck on, because it damages other tomatoes in the buckets and bins. Some crews also reported being required to move tomato vines out of the furrows as they picked. Thus, workers are not only picking tomatoes, but also cleaning and sorting them, and possibly also cultivating the field. These latter tasks significantly slow the work. They have probably assumed even more importance, however, as the packing houses have become automated.

The workers are endlessly harangued while they are working, because it is assumed that the workers are basically trying to cheat and must be threatened with adverse consequences. The permanent threat is that they will not be given a *ficha* (a token to account for buckets picked) for some failing: wrong color, wrong size, three misshapen or bad tomatoes, buckets not full enough, and so forth. For this threat to be credible, the penalty must actually be imposed on occasion. Although management denied withholding *fichas*, every worker interviewed reported its occurrence. Several workers said it happened to them on average once a week.

While the number of dumpers varies among crews and shippers, the number of dumpers on at least some Stockton crews has been reduced since 1989. A full complement would be four dumpers to a truck, but some crews operate with only two to a truck, which means that workers must lift the buckets higher up, rather than pass them to an intermediate dumper. This maneuver is particularly difficult for women.

Workers in Stockton are now expected to buy their own tools, which in tomato harvesting consists of buckets and gloves. Combining these items, a reasonable estimate of tool cost would be $5 a week for each worker. Long-time workers report that they were given the buckets free until 1983–85, but that since 1985 they have had to buy them. This change thus occurred before IRCA. By contrast, tools are provided for workers in Baja and San Diego, and buckets also are generally provided in Fresno.

Whereas formerly contractors curried the favor of workers and went to the camps to recruit, now the situation is reversed — workers must look for the contractor. In the entire survey, of the workers who had some contact before the season started with the employer, 74 percent said the worker contacts the *mayordomo*, contractor, or grower.

Essentially, the only form of recruitment in Stockton now is walk-in by word of mouth. Some contractors or shippers register the workers at a central office a month before the season, but more people usually appear at every field each day than are allowed to work. Thus, people are turned away daily, and it is because of this system that problems arise in limiting the numbers in the field. This system serves as a constant reminder to workers that there is a surplus of labor at present. To the extent that it allows crew sizes to increase, however, it frustrates workers with excessive waiting and lowers their incomes, which causes resentment.

Fresh tomato work, as with much agricultural labor, is replete with unpaid waiting time. It is common for workers in Stockton to arrive at the fields and wait four or five hours to begin harvesting. If a truck gets stuck in the mud, workers wait. If there are too many workers for the number of trucks and dumpers, then they wait in line to dump their buckets.

These complaints of the Stockton workers effectively demonstrate the extent to which working conditions and incomes are being affected by the excess labor supply and the competition among contractors. Essentially, the workers are being nickeled and dimed and asked to bear more and more costs themselves. They must spend their time sorting, cleaning, and cultivating, and they must heap the buckets. They lose $1 a week in "fines," maybe $.50 a week to cash their checks, and pay $5 a week for tools. They pay their own transport to the fields, are overcharged at the lunch wagon, and suffer direct physical pain to save on dumpers. In addition, they wait long hours at times to start harvesting; they must often wait in line to dump their buckets.

Some of these cost savings accrue to the shippers, some to the contractors. The contractors in Stockton are small family businesses. The dumpers, *ficheras*, supervisors, and lunch wagon are usually all staffed by relatives of the contractor. It is a group enterprise that is proving highly efficient at cost minimization — with many impacts on working conditions.

Florida

Foremost among south Florida vegetable producers' production practices is their use of farm labor contractors (FLCs). Without exception, all growers interviewed use FLCs, although the character of their use varies: Some growers allowed FLCs to handle farm labor relations entirely, while others worked alongside FLCs in partnership-type arrangements.

Payment systems have changed on some farms, with more growers paying weekly instead of daily for most of the year. These changes seem related to the growing practice among growers to keep records on individual farm workers as opposed to leaving this responsibility to crew leaders. It is likely that the shift from paying FLCs a single check for the entire crew to paying workers individually occurred because of questions concerning tax-withholding practices by FLCs. Under the old system, FLCs could (and did, judging from lawsuits) claim that they were withholding taxes from workers and keeping the amount withheld, or they could negotiate different wage rates with growers than those offered workers. Increased litigation over this problem, combined with joint liability questions, led to this shift.

This change in payment involves a significantly altered relationship between growers and FLCs, effectively reducing the FLC's role as contractor to more of a supervisor/labor manager and employee. Only one of the 18 growers interviewed reported moving in the opposite direction — from having FLCs as foremen to subcontracting more of the work to them. This, of course, affects the overall business of farm labor contracting.

FLCs occupy a somewhat unique and often precarious position in the farm labor market. Florida independent farm labor contracting is often a family tradition, much of it originating in the Mexican–American and Chicano farm working families of the lower Rio Grande Valley. The migrations into Florida of Spanish-speaking farm workers during the 1950s and 1960s provided opportunities for enterprising farm workers who possessed the cultural, linguistic, and social skills to establish themselves as small "nickel" contractors and gradually build their businesses into the complex transporting, labor supervising, and housing operations many of them are today. It was common for FLCs to respond to questions about their backgrounds in labor contracting by citing some family involvement. "My father, uncle, and grandfather were all crew leaders," one said. "I was born into it. The whole family (four men) works in the business."

More commonly, however, crew leaders and FLCs mention backgrounds as farm workers, working their way into labor contracting and supervisory positions in the fields. The opportunities for becoming an FLC still exist, and one of the complaints of labor contractors in south Florida today is that there are too many of them, competing against one another for workers.

Crew sizes are normally around 30 to 40 workers, but can go as high as 60 or 70. Crew sizes have decreased over the last 5 years — once being as high as 250 to 300 per crew. The shrinkage in crew size may underlie the growth of crew leadership positions that current FLCs complain about, since more crews of smaller size create opportunities for farm workers to become field walkers, crew supervisors, or "nickel contractors." This gives farm workers supervisory experience, a prerequisite to becoming a crew leader.

With daily supervision of workers and other tasks come several other responsibilities and legal requirements. Some responsibilities are shared with growers; ultimately, both growers and FLCs are jointly responsible for serious problems such as occupational injury. In a practical sense, the FLC and his helpers assume full responsibility for daily and hourly problems that arise with workers. In the field, a typical method of disciplining workers, for example, is simply to order the worker to leave the field, an action that is particularly harsh when the field may be located as far as 20 miles from the farm worker's living quarters. A less harsh means of reprimanding workers is to refuse to "punch their ticket" (punch a hole in a card that is used to keep track of a worker's productivity and reckon his or her pay) when the worker brings bruised fruit to the truck.

Both of these disciplinary actions involve direct confrontations with farm workers, which, witnessed by other farm workers, can either strain relations with the entire crew or further enhance the crew leader's power. In either case, crew leaders, FLCs, and field walkers are responsible for disciplining workers, as well as offering workers advice on how to improve productivity and rewarding workers with tasks or assignments that are perceived to be "better" than others (assigning a worker rows nearer the truck, for example, so that he or she can work faster, pick more, and earn more than workers further from the truck).

Recruitment varies little between the eastern and southwestern regions: Both regions have well-known "shape-up" areas where large numbers of workers gather, waiting for FLC buses, on a daily basis. A few of the FLCs interviewed, however, tended to use a more direct recruiting method, trying to keep the crew together on a daily basis by picking up workers at their living quarters. Others use housing as a recruiting tool, renting units on the condition that the occupants work for them; on the East Coast, many growers have their own housing for seasonal workers, again occupied on the condition that workers work for the grower who owns the housing.

Most commonly, FLCs use a combination of getting workers at the local shape-ups and picking up workers at their homes, and the latter seems to be emerging as a common method of labor control, reducing the amount of negotiation workers can do, as well as reducing the potential for workers to move into nonagricultural jobs. Overall, recruitment relies heavily on traditional and informal techniques (shape-ups and networks) and almost not at all on formal agencies such as the U.S. Employment Service.

Field supervision consists of staying with workers throughout the day, transporting them between activities and fields. Field walkers and crew leaders, many of whom have close kinship and friendship ties with a core of crew members, tend to maintain the pace of work as well as order, through direct interactions with workers, using various supervisory styles. Styles range from abusive to affective; workers are sensitive to stylistic differences in supervision, and often base decisions to quit on field walkers' poor supervisory skills. There are no formal grievance procedures or channels for appealing a crew leader's decision; workers' abilities to complain rest on their personal relations with their crew leader. FLCs' abilities to enforce rules or have their orders followed rest, in large part, on labor surpluses and the ultimate fact that workers are interchangeable.

Wages

Piece rates for picking tomatoes have risen in nominal terms since IRCA, but have fallen slightly after adjusting for inflation. Piece rates in Stockton were $.475 a bucket in 1991, compared to $.40 in 1986; in real terms, this represents no increase. Piece rates in Fresno averaged $.375 a bucket in 1991, compared to estimates of $.33 to $.36 cents in 1986; thus, in real terms, piece rates fell by 6 to 15 percent.

Wages for tomato picking not paid by piece rate are at or near the minimum wage. Of 14 tomato jobs in San Diego and the San Joaquin Valley where workers were not paid by the bucket, ten were paid $4.25 an hour, one was paid $3.44, and the other three were paid $4.35 or $4.50. The mean wage was $4.24 for this group.

Hourly equivalent wages for piece-rate tomato harvesting averaged more than twice these hourly wages. In fact, in every case, the piece-rate jobs paid more than the jobs paid on an hourly basis. Piece-rate pay for harvesting tomatoes ranged from $5.20 to $16.84 an hour in California, with a mean of $9.30 an hour. Combining hourly and piece-rate harvesting jobs, the mean wage was $7.91 an hour in California and the median was $7.14.

Hourly equivalent wages in the survey averaged $8.20 in Stockton and $8.11 in Fresno, where workers received piece rates. They averaged $6.53 in San Diego, where most workers were being paid by the hour, and they averaged $.88 in Baja, where most workers were paid by the day. While the hourly wages were higher than in many agricultural jobs, workers complained that they could not work enough hours a day or days a week. While workers in Baja averaged 58 hours a week, the average in San Diego was 44 hours, in Fresno 37 hours, and in Stockton only 27 hours. Thus, while Stockton workers received the highest piece rates and the highest hourly wages, they averaged the lowest daily and weekly incomes of all California tomato workers.

The least variable aspects of south Florida vegetable production are piece rates, methods of payment, and hourly wage rates for seasonal workers. There exist two basic methods: (1) workers are paid a standard piece rate that varies little, if at all, from farm to farm or (2) workers are paid an hourly wage, in almost all cases the minimum wage, plus a piece rate one-third to one-quarter of the standard piece rate. A minority of growers pay only an hourly wage.

Without exception, all growers in our sample reported that wage rates had increased along with the minimum wage, yet disagreed over the issue of the proportionate cost of labor. Budgets compiled by

University of Florida production scientists suggest that there has been little change in the proportionate cost of labor. For tomatoes, labor costs fluctuated between 17 percent and 24 percent of total production costs, failing to demonstrate either a falling or rising trend over the past five years.

Similar to accounts of growers, FLCs, crew leaders, and foremen reported that wages had increased along with the minimum wage, although pay records from court documents from the early 1980s, as well as other sources (Griffith et al. 1991, Heppel and Amendola 1991), indicate that piece rates have remained constant at least for the past 10 to 12 years. Most of the work is paid by the piece, although a few reported combining piece rates and hourly or daily rates. Piece rates for the tomato harvest are still between $.35 to $.60 a bucket, with $.40 a bucket being the most common. Wages rise and fall during the season and from field to field, depending on supplies of labor in the first case and yields in the second. In terms of benefits, most FLCs reported, again like growers, that they paid only workers compensation.

Data about piece rates suggest for some that farm work is a potentially high-paying field relative to other unskilled occupations, but for others that it is an abysmally low-paying field. Numbers of buckets picked per day range from a high of 260 ($91 to $104, depending on the piece rate) to a low of 12 ($4.20 to $4.80), with the modal number being 150 ($52.50 to $60.00), the median 115 ($40.25 to $46.00), and the mean 117.11 (sd=51.87; $40.98 to $46.84). With the exception of those at the low extremes (one in five workers makes less than minimum wage), most workers seem to make more than minimum wage while working in south Florida vegetables; again, however, the variation is fairly large.

The wide ranges in each of the categories presented above encourage further, more discriminating analysis of these data. When we examine the sample closely, we find that systematic differences exist between workers grouped according to variables that reflect their experience and relative power in the U.S. labor market. The foreign born, who are among the most migratory, vary by their experience in the U.S. and in the farm labor market, which in turn affects their earnings. New immigrants, in short, earn less than earlier arrivals.

Nominal wages have increased in Baja as growers try to retain laborers. The proximity to the United States is a constant temptation for workers, and now that the Mixtec have established migration networks in the United States, there is real competition for this labor force. In fact, to avoid this competition, growers in Baja have attempted to recruit in villages in Oaxaca that do not send migrants to the United States. In our interviews in Baja, we found a number of villages and regions represented that were not among the 180 villages found in a 1991 Mixtec census in California.

A significant proportion of tomato workers in Baja are women — more so than in any region of California. This is undoubtedly because of both the difficulty women have in migrating and working in the U.S. farm labor market and the more flexible work environment in Baja. Women can take days off in Baja to attend to their children without repercussions. In addition, the convention of paying workers on a daily basis has permitted more women to participate, as the daily minimum number of pieces required has been set at relatively low levels. Finally, the ability of families to self-construct small shacks allows them to live as a group much more cheaply than would be possible anywhere in California.

Although real wage rates are not yet increasing in Baja, the gap in real wage rates between Baja and California has declined recently. In the San Quintin Valley, daily wage rates as of December 1991 were approximately 20,000 to 25,000 pesos ($6.51 to $8.14 a day) compared with 13,000 pesos a day ($5.28) in 1989. The official minimum wage in Baja as of November 11, 1991 was 13,300 pesos, so it is clearly no longer relevant to actual wage rates. Although the current labor rate is still extremely low relative to typical California wages of $4.25 to $6.00 an hour, labor is generally less productive. This partly offsets Mexico's advantage in lower wage *rates*, which makes labor *costs* not as low as would be expected given the differential in wage rates (see the discussion on productivity).

Furthermore, because growers often must provide worker housing (no longer a common practice in California), incur significant transportation costs to attract workers from distant regions, provide local transport, and offer social services for large worker families, there are many labor costs for which the wage rate does not account. Increasing grower concern about labor availability is sparking some to pay greater attention to labor management, and more producers are introducing at least partial piece wage rates as a means to increase productivity. Piece wages can double daily wages. In addition, some areas of northern Baja, with more seasonal work than the San Quintin Valley, pay closer to 30,000 pesos for the basic daily wage rate.

Migration Patterns

Because some of the peak seasons were missed in surveying the workers, it is likely that the workers interviewed are less migratory than the population as a whole. With that caveat, of the workers interviewed, 46 percent reported not leaving California in 1990, 48 percent went to Mexico at some point during the year, and only 6 percent migrated to another U.S. state without also going to Mexico. About 14 percent of the California tomato workers migrated to another U.S. state; the only states reported were Oregon, Washington, Florida, and Arizona, in that order of importance.

There were regional differences. In Stockton, 50 percent said they stayed all year in California, 36 percent stayed in the Fresno area, and 31 percent stayed in San Diego. Viewed from the other side, 42 percent of Stockton workers went to Mexico, while 57 percent of Fresno workers and 69 percent of San Diego workers went to Mexico. This evidence of greater settlement by the Stockton workers confirms other information presented here.

Tomato workers in California tended to specialize in a certain set of tasks. In Stockton, workers harvest cherries, apricots, or asparagus, in addition to tomatoes, but they do not prune. In the Fresno area, most of the alternative work is in grapes, which includes pruning among other tasks, but Fresno tomato workers do not work in tree fruit or citrus. The Mixtec also travel to such places as Oregon or Florida to harvest tomatoes, berries, or vegetables. In San Diego and Baja, growers complement tomatoes with plantings of other vegetables and strawberries, which accounts for most of the additional employment in those regions.

Most of the workers in our sample migrate during the year to other farm work jobs in the United States or back to their home countries every so often. First, among the foreign born, only 45 percent report never returning home, while 12 percent return home every four or more years and 21 percent every two to three years. A significant proportion, 21 percent, are what we consider cyclical migrants, returning home seasonally every year; most cyclical migrants (69 percent) are Mexican workers, while the remainder are Haitians (31 percent cyclical).

In addition to such high international mobility, fully 83 percent of the farm workers were migrants within the United States, working in other areas within Florida or other states. One-quarter of the population work only in Florida, while 35 percent work in two states, 25 percent in three, and 12.5 percent in four or more states during a typical annual round. We elicited 78 different work locations in and out of Florida, 18 different states, and 4 foreign countries where respondents had worked or lived during the year prior to our survey, again testifying to a highly mobile work force and confirming the idea that south Florida is both a labor-importing and labor-exporting region for the U.S. farm labor force. Most of the states were in the southeastern, eastern, and midwestern United States, although we encountered farm workers who had worked in California, Arizona, and Oregon in our sample. Among the migrants, farm work locations immediately prior to working in south Florida vegetables were: other parts of Florida (62%), other southeastern states (11%), the West/Southwest (9%), the mid-Atlantic states (9%), the Midwest (4%), and the Northeast (4%).

Given the low-skill, varied nature of work in the fields, which makes it easy to move workers from task to task or from crop to crop, farm workers rarely specialize in a single crop or task. In our sample, the farm workers named five crops in which they were currently working and listed ten tasks. Over the course of the previous year, however, the farm workers in our sample had gained experience in over 25 different crops and listed over 20 tasks. A typical annual round among south Florida vegetable workers is to work in vegetables through the year, moving from south Florida to the Palmetto–Ruskin area of Florida in late spring or early summer, then into Georgia or the Carolinas and on up into the vegetable harvests of the mid-Atlantic states, the Northeast, or midwestern locations such as Ohio or Michigan. The link between Florida and Texas, especially among Mexican nationals and Mexican–Americans, remains durable as well.

International Competitiveness and Productivity

To the extent that IRCA increased the supply of labor in U.S. agriculture, it has clearly improved the international competitiveness of U.S. fresh market tomato production. Comparisons of costs in California and Mexico demonstrate that much of the movement to Mexico in the 1980s was due to an undervalued peso and various input subsidies in Mexico. The revaluation of the peso since 1987 and the gradual removal of those subsidies has eliminated many of the cost advantages that Mexico offered. Export tomato acreage in northern Mexico has actually declined in recent years.

While labor costs are still much lower in Mexico than in California, when one factors in productivity, the differences are much smaller. Workers in Mexico pick relatively few buckets of tomatoes for their daily wage, and this makes unit labor costs higher than they at first appear. While hourly tomato wage differentials between Baja and California are approximately nine to one, unit labor cost differentials were calculated at about three to one.

Assuming that the red bucket (used in Stockton and Florida) holds 28 pounds of tomatoes (and it could hold more, depending on how much one wanted to supervise the workers), then at $.475 a bucket in Stockton, the unit cost is $.017 a pound. If the white bucket (used in Fresno, Salinas, and Baja) holds 25 pounds of tomatoes, then at $.375 a bucket in Fresno, the unit cost is $.015 cents a pound. The unit cost of picking tomatoes would be only 13 percent higher in Stockton, although the piece rate is 27 percent higher.

In the Salinas Valley, workers are paid about $.43 for a 25-pound bucket working for a farm labor contractor. The one firm operating under a union contract in 1991 paid $.495 for a similar bucket *copeteado*, which worked out to 26 pounds on average, according to the firm. At these rates, contractors in Salinas were paying $.017 a pound and the unionized firm $.019 cents a pound.

Similarly, in Florida, if a bucket is assumed to weigh 28 pounds, then, at the most common piece rate of $.40 a bucket, the unit labor cost is $.014 cents per pound to harvest, or slightly lower than the lowest cost found in California.

In San Quintin, workers are picking 25-pound buckets. With extensive data provided by some firms in Baja on their costs, we calculated that in the fall of 1991, the average cost to pick a bucket was 448 pesos, or $.146, at that time. That works out to about $.58 per pound, or 39 percent of the Fresno unit cost of picking ($.015 cents), which is the lowest we found in California. It is clearly much less expensive to pick tomatoes in Baja, but the unit harvest cost gap (2.5:1) is not so great as the hourly wage differential (9:1) would seem to imply.

California tomato growers also benefit from other advantages, such as higher yields and better infrastructure, as do growers in Florida. We conclude that the fresh tomato industry is already largely restructured along the lines that would emerge with completely free trade with Mexico, and that the various regions are much more complementary than competitive. The continued large supply of immigrant labor, which has been further encouraged by IRCA, has allowed California to maintain its share of the U.S. market and even to begin to ship fresh tomatoes to Mexico (See Runsten et al. 1993, Cook et al. 1992).

Conclusions

The similarities and differences between Florida and California reveal some of the more essential features of the farm labor market. Both regions have experienced a long history of ethnic succession, with differences within each of the regions regarding the stability and "reproductive" capability of some groups over others. Some of the earlier Mexican–Americans have been able to work into positions as FLCs in south Florida, while some Mexican families in Stockton, largely because of access to subsidized housing, have achieved some stability across generations in farm work and migration patterns.

In contrast to these proportionately small instances of stability is the more dominant reality of a highly transient, predominantly young male work force supplied through constant waves of immigration, cyclical migration, and return migration. Baja seems to be the only region with large numbers of women workers; in other production areas, fewer women work alongside their husbands or male relatives.

As part of the continued reliance on an immigrant work force, new labor supplies have been tapped in Mexico and Central America, primarily among indigenous peoples such as the Mixtec from Oaxaca or the Kanjobal or Chuj from Guatemala. It is remarkable that all regions studied (Florida, California, and Baja California) are increasingly drawing on this same group of immigrant workers. One consequence of this development will be that the U.S. tomato labor force will be increasingly composed of undocumented workers.

Wages were also found to be similar in California and Florida. This is not surprising given the similarity of the labor force and the number of shippers who operate in both regions. While wages are much lower in Mexico, the lower labor productivity there implies that unit labor costs are much higher than they at first appear. Given the much higher productivity of Mexican workers in the United States than in Mexico — and the declining real wages in California — there is a clear tendency toward convergence of unit labor costs in U.S. and Mexican tomatoes, which is a tendency that is likely to continue as the two economies become more integrated and actually utilize the same workers to pick tomatoes in both countries.

Notable differences between California and Florida are also quite telling. First, California's farm worker housing picture is much more variable than Florida's. Florida has little government-subsidized housing and that which is available (for example, the Farm Worker Village in Immokalee) has been "colonized" by many families who have left the farm labor force since occupying this housing. A cyclical migration pattern between Florida and Mexico, as is common in California's state camps, has not been established with this housing. Second, Florida has seen considerably less union activity than California. Third, whereas the use of farm labor contractors is universal among Florida tomato growers, some California growers rely on FLCs while others do not. The need for FLCs in Florida may derive, in part, from the distance to the border. In this context, FLCs team up with *coyotes* and others who provide transportation from border states to Florida. In California, the use of FLCs in tomatoes is much more a managerial decision — or an effort to circumvent union organizing — as many growers and shippers hire directly at least part of the work force.

Because tomato employers uniformly expressed the opinion that IRCA had no real effect in terms of constraining their operations, one must turn to the workers to evaluate the real effects of the law. The workers' view of IRCA can best be summarized by this quote from a Stockton worker: "The amnesty was an agreement with the growers so that many new people would come [from Mexico] to work." By legalizing over one million Mexican immigrants under the SAW program, there can be little doubt that IRCA encouraged more migration from Mexico as it facilitated settlement in the United States. In this sense, it made U.S. tomato production more competitive with Mexico as real wages declined with continued immigration.

Comparisons of costs in Mexico and the United States indicate that free trade agreements are unlikely to significantly restructure the U.S. fresh tomato industry — that, indeed, it was already largely restructured by the mid-1970s and that market shares have subsequently changed little. The key element in competitiveness, apart from marketing considerations, appears to be productivity. California and Florida have remained competitive primarily because of increased yields and higher labor productivity.

NOTES

1. This chapter is based on two case studies conducted for the U.S. Commission on Agricultural Workers. See Runsten et al. 1993 and Griffith and Camposeco 1993 for more detail.

2. Our information on tomato workers in California is mainly based on interviews with 86 such workers spread over a period of two years. Most of the data, however, refer to a group of 56 tomato harvest workers interviewed in three regions of California and 14 tomato harvest workers interviewed in Baja California. The information presented here also relies heavily on open-

ended interviews with participants in the fresh tomato industry. Interviews were conducted with a dozen grower-shippers of varying sizes in California and Baja, who collectively operated in all of the fresh tomato regions of California and northwestern Mexico. Several contracted growers were interviewed in three distinct regions. Information on the operations of farm labor contractors was obtained through an earlier study conducted by Suzanne Vaupel in the Fresno area (Vaupel 1991), and through interviews conducted by the California Institute for Rural Studies as part of a statewide survey of farm labor contractors (California Employment Development Department 1992). Finally, in-depth interviews of tomato workers and former tomato workers were conducted by Anna Garcia in Stockton, Fresno, and San Diego (Runsten et al. 1992).

In Florida, we conducted 36 interviews (18 growers and 18 labor intermediaries) that reflect the labor- demand side of the south Florida labor market for vegetable production. In addition, a large number of interviews were conducted with other informants and observers of south Florida agriculture. Information on Florida tomato workers is based on interviews with 88 such workers sampled somewhat purposively to provide adequate representation of various ethnic groups (Griffith and Camposeco 1993).

3. There are two basic types of tomato marketed in the United States: vine-ripe or "pinks" and mature-green or "gassed." Vine-ripe tomatoes are simply allowed to ripen on the vine until they turn pink, and are shipped to market as they are picked. Most vine-ripe tomatoes are grown with staked culture on wires strung between the stakes. An alternative culture is to grow bush tomatoes on the ground, pick them green, and then gas them so that they continue to ripen. One can also grow mature-green tomatoes with staked culture, as they do in Florida or at times in Mexico.

19

IRCA and Raisin Grapes[1]

ANDREW J. ALVARADO, HERBERT O. MASON, AND GARY L. RILEY

Introduction

This chapter describes one of the most labor intensive of the approximately 250 crops grown in California's central San Joaquin Valley: raisin grapes, whose production relies on a peak labor force of 40,000 to 50,000 workers during a 3- to 4-week period. All of the workers studied were Hispanic, most were Mexican born (94%), and they worked under some of the most marginal conditions.

The focus of this study was to determine the consequences or impacts of the Immigration Reform and Control Act of 1986 (IRCA) on the raisin industry's labor force, labor practices, employee-employer relationships, labor costs, and shifts toward mechanization. The raisin grape harvest is an ideal case study to assess the effects of attempts to manipulate agricultural labor supplies due to the sudden demand for harvest labor, which lasts for only three to four weeks. That is, if legislation has indeed affected the supply of labor, it would most likely affect this commodity, which is considered to be one of the points of entry for workers into the U.S. agricultural labor market (Alvarado, Riley, and Mason 1990).

Small-acreage farms dominate the industry and as many as 5,500 raisin growers can be found in the Central Valley. These farm units are typically family owned and operated (U.S. Census 1989). The most common variety of grapes grown for raisin production in this region are the Thompson seedless. In the entire state, 95 percent of all raisins are made from this variety as well, and total raisin production in recent years has reached between 135,000 and 140,000 tons. Conversely, only about 40 percent of the Thompson seedless grapes are made into raisins. The remainder of this variety's production is sold in fresh fruit, wine, or juice markets.

The raisin industry experiences economic variability associated with market and production conditions. Raisin grapes are particularly vulnerable to rain during the three-week drying period in

[379]

September, when the harvested grapes are laid on paper trays to dry. Rainfall during this time can cause substantial crop loss.

Ninety-five percent of all raisins produced in the United States are grown within a 60-mile radius of the city of Fresno. This geographically centralized commodity is typically grown on small farms averaging 45 to 50 acres each, with a total annual production of between 325,000 and 350,000 tons of raisins.

The most salient findings of this case study are:

- the dominance of farm labor contractors utilized by employers during the harvest

- the continued and unabated influx of undocumented or fraudulently documented workers

- an abundant labor supply

- the near nonexistence of the border patrol in the region

- the availabilty of mechanization as an alternative to reduce labor demands, but that does not appear to be economically viable under current labor market conditions

- the benefits to workers who were legalized under the SAW program

- the apparent exiting of SAW workers from the raisin harvest but not from farm work.

Data

To obtain the information needed to accomplish the objectives of this study, three sample groups were identified. The first was a random sample of 125 raisin harvest workers drawn from among the crews of 12 Fresno County growers. Interviews were conducted on the job sites, with employer permission. Large, medium, and small employers (growers and farm labor contractors) were equally represented.

The second study sample consisted of 1,500 raisin grape growers randomly selected from a comprehensive listing of 4,500 growers maintained by the Raisin Administrative Committee of Fresno County. A total of 323 questionnaires were completed, resulting in a response rate of 21.5 percent.

The third source of study information included 12 raisin industry experts purposefully selected by the researchers from among central California agency and industry leaders. They are considered to reflect the highest levels of decision making and experience regarding the raisin industry.

Findings

Farm Labor Contractors

A clear trend that has emerged in the raisin industry is the increasing use of farm labor contractors (FLCs), particularly since IRCA was passed in 1986. As indicated by the data presented in Table 19.1, growers relied on farm labor contractors to a much greater extent in 1991 than in 1985. Three-fifths of the 323 growers who responded to our mail survey indicated that they used FLCs in 1991. This same sample of employers stated that only 35 percent of them used FLCs in 1985, the year prior to passage of IRCA.

Table 19.1 Use of Farm Labor Contractors in the Raisin Harvest Since 1985, as Reported by Employers

Year	Percent	Year	Percent
1985	35.1	1989	52.0
1986	37.0	1990	57.1
1987	41.4		
1988	44.5		

The 1991 data are similar to the results found by others. For example, Heppel and Amendola (1991) found that slightly over 60 percent of Fresno County raisin employers used FLCs in the 1989 harvest season. Our interviews with industry leaders, growers, and FLCs indicate that these are conservative estimates of the importance of farm labor contractors in the raisin harvest. Typical estimates given by these sources were that 70 to 80 percent of growers are now using farm labor contractors for the raisin harvest.

The major reasons given by employers for the shift to farm labor contractors are:

1. *Too much paperwork in hiring workers directly.* Ninety percent of the responding employers who used FLCs in 1991 indicated that they did so in part because of difficulties in complying with all the paperwork requirements involved, even for short-term employees. IRCA is only one of several laws requiring that employers create a paper trail. Examples beyond the I-9 form required by IRCA include forms for withholding and documenting taxes, demonstrating health and safety standards, and paying workers' compensation, unemployment, and state disability insurance. It is clear that many employers are willing to pay the increased costs associated with using a farm labor contractor in order to avoid having to create a payroll and documentation system that is required for only a short season.

2. *Too many regulations imposed on employers.* Another major concern expressed by growers is related to the sanctions and liability associated with noncompliance with the variety of laws and regulations that face them. Although it is evident that many growers have become less apprehensive about failing to comply with IRCA, other liability concerns, such as work-related injuries, reporting requirements, mandatory workers' compensation insurance, and other regulations are worrisome issues. Approximately 84 percent of the employers in our survey who used farm labor contractors in 1991 indicated that one reason for using them was to avoid dealing with various government regulations.

3. *Labor supply concerns and difficulty in recruiting workers.* Approximately 58 percent of employers who used FLCs in 1991 indicated that difficulty in recruiting workers for the raisin harvest was a reason for electing to use an FLC. Although this reason was checked by the majority of those who used FLCs, it appears to be significantly less important than compliance with laws and regulations in terms of motivating growers to rely on FLCs to recruit and supervise their harvest crews.

Reducing labor costs apparently is not a motivating factor in using farm labor contractors. Only about 7 percent of the raisin growers who used FLCs in 1991 indicated that such costs were reduced by using a farm labor contractor. In fact, 86 percent of the total sample responded that their total labor costs had increased since 1986, even though harvest piece rates have remained relatively constant during this period. Fifty-six percent of the growers indicated that the increase in labor costs was due to a combination of increases in minimum wage and increases in piece rates; 30 percent indicated that it was due to increased piece rates alone, and 8 percent cited increased fees paid to farm labor contractors. These responses suggest a general sense among raisin growers that labor costs have increased since 1986, but this only seems a vague notion that it costs more to do business every year, rather than a firm grasp of where the monies are going.

Grower-FLC Arrangements

Typically, the grower-FLC relationship is an informal agreement. Ninety-three percent of employers who used FLCs in 1991 said that they worked under a verbal agreement with the FLC. Three general types of FLC payment methods are used in the industry:

1. The most common method is to pay the FLC a flat rate for each tray picked. Approximately 53 percent of the employers who used FLCs in 1991 responded that they used this method of paying FLCs. The median average paid to FLCs via the flat rate was $.21 per tray.

2. Another method used to pay FLCs is a percent commission on payroll. The median response from the mail survey was that growers paid FLCs 31 percent of payroll for their services in 1991. About 41 percent of respondents who employed FLCs indicated that they used a percentage commission arrangement with their farm labor contractors.

3. The final method used by a small portion of growers is the "penny contractor." In this type of arrangement, the "farm labor contractor" recruits the workers and sometimes supervises the crews. The penny contractor is paid $.01 or $.02 per tray for recruiting the workers, and then is paid on an hourly basis for supervising the harvest crews. The grower is responsible for the payroll and related costs. It appears that many growers who use penny contractors do not consider them to be FLCs, although many governmental agencies would.

According to the results from our mail survey of employers, workers who were employed by farm labor contractors in 1991 earned about $.16 per tray. This leaves a commission of about 31 percent for the farm labor contractor. When one accounts for mandatory employer-paid payroll taxes and insurance, including OASDI (7.65 percent), unemployment insurance (as high as 5.6 percent), and workers' compensation insurance (base rate for grapes was 8 percent in 1991), it is evident that the FLC works on a slim profit margin in the raisin harvest. The minimum commission figure estimated by farm labor contractors and industry experts we interviewed that would allow an FLC to operate legally by complying with all regulations and required payroll expenses would equal a 32 percent markup on direct labor charges. One farm labor contractor stated that he would not negotiate a commission below 36 percent. Consequently, he has gradually reduced the number of growers he works for in the raisin harvest and notes that the competition among FLCs engaged in the raisin harvest is fierce.

There are a variety of ways that an FLC can make a profit, even with these small commissions. Unscrupulous farm labor contractors may not pay taxes they withhold from workers or may report a reduced wage bill to the IRS and EDD. They may not pay workers' compensation insurance or other insurance premiums, or they may not carry required liability insurance. In some instances, the workers are charged for services provided by FLCs. Sixty-one percent of the 125 workers we interviewed indicated that they paid someone to provide a ride to work. FLCs, foremen, and *raiteros* (drivers of privately owned vehicles who may be FLCs, a foremen, or fellow workers) accounted for 60 percent of those who were paid for transportation to the fields. The cost of the ride ranged between $3.50 and $4.00 round trip each day in 1991. Another charge or cost imposed by both FLCs and growers is for equipment use. Virtually all (124 of 125) of the workers interviewed indicated they had to pay for both gloves and knives, even though required equipment is supposed to be paid for by employers in California if the worker earns less than twice the minimum hourly wage ($4.25 per hour in 1991).

Crew Bosses and Foremen

Previous studies in a variety of California agricultural crops have found that crew bosses, supervisors, or *mayordomos* are often the key link between growers and seasonal farm workers (Mason, Alvarado, and Riley 1993; Mines and Anzaldua 1982). Supervisors are typically bilingual and often are responsible for recruiting new workers as seasonal demands increase. Workers are usually found through family and friendship networks, quite often based in rural villages in Mexico.

Among the growers who responded to our survey, 13.5 percent indicated that they used their foremen to recruit crews to harvest raisins in 1991. While this is a relatively small proportion of the total sample, it represents about 29 percent of the growers who did not use farm labor contractors in 1991. Farm labor contractors were not included as employers in the mail survey. It is likely that crew bosses and supervisors perform important roles in recruiting and hiring workers employed by FLCs, as well.

Among the raisin growers who used foremen to recruit workers in 1991, the dominant method of payment was a tray rate. Of employers who paid foremen for recruiting, 92 percent indicated that they paid them on a per tray basis. This arrangement probably reflects the proportion of growers who use the penny contractor arrangement. That is, about 11.7 percent (.127 x .92) of the growers used their crew boss in a penny contractor type of arrangement to recruit workers.

How Workers Find Jobs

While employers recruit workers, workers search for jobs. It is clear from the questions we asked farm workers that the job search among raisin harvest workers relies on an informal network of friends and family members. This is consistent with our previous findings among the general farm worker population in the Central Valley region. In the current study, we found that only 12 percent of workers were recruited by a grower or his foreman, and slightly fewer than 10 percent were recruited by an FLC or his foreman. The majority of the workers interviewed (54 percent) indicated that they found their current job through a personal reference from a friend or relative.

Labor relations are casual, with farm labor contractors or crew bosses often acting as intermediaries between growers and workers. Employment with an individual grower typically does not last for more than two or three days. In our interviews with workers, we found that many workers did not know who the farm owner/operator was at the specific work site. In several instances, workers did not know who the farm labor contractor they were working for was. However, in most instances they could identify by name either the crew boss or foreman.

Slightly more than half of the worker sample stated that one or both of their parents had worked in U.S. agriculture. Since 99 percent of the workers interviewed were born outside of the United States, this suggests that there is an intergenerational migration pattern that relies heavily on friends or family members already in the farm labor force for job information.

This international network, which links rural villages in Mexico with agricultural jobs in California, is clearly an important connection for migrants seeking work. Fifty-three percent of the raisin workers interviewed return to Mexico each year, a way of maintaining the friendship and kinship networks that communicate information about employment opportunities. Seventy-nine percent of the workers who responded to this question indicated that they knew someone in Fresno County who helped them get their job when they migrated from Mexico. Eighty-three percent of the respondents knew they would be working in agriculture, and 65 percent planned to work in the raisin harvest when they left Mexico.

Harvest Mechanization

With a few exceptions, harvesting, tying, and pruning tasks in the production of raisins have not been affected by improved technology or mechanization for the past 50 years. According to University of California Cooperative Extension estimates, approximately 103 hours of labor per acre are required each year to produce raisins (Mamer and Wilkie 1991). Despite the labor-intensive nature of growing raisins, there has been little sustained interest in adopting new technologies that would reduce these labor requirements.

During recent years, there has been experimentation with mechanized harvesting in the region, but such practices are not yet deemed to be cost effective by most growers, whose average raisin grape farm does not exceed 50 acres. Key informants interviewed for this study estimated that a grower should have at least 200 acres of raisin grapes in production for current mechanization technologies to be cost effective. Local growers first began to experiment with mechanized raisin harvest systems approximately 20 years ago. Presently, only a few growers in the region have invested as much as $120,000 to develop or purchase mechanized systems that harvest, lay continuous paper trays on the ground, and later retrieve the raisins.

Five basic systems are currently operational. Two of the systems mechanically harvest the grapes. About one week prior to harvest, fruit canes are cut with hand shears, allowing the grapes to partially dry while still attached to the grape stem. When the mechanical harvester comes in, the grapes fall off the stem as single berries. The berries are then conveyed to continuous paper trays that have been mechanically laid down in the middle of the rows. Neither of these two systems turns or rolls the trays prior to retrieval, but they do utilize mechanical retrievers to pick up the raisins when dried.

The other three systems rely on manual harvesting of the grapes onto continuous paper trays that are laid out by hand in the middle of the grape rows. According to the raisin growers who use the continuous paper trays, the speed of harvesting with this method is approximately 20 percent faster than it is with the traditional individual trays. Next, the continuous paper trays are turned mechanically by a tractor-drawn machine about a week to 10 days after the grapes are first laid on the ground. The final step in the process involves the mechanical retrieval by a tractor-drawn machine of raisins that are sufficiently dry. The raisins are then loaded into bins for transport to the packer.

The key issue in the use of any of the mechanized systems available today for harvesting raisins appears to be cost effectiveness. When we asked our 1991 survey sample of 323 raisin growers whether they would consider using mechanized systems in the event of a labor shortage, only 28 percent answered affirmatively. As one grower we interviewed commented, "It doesn't make any sense that they make so much fuss about these expensive machines when workers do a better job." What this grower really meant was that workers do a better job cheaper.

Raisin Workers

The 125 raisin harvest workers interviewed as part of this study were entirely Hispanic, with 94 percent born in Mexico. Only one worker interviewed was born in the United States and seven workers (5.6 percent) were born in Central America. Of the Mexican workers, 32 percent were from the state of Michoacan and 30 percent were from Guanajuato.

Of the workers interviewed, 92 percent were male and 41 percent were single. The median age of the workers was 28 years, and 93 percent of the sample were citizens of Mexico. The median year that the workers first came to the United States was 1983, but 35 percent indicated that they had come here initially in 1986 or later. About half of the sample planned to remain in the United States permanently, with the remainder planning to return to Mexico or unsure of their plans.

The mean average of experience in farm work in the United States was about nine years, while the median was only six years. Nineteen workers indicated that 1991 was the first year they had done any farm work.

About 38 percent of the farm workers had done some nonfarm work in the United States. Gardening, as well as work in construction, restaurants, and factories, were listed as the most frequent types of this employment.

Formal levels of schooling among this population were low, but consistent with those among farm workers employed in other crops. The average (mean) years of schooling completed in Mexico was six years, and only seven of the workers had completed any school in the United States. Considering that most workers had been in this country for fewer than ten years (often for only part of each year) and had almost no exposure to English in school, it is not surprising that the language spoken by 99 percent of the sample was Spanish. Ninety-four percent of the workers said they were able to speak very little or no English. Eighty-seven percent understood little or no English.

Slightly more than 35 percent of the workers interviewed indicated that they were working in the United States without legal work documents. Forty-two percent had applied for legal status through the SAW program, with the remaining workers legalized through general amnesty or prior to IRCA. Less than 2 percent of the raisin work force were found to be U.S.-born citizens. The industry experts we interviewed estimated that at least one-half of the current work force in the raisin harvest was undocumented or using fraudulent documents, which indicates that the self-reported 35 percent proportion of the sample represents the lower limit of the importance of illegal workers in the raisin harvest.

Seventy percent of the workers interviewed responded that it was not hard to obtain documents to work in the United States. This response was verified by project researchers, who were able to openly purchase a social security card and driver's license for $25 at a local flea market that caters to farm workers.

Employment

For most of the workers, the raisin harvest is not the only source of employment. Seventy-two percent of the workers indicated that they had worked in other crops in California in the past year; 37 percent had worked in agriculture outside of California during the past five years. (These subsets are not mutually exclusive.) The "typical" worker had been employed in one to two other crops during the previous year. The task and crops most often cited were harvesting vegetables, tree fruit, citrus, and olives. Eighty-two percent of the workers indicated that they had worked in only one county in the past year, which suggests that the primary form of migration is from Mexico to one area in California and that raisin workers do not "follow the crops" within the state or the West Coast region.

Despite their pursuit of work in other crops, the raisin workers were unemployed for five to five-and-a-half months during the past year. Forty-eight percent of the workers applied for unemployment insurance during these periods of unemployment. The most frequent reason not to apply for unemployment insurance was that they thought they would not qualify. We also asked the workers if they knew they were covered by unemployment insurance. Thirty-nine percent said they were covered, 41 percent said they weren't covered, and 20 percent did not know their status.

Although growers reported that labor costs have increased in the raisin industry since the implementation of IRCA, these increased costs do not include either higher earnings or increased benefits for workers. Indeed, according to the 125 workers interviewed for this study, benefits for some workers have been reduced since 1985, as have their piece-rate earnings.

In studies of central California agricultural labor (Mason, Alvarado, and Riley 1993; Alvarado, Riley, and Mason 1990, 1992), the point is frequently made that raisin workers seem to fall at the lowest end of the "wage-and-benefit" continuum. We have seen in previous studies that raisin workers report the most meager and fewest of employee benefits, the highest seasonal rates of unemployment, and the lowest overall or annual earnings, compared to workers in other commodities, such as table grapes, citrus, and melons. These previous findings were echoed in the data obtained from workers and growers in this 1991 study, but with greater resonance than before, due to the diversity of data sources and the level of detail allowed by this current study's exclusive focus on the raisin industry.

Worker Supervision and Compensation

A critical factor among an overwhelming majority of the 1991 raisin harvest workers interviewed for this study was their inability to understand English. When asked how well they understood their English-speaking employer, 87 percent responded "not at all" or "very little." As a result, these individuals are dependent on others who are able to communicate in English for most, if not all, of their employment-related transactions. This function is commonly provided by either farm labor contractors or by foremen employed directly by the growers. Because raisin harvest workers usually work independently rather than as part of a crew, once they begin their work routines at the farm site, direct supervision of workers is often lax or nonexistent.

We found that all the raisin harvest workers were paid on a piece-rate basis, averaging $.16 per tray within a range of $.15 to $.17 cents per tray. Workers reported an average of nine hours of work per day, and over the total harvest season they reported earning an average of $6.25 per hour. About 10 percent reported earning less than $4.25 per hour and another 10 percent reported earning over $8.00 per hour. Since the raisin harvest lasts for only three to four weeks, most workers earn less than $1,000 during the harvest season. Worker earnings determined by a piece-rate basis are sensitive to several factors that ultimately affect individual earnings. Among these factors are worker skill, vine and crop conditions, and the piece rate paid by employers. Workers employed by FLCs averaged $.01 less than those employed directly by the grower.

There is a connection between the low prevailing rate of pay among raisin harvest workers and the increasing prevalence of illegal workers. It is not merely that undocumented workers are willing to work for less. Because of their lack of English skills, their inability to arrange for employment directly with the growers, and their need to minimize the possibility that their illegal status be disclosed, most workers with fraudulent documents are forced to seek employment through a middleman.

In some cases, the middleman is a registered farm labor contractor. But in many other cases, the middleman is simply a foreman who works on a salary for the grower and arranges for a harvest crew through personal contacts, often reaching directly into Mexico and small, rural feeder villages there. Whether working for a registered farm labor contractor, an unregistered farm labor contractor, or a foreman functioning as a farm labor contractor, the raisin harvest worker must sacrifice in earnings the difference between what the grower pays the contractor and what the contractor pays the worker. This difference is often as much as $.02 or $.03 cents per tray.

In order for a worker to earn the minimum wage of $4.25 per hour, it is necessary to harvest 27 trays hourly at $.16 per tray. On numerous occasions, workers reported harvesting in excess of 400 trays per day, a number that is difficult to attain, according to experienced workers and supervisors. During a 10-hour work day, a worker would need to harvest an average of 40 trays per hour. However, if the average wage reported by workers was $6.25 per hour and the average rate per tray was $.16, and workers average nine hours of work per day, then the average number of trays was about 350. Using these averages, the daily pay would be about $56.

In the 1991 raisin harvest, EDD staff conducted tray counts at selected farm sites in Fresno County where some workers harvested 50 trays per hour (interview with Ms. Socorro Davila, EDD, Sacramento, Calif., March 17, 1992). However, these observations were made in the early hours of the work day before the summer heat exceeded 80° F. It is doubtful that a worker could maintain such a frenzied pace through the entire day. In instances where workers reported such high piece-rate production, we observed trays that were only half full. A full tray of freshly harvested raisin grapes averages about 22 pounds. If an employer or supervisor will accept trays weighing 15 pounds or less, the workers are then able to increase their average hourly wage by picking the same amount of fruit at the average piece rate for 400 half-trays as those who are compensated at the same rate for 270 full trays.

Rising Labor Costs and Declining Worker Earnings

As noted, there is a clear trend toward the use of FLCs by employers for the harvest of raisin grapes. We found that 40 percent of those interviewed were working for an FLC at the time this study was conducted. (Because of difficulties in locating FLCs and their employees, this 40 percent figure should not be considered representative of the importance of FLCs in the raisin harvest.) Although growers reported that labor costs were increasing, the increases were not reflected in higher wages to workers, but rather in commissions and fees paid to the FLCs. The increases in labor costs reported by employers were also probably due to increased payroll taxes, social security insurance, workers' compensation insurance, and other mandated fees. When asked whether the piece rate for the raisin harvest had changed during the past five years, 62 percent of the workers indicated that the rates had not increased or decreased. Nearly all of the employers we surveyed indicated that labor costs had increased during the same period. They attributed the increase to a combination of factors, including piece-rate hikes, commissions or fees paid to farm labor contractors, and the increase in the minimum wage.

Industry leaders we interviewed agreed that there has been a slight increase in the piece rates for the raisin harvest, but that such increases have been minimal and not commensurate with increases in the cost of living. In fact, the prevailing piece rate reported by most sources — workers, employers, EDD — was the same in 1986 as what was reported in 1991. This suggests that most workers employed in the raisin harvest have experienced a real decline in wages since the passage of IRCA in 1986.

Housing Conditions

The lack of adequate housing is one of the most critical problems experienced by farm workers in the county. Scarcity of housing is even more acute during the raisin harvest, with a large number of workers coming into the region for a relatively brief period of time. Such housing often consists of cramped, substandard structures lacking basic necessities.

According to the grower survey, about 17 percent of growers provided housing opportunities for their raisin harvest crews in both 1985 and 1991. Although these percentages coincided, further analysis of the data suggested that there has been about a 25 percent turnover in growers providing housing between 1985 and 1991. That is, about 25 percent of the raisin farmers who provided housing in 1985 had discontinued that practice by 1991. When asked why they discontinued making farm labor camp housing available to their workers, 40.7 percent said that the laws and regulations had become too restrictive. Others noted the poor condition of existing housing and the high maintenance expenses.

In our 1989 survey of farm workers in central California, we found that 85 percent of the general farm worker population was living in rented nonemployer-provided housing, mostly in single-family dwellings (Alvarado, Riley, and Mason 1990). Among our 1991 raisin worker sample, 16 percent indicated they were renting from their raisin employer and living on the farm premises (Alvarado, Mason, and Riley

1992). Almost 38 percent (37.6 percent) reported living in boarding houses or labor camps in this study. Slightly more than 6 percent (6.4 percent) were found to be homeless, sometimes living in vehicles. Seven percent lived in apartments, and nearly 9 percent lived in trailer homes. When asked if housing was an important factor in deciding which raisin employers to work for, 32 percent said yes.

In conducting the field research for this study, researchers observed a variety of housing situations for raisin workers. The most prevalent was a number of families occupying what appeared to be substandard single-family houses and apartment suites. In many instances, this housing was occupied by several single men. The worst conditions observed included workers sleeping in cars, under trees, and under a bridge. The best housing situations were often employer-provided housing or housing provided and subsidized by public agencies.

Overall, housing conditions for raisin harvest workers are among the most deplorable of all farm workers, and the prospects appear dim that either governmental agencies or employers will address the problem. Employers repeatedly expressed their frustration with stringent government regulations that serve to discourage them from providing housing for their workers, often citing experiences where compliance with building codes for farm worker housing would exceed building standards for their own homes.

IRCA's Effects

Our observations and the data collected for this study showed that the direct effects of IRCA have been minimal. We found no serious disagreement with the notion that there exists an overabundant supply of farm workers in the region's raisin grape harvest. Employers, state employment officials, industry experts, and grower representatives all agreed that IRCA did not diminish the size of the labor force for the raisin grape harvest. On the contrary, most agreed that since the enactment of the law, and even more so after 1988, an oversupply of workers was evident.

This increase in the size of the labor force can be attributed to numerous factors. First, more workers were legalized as SAWs than anticipated. Second, others, albeit a smaller number, sought legalization under IRCA's amnesty provision. And finally, the number of workers joining the labor force after 1986, and hence not qualifying for legal status under this law, continues to grow. Indeed, our data show that those in the latter group now constitute the majority of the raisin grape harvest workers in the region.

Undoubtedly, IRCA served to stabilize a previously existing undocumented agricultural labor force: legalized workers who qualified under the SAW or amnesty provisions, and who were routinely and frequently apprehended and deported by the border patrol. Our findings show that, until now, there has been minimal leakage among these previously undocumented workers out of agriculture and into other labor markets. Moreover, the undocumented labor force is growing because of various shifts in policy and enforcement of immigration law. We refer here to reductions in the number of border patrol agents in the region, the termination of "field inspections" by the border patrol, and the nonenforcement of employer sanctions for hiring undocumented workers. Essentially, once an undocumented worker crosses the United States–Mexican border and arrives in California's San Joaquin Valley, there is little likelihood he or she will be apprehended and deported. The net result is obvious: a larger labor pool.

All other effects seem to be secondary. Increases in the labor force during the raisin grape harvest place greater stress on the availability of adequate housing in an already depleted farm worker housing market. Changes in wages and working conditions are similarly affected, since improvements have been negligible for about a decade. Reported increased labor costs by employers generally do not translate into increased earnings for the farm workers; rather, such costs seem to be siphoned off by farm labor contractors. Yet, in most instances, we found that farm labor contractors provide a valuable service to both employer and worker. Employers evidently consider the increased costs associated with using a contractor to be a fair exchange for the decreased paper work and for contractors to function as a buffer between the grower and the border patrol and other regulatory government agents. On the other hand, raisin workers are generally able to obtain a greater number of work days when they work for farm labor contractors than when they work directly for the growers.

Conclusions

The direct impacts of federal immigration legislation on the raisin industry in central California appear to be minimal. Farm workers who qualified for legal status under the SAW provisions of the 1986 law have benefited largely from the ease with which they can travel to and from Mexico during periods of employment. The increasing number of undocumented or fraudulently documented workers employed in the harvest of this crop indicates that employer sanctions intended to remove such persons from the work force have been ineffective. Here, it is apparent that there has been little enforcement of the law by the border patrol.

The increased use of farm labor contractors during the past decade is frequently attributed by employers to greater scrutiny by governmental regulatory agencies of agricultural operations in general and to labor-intensive crop commodities specifically. In addition, the seasonality of the harvest, the reliance by farm workers on others for transportation, and workers' relative unfamiliarity with specific employers are all part of a tenuous work environment. For these reasons, workers prefer to work directly for employers rather than for FLCs, if given a choice. And even though nearly 80 percent of the workers in our sample indicated that they earned lower piece rates while working for FLCs, they still were more likely to maximize their seasonal earnings by working for contractors who were able to place the workers in more jobs.

Labor-organizing activities during the past decade have been almost nonexisten, and there is little evidence that efforts to unionize raisin harvest workers have ever been successful. The overabundant labor pool makes it even more difficult for union organizers to win labor contracts and obtain better wages and fringe benefits for workers.

The 1986 IRCA legislation has had some effect on the raisin labor force, particularly on those workers legalized through the SAW provision of the law. These newly legalized workers are now able to travel to and from Mexico without the risks associated with illegal entry into this country. During the off-season winter months, thousands of recently legalized SAW workers return to Mexico for two or three months until the spring employment surge begins. The border patrol no longer conducts field inspections or raids in the work place, a practice contemptuous to both employers and workers. On the other hand, illegal entry into the region continues unabated; upon arrival in Fresno County, undocumented raisin harvest workers are able to obtain employment without much difficulty. Fraudulent documents, readily and inexpensively obtainable, are routinely presented by workers to employers as evidence of their legal status in this country.

Despite their legal right to work in the United States, workers legalized under IRCA's special agricultural provisions in general do not seem to be exiting from farm work. With limited English-speaking skills, limited nonfarm employment experience, and few personal contacts in the urban areas, there appears to be minimal movement from farm employment to urban jobs within this group. Participation by SAWs in the raisin harvest does, however, appear to be decreasing, probably due to their ability to move more freely within agriculture and seek the more desirable jobs among the various crops. The combination of the workers' exit from the raisin harvest and a continuous influx of undocumented workers creates an increasing reliance on the latter by employers.

It is likely that the raisin industry will continue to depend on immigrant workers in the future. To the chagrin of the farm workers, the buffer provided by FLCs between growers and workers increasingly dominates hiring practices. Raisin acreage in the region remains stable, with most of the employers growing other crops along with raisins and grapes. There is no indication that raisin growers have reduced their raisin acreage due to labor shortages or that IRCA has had any effects on prevailing wages or personnel practices in the industry.

Abundant labor supplies since 1986 have allowed the raisin harvest to continue unfettered and with little concern about future labor needs. Yet, this solution is unsatisfactory to most involved. Growers and industry leaders readily admit their continuing dependence on undocumented workers, and are keenly aware that their crops would be jeopardized if the replacement workers stopped crossing the border. These abundant supplies of labor have also thwarted most of the salutary effects anticipated by IRCA proponents. The work force has not been stabilized, employment opportunities have not expanded for legalized workers, and the economic position of farm workers has not improved. While the current situation is workable, in the sense that crops are being harvested and workers are finding jobs, it is an uneasy solution that satisfies neither the growers' desire to have a stable and predictable work force nor the goals of those seeking to improve the economic lot of farm workers.

In most instances, growers continue to operate with little concern about the supply of labor or field inspections by immigration authorities. Economic incentives for workers from Mexico to migrate to California to work in the raisin harvest remain strong. For example, the daily minimum wage of farm workers from one primary feeder region in Mexico is about one-tenth of the minimum earnings among raisin harvest workers in central California. Thus, with lax screening by employers and farm labor contractors for workers' legal status, the ability of newly arrived workers to find and perform the harvest tasks, the sudden demand for a large labor force, and the absence of the border patrol at the farm sites make this crop a logical point of entry into farm labor by illegal and inexperienced workers. Worker turnover in raisins is among the highest of all crops in the region, with as many as one-third of the workers reporting that they would not work in the raisin harvest during the next season. In comparison to other crops in the region, raisin harvest workers report the lowest annual earnings, the shortest work seasons, and the fewest work-related benefits.

NOTES

1. Martina Acevedo, research assistant, provided valuable assistance in interviewing farm workers and collecting other data. Research support was provided by the Commission on Agricultural Workers and the California Employment Development Department.

20

Immigration Reform and Florida's Nursery Industry

JOHN J. HAYDU AND ALAN W. HODGES

Introduction

The purpose of this chapter is to assess the effects of the Immigration Reform and Control Act of 1986 (IRCA) on the supply and availability of labor for Florida's ornamental nursery industry. Like many horticultural commodity sectors, the nursery industry relies heavily on hand labor to perform numerous tasks in the production, care, assembly, and distribution of its products. However, a continuing problem facing researchers who examine the nursery industry is the lack of consistent and reliable time-series data describing key aspects of the production and marketing process. This problem is national in scope and is due in part to two factors. First, the nursery industry has not been included in state and federal data collection efforts, as have some of the more traditional commodities, such as citrus, wheat, cotton, corn, and soybeans. Second, the diversity and number of nursery crops produced annually runs into the thousands. Consequently, to address researchable topics in a meaningful way, investigators must often rely on periodic and costly primary data collection studies. Addressing the impact of IRCA on the supply and availability of labor in Florida illustrates this point.

This chapter utilizes three separate studies to evaluate the labor supply issue. First, a wage-and-benefits study conducted for the Florida Foliage Association in 1987 is summarized. The major purpose is to provide a benchmark from which comparisons can be made with the other two sources. A second source of labor information was derived from the Nursery Business Analysis Program (NBA) at the University of Florida. This program collects detailed financial information annually from a sample of nursery firms. From this data base, labor information was extracted for the period 1984–91. Finally, a third source of data was generated by replicating key portions of a labor study conducted by Polopolous, Moon, and Chunkasut in 1988 (1990).

Questions addressing labor availability and the impact of IRCA were repeated to a sample of 54 nursery firms. Results were contrasted directly with the 1988 study.

Background on Florida's Nursery Industry

Nursery and greenhouse crops represent the seventh largest agricultural industry in the United States, with a total value of $8.4 billion at the wholesale level in 1991 (USDA/ERS 1993). Florida was the second leading state in terms of production, with an industry value of $957.8 million in 1988. This compared with California's $2.15 billion in sales for the same year. Florida's nursery industry grew rapidly throughout the 1970s and early 1980s. With the eventual maturation of the industry came the attendant problems of overproduction, depressed prices, low profitability, and a growing rate of business failure (Haydu 1989; Hodges and Haydu 1989; Strain and Hodges 1989). Table 20.1 illustrates this maturation process for one industry sector.[1] Based on first-hand knowledge, the authors believe that similar changes have occurred in other nursery sectors, as well. Declining numbers of firms (down 25 percent), increased production area on a per firm basis (up 19 percent), and higher unit sales (up 25 percent) represent responses to increasingly competitive market forces. A general erosion in the financial health of many nurseries also occurred, as is evident from results of the University of Florida's NBA. Three performance measures — productivity, efficiency, and profitability — indicated that in general nurseries fared best during 1986 and 1987, with financial performance declining thereafter. The last year examined, 1990, appeared to be the most difficult for the sample of firms. The rate of return to capital for foliage producers, for instance, fell from nearly 20 percent in 1984 to less than 6 percent by 1989. Net nursery income dropped by nearly two-thirds for the same period, from $110,800 to $41,700 (Haydu 1991).

Reasons for deteriorating financial performance are not limited to falling demand or the recent economic downturn. Industrywide overproduction has also played a crucial role. Lack of economies of scale in production and relatively low capital requirements due to Florida's warm winter climate pose very low barriers to the entry of new firms. In fact, for the industry overall, the population of registered nursery firms has grown more rapidly than aggregate sales, resulting in widespread sales of products below costs (Hodges and

Table 20.1. Changes in Number of Firms, Production Area, Net Value of
Sales, Net Sales Per Firm, and Production Area Per Firm, for Florida
Foliage Growers, 1984–91[a]

Year	Number of Firms[b]	Production Area (ft²) (million)	Net Sales Value (million)	Average Prod Area (Ft²)	Average Net Sales
1991	736	174.8	$269	2,370	$3,600
1990	816	169.0	$280	2,070	$3,400
1988	920	179.9	$279	1,950	$3,000
1986	967	193.3	$309	1,990	$3,100
1984	985	192.3	$269	1,940	$2,700

[a] Source: Florida Department of Agriculture, Division of Marketing
Florida Crop & Livestock Reporting Service, Orlando.
[b] Commercial-sized businesses with a net value of sales exceeding
$10,000.

Haydu 1989; USDA/ERS 1991). This oversupply problem has been
exacerbated with the advent of tissue culture. This propagative technology allows many marginal businesses to produce large numbers
of uniform plants quickly. Prior to tissue culture, propagation was a
natural filtering device eliminating many less capable firms.

Recently, three separate disasters have impacted Florida's nursery industry considerably, disrupting product supply and perhaps
labor availability. First, in 1990 Du Pont's popular fungicide, Benlate,
caused widespread damage to the state's vegetable, strawberry, and
foliage plants. Major economic losses were incurred and, as of May

1992, Du Pont has voluntarily paid $330 million in damages, most of it to Florida growers (Jackson 1992). Second, in March 1992 a severe hailstorm struck central Florida, causing $30 million of property and plant damage to foliage growers (Jackson and Pankowski 1992). Finally, in August 1992 hurricane Andrew caused an estimated $329 million in damage to south Florida nurseries (Hodges 1992). Because the scope and severity of destruction was so pronounced, rebuilding efforts may continue for several years. Currently this "economic vacuum" is drawing large numbers of both skilled and unskilled labor into the area. This locational shift in labor is expected to affect the nursery industry, as well as many other labor markets.

Labor in Florida's Nursery Industry

Summary of 1987 Wage-and-Benefits Study

A wage-and-benefits survey of Florida Foliage Association (FFA) members was completed in July 1987 (Brown 1987). The purpose of the survey was to inform members of current wage-and-benefits practices within the industry. Approximately 450 questionnaires were mailed to FFA members. The survey included a description of 18 common jobs typically found in nurseries, as well as information regarding wage rates paid and data on current fringe benefits provided to employees.

The survey revealed a wide range of wage-and-benefits practices among nursery operators. Some nurseries appeared progressive and innovative; others were new and just beginning to formulate their compensation policies. The study indicated that owner/managers were strongly concerned with the impact of wages on profitability. Still, many employers claimed to use compensation as a tool to provide quality products and to retain an effective work force. The survey disclosed that the average wages actually paid for similar positions varied widely by geographic area. Southeast Florida typically paid about 10 to 16 percent more for nursery workers than employers in central Florida and 15 to 20 percent more than employers in the western part of the state.

The survey provided useful data on wage classification structures. For example, in some nurseries workers started at $3.35 per hour (the minimum wage at the time), while others in the same geographic area began workers at $4.40 per hour. Maximum rates paid for similar jobs also varied widely. For instance, maximum shipping worker wages varied from a low of $3.80 an hour to a high of $9.03 an hour. These variables could reflect cost-of-living differentials across areas, the tightness of the local labor markets at the time, the number of long-term employees, turnover rates, or unique aspects of job assignments.

No consistent pattern emerged indicating that business size had any strong influence on wage patterns. In fact, in many cases, top pay was found in smaller units. This may suggest more careful compensation practices in large units or the possibility of broader and more varied responsibilities in smaller units. The survey also found that some employers paid below the minimum wage. While this may have been due to special circumstances or to clerical errors in completing the survey form, it warrants attention. A review of turnover statistics suggests that some nurseries experienced significant turnover (in excess of 200% annually). The higher turnover appeared to be in the medium- or large-sized organizations. Pay levels may have been a factor in employee retention. Some nurseries adopted practices for retaining long-term employees, such as special raises, one-time bonuses, profit sharing, service awards, added health benefits, and vacation trips.

The 1987 wage-and-benefits study provides a rough picture of some labor issues for an important sector of Florida's nursery industry. Labor represents the single greatest cost for most nurseries. How this cost was dealt with varied greatly from firm to firm, based on wage differentials, incentive programs, and the strong concern of managers with the impact of wages on profitability. From this information we conclude that, for some nursery sectors, IRCA legislation *could* have posed a significant problem. How it eventually impacted the industry is discussed in the two sections that follow.

Labor Trends from Nursery Business Analysis

Labor trends in Florida's nursery industry were evaluated with data from the University of Florida's NBA. This ongoing program gathers confidential business records from wholesale ornamental nursery firms in order to develop industry standards for business performance and to analyze individual firm performance. A total of 707 records were collected from 234 firms from 1984 through 1991. This timeframe allowed for a comparison of business performance before and after the implementation of IRCA. Because participating firms were accepted on a voluntary basis, the data set was *not* a statistically designed sample of firms, although it is probably representative of firms with above-average management quality, by virtue of their willingness to participate in such quality improvement programs.

Data were separately compiled and analyzed for firms classified in five different sectors of the ornamental nursery industry: containerized woody ornamental plants, field-grown woody ornamental plants, flowering plants, and tropical foliage plants in subregions of central and south Florida. Data included 252 records for container firms, 60 for field firms, 163 for central Florida foliage firms, 199 for south Florida foliage firms, and 33 for flowering plant firms. Within each industry grouping, data were also analyzed for highly profitable firms — those with profit margins of more than 30 percent. Results were weighted by firm size in each group in order to represent the greater impact of large firms. Dollar values for these time-series data were adjusted for inflation by deflating with the Consumer Price Index (base period of 1982–84).

Labor productivity reflects the changing economies of labor for the nursery industry. Productivity was measured in terms of value produced per full-time equivalent worker (2,080 hours per year). Total annual production of firms was measured as annual sales plus change in plant inventory value. Data collected on total employee hours worked were expressed as percentages of full-time equivalents.

Figure 20.1 shows that labor productivity generally declined over the 1984–91 period for all firms and for profitable firms. For all firms, average labor productivity declined from $60,000 per FTE in 1984 to $46,000 per FTE in 1989, then rebounded to $48,000 per FTE in 1991. Overall, this represented a 20.1 percent decrease in labor productivity. Profitable firms had a generally higher level of labor productivity and showed a similar steady decline from $80,000 per FTE in 1984 to $55,000 per FTE in 1991. Among industry groups there were substantial differences in overall levels of labor productivity; however, only flowering plant nursery firms showed a different trend, with value produced per FTE increasing between 1987 and 1991. The rate of decline in labor productivity after IRCA (1988–91) did not appear to differ from the rate prior to IRCA (1984–87). Therefore, we conclude that IRCA probably did not influence the trend already underway. The decline in labor productivity is attributed to increasing competition leading to lower product prices in the industry.

Figure 20.1. Labor Productivity Per FTE (Including Production, Sales, Administrative, and Management Personnel)

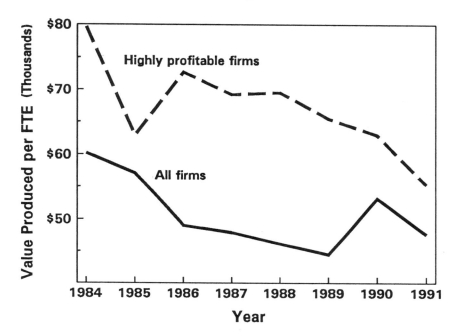

Capital-labor intensity is an indicator of the balance of labor and capital resources utilized. It was measured as the ratio of total managed capital (owned plus leased) to employment (FTE). Assets and liabilities were evaluated at the book value stated on company financial statements, including investments in land, buildings, improvements, equipment, and plant inventories. Leased capital assets in land, buildings, and equipment were evaluated at current market value.

As shown in Figure 20.2, capital-labor intensity varied irregularly from year to year, which may reflect the changing sample of firms. All firms and profitable firms had similar levels for most years of the study period. For all firms, capital managed per employee increased slightly, from $74,000 in 1984 to $80,000 in 1987, then declined to $62,000 in 1990 before rebounding to $83,000 in 1991. Profitable firms also showed a significant rebound in capital managed per employee in 1991, following a period of general decline during 1988 through 1990. The pattern of trends was apparent for all industry groups. Field nurseries had higher levels of capital managed per FTE,

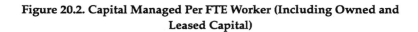

**Figure 20.2. Capital Managed Per FTE Worker (Including Owned and
Leased Capital)**

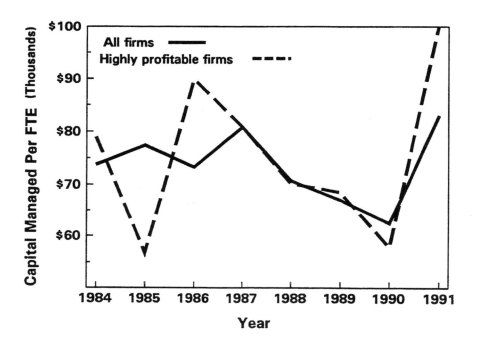

reflecting their greater mechanization and lower labor intensity with
respect to capital as well as production space.

  Labor costs are perhaps the most important indicator of trends in
labor economics. In this analysis, labor costs included all wages, salaries, payroll taxes, and fringe benefits for all employees. To remove
the effects of varying firm sizes, labor costs were evaluated as a share
of value produced. As shown in Figure 20.3, labor costs generally
followed an increasing trend from 1984 through 1991. Labor costs for
all firms increased by 39 percent during this period, from 25 percent
of value produced in 1984 to 35 percent in 1991. Profitable firms had
substantially lower labor costs overall, but had even more markedly
increased costs, from 17 percent of value produced in 1984 to 29 percent in 1991. In contrast to the general trends, flowering plant nurser-

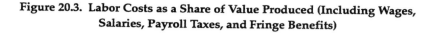

Figure 20.3. Labor Costs as a Share of Value Produced (Including Wages, Salaries, Payroll Taxes, and Fringe Benefits)

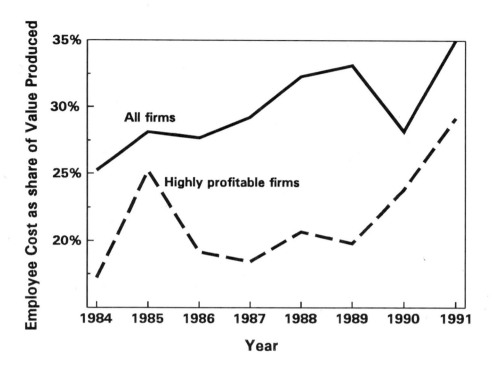

ies showed a decrease in labor costs from 42 percent of value produced in 1984 to 29 percent in 1991, reflecting their improved productivity.

The rate of increase in labor costs was not discernibly different in the pre-IRCA and post-IRCA periods (1984–87, 1988–91, respectively). This does not support the hypothesis that immigration reform caused a tightening of labor supply that would increase labor costs. However, neither does this rule out the possibility that changes in labor supply brought about by change in immigration policy reinforced labor market trends that were already underway.

From the three indicators examined (labor productivity, capital-labor intensity, and labor costs), there is no strong evidence to conclude that IRCA legislation acted as an impediment to the supply and availability of labor to the nursery industry. At the same time, neither do the data indicate what measures or actions employers may have taken to circumvent or blunt the impact of this legislation.

In the spring and summer of 1988, a study was conducted for the purpose of determining wage rates to seasonal workers, prospective labor supplies, and the likely impact of IRCA on the availability of seasonal workers (Polopolous, Moon, and Chunkasut 1990). Although the full impact of the immigration legislation may not have been felt in 1987, the information obtained provides a solid benchmark for labor conditions and producer concerns at the time. With this in mind, we selected several key issues from the 1987–88 study and examined them again for the 1992 calendar year. Major questions asked include: (1) the number of firms hiring temporary or seasonal workers directly or through contractors, (2) average hourly wage rates for selected tasks, (3) attitudes about supply of seasonal workers in peak week, and (4) producers' views regarding the impact of IRCA on labor supply.

Nursery Labor — 1988 versus 1993 Studies

The sample of respondents for the 1993 survey was obtained from trade association member lists for foliage, woody ornamental, and cut fern growers. Fern growers were chosen randomly by selecting every n^{th} producer. Foliage and woody growers were selected semirandomly, in that the sample focused on larger firms. The rationale for this approach was that more information can generally be obtained from larger businesses and larger businesses also comprise the greater share of the industry in volume of sales. A survey instrument was developed based on questions from the 1987–88 study. Due to the small amount of time available, a decision was made to conduct the survey by phone, utilizing a professional market survey organization. From a target population of 183 potential respondents, 54 completed questionnaires were obtained, representing a 30 percent response rate (Table 20.2).

Table 20.2. Number of Respondents Using Temporary or Seasonal Labor,
by Nursery Sector, Florida, 1988 and 1993

Year	Foliage	Fern	Woody	Total
		——————Nursery Sector——————		
		——————Number——————		
1988*	71	17	30	118
1993	11	19	24	54

*From Polopolous, Moon, and Chunkasut 1990.

Employment and Wage-Rate Information

In contrast to fruit and vegetable industries, year-round employment is much more common for ornamental producers, except ferneries (Polopolous, Moon, and Chunkasut 1990). In 1992, roughly 60 percent of the ornamental nurseries interviewed claimed to utilize direct hire only (Table 20.3). This represents a considerable decline since 1987 when 94 percent reported hiring labor directly. Similarly, the number of respondents claiming to hire indirectly through labor contractors increased from 3 to 13 percent, and those indicating some form of both methods grew from 3 percent to 28 percent between 1987 and 1992. In other words, based on results of these two studies, there was a discernible shift toward *greater* use of indirect labor. How much of this is due to IRCA and how much can be attributed to other factors is unclear. For instance, one response to increased competition and narrower profit margins, which has been a common experience for Florida ornamental producers, is to reduce the permanent labor force. This would be particularly true in locations where seasonal labor is readily available.

Average hourly wage rates were also examined to determine whether changes occurred during the past five years. One might expect, for example, that if the impact of IRCA *did* reduce the supply and availability of labor for the nursery industry, wage rates would increase to offset this trend. Three types of hand skills were identified for wage rate estimation: potting/planting, hand harvesting, and packing/shipping (Table 20.4). Although the latter activity was not

Table 20.3. Distribution of Firms Hiring Temporary or Seasonal Workers, either Directly or through Labor Contractors

			—————Employment Category—————		
				Hired &	
Year	Industry	Hired	Contract	Contract	Total
			————————percent————————		
1988[a]	Foliage	88	6	6	100
	Fern	100	0	0	100
	Woody	94	4	2	100
	Total	**94**	**3**	**3**	**100**
1993	Foliage	64	0	36	100
	Fern	58	31	11	100
	Woody	56	8	36	100
	Total	**59**	**13**	**28**	**100**

[a] From Polopolous, et al, 1990.

Table 20.4. Average Hourly Wage Rates for Selected Tasks

Labor Activity	——————Industry——————			
	Foliage	Fern	Woody	Average
Pot/Planting	——————Dollars/hr.——————			
1993	4.85	3.72	4.42	4.39
1987/88[*]	4.29	4.83	4.52	4.54
Harvesting				
1993	4.69	5.42	7.35	6.18
1987/88[*]	4.59	5.50	4.55	4.88
Packing/Shipping				
1993	5.29	5.67	8.15	6.40
1987/88[*]	N/A	N/A	N/A	——
Machine Operation				
1993	5.08	4.88	6.97	6.07
1987/88[*]	5.80	6.01	5.64	5.81
Permanent Field Workers				
1993	5.53	5.68	5.45	5.55
1987/88[*]	4.92	5.16	4.82	4.96

[*] From Polopolous et al, 1990

examined in the 1987 study, it was included for 1993 because it represents an essential task and can be used as a benchmark for future studies. Two additional wage categories were also compared with the 1987 study: machine operation and average hourly wages for permanent employees.

With the exception of potting and planting, only modest wage increases occurred over the five years for the tasks identified. Wage increases were found for hand harvesting (20%), machine operation (5%), and permanent employees (10%). For potting activities, the 1993 survey indicates that hourly wage rates declined in both the fern and woody sectors, whereas they increased for foliage. The drop in wages was considerable in the fern industry, nearly 25 percent. Interestingly, for 1987, the fern industry reported the highest wages paid for all activities, including for machine operation and permanent employees. Why this trend reversed is unclear. However, it should be noted that the fern industry is typically the most likely sector to utilize seasonal labor.

Labor Supplies and the Impact of IRCA

Questions were asked about employer attitudes concerning the supply of seasonal workers during the peak week (Table 20.5). In 1987-88, 74 percent of respondents deemed the supply adequate and 8 percent abundant. Conversely, nearly 20 percent believed seasonal labor supplies were inadequate. Of the three producer groups, fern growers were the *least* likely to claim that labor supplies were insufficient. By 1993 attitudes had shifted discernibly. For the present study, nearly 90 percent of respondents believed supplies of seasonal labor were either adequate or abundant. Similarly, the number claiming they were inadequate dropped from 18 to 11 percent.

Finally, changes can also be seen regarding attitudes on the impact of IRCA on labor availability (Table 20.6). During 1987–88, 53 percent of respondents felt that "immigration laws" affected labor supplies adversely. Fern growers were most likely (70% affirmed) to make this claim. In contrast to this, five years later only 40 percent of respondents believed immigration hurt labor supplies. However, fern growers as a group remained unchanged in their views regarding IRCA, that is, 69 percent believed it was a problem.

Table 20.5. Attitudes about Supply of Seasonal Workers in Peak Week

Year	Industry	Response Too Few	Adequate	Abundant	Total
1993	Foliage	1	4	4	9
	Fern	3	8	1	12
	Woody	0	10	5	15
	Total	4	22	10	36
Column Percent		11%	61%	28%	100%
1988[a]	Foliage	11	46	4	61
	Fern	4	11	2	17
	Woody	15	66	7	88
	Total	30	123	13	166
Column Percent		18%	74%	8%	100%

[a] From Polopolous et al, 1990

Table 20.6. Have Immigration Laws Affected Your Labor Supply?

Industry	1993 Yes	No	Total	1987/88[a] Yes	No	Total
Foliage	2	6	8	32	32	64
Fern	9	4	13	12	5	17
Woody	3	11	14	46	42	88
Total	14	21	35	90	79	169
	40%	60%	100%	53%	47%	100%

[a] From Polopolous et al, 1990

Conclusions

Three separate methods were used to assess the effects of IRCA on the supply and availability of labor on Florida's nursery industry. Basically, all three led us to the same conclusion: that the IRCA legislation has had little or no impact on ornamental nurseries. In the 1987 wage-and-benefits study, considerable variability in wages was found for similar types of activities. Some employers appeared concerned over employee retention, while others did not. One section of the study mentioned that some nurseries experienced significant turnover rates, suggesting low pay and relatively easy access to labor.

For the nursery business analysis, three indicators — labor productivity, capital-labor intensity, and employee labor costs — were examined for the 1984–91 period. If immigration reform had restricted labor supplies significantly, compensating actions by nurseries would be expected, such as a substitution of capital for labor, higher labor costs, and greater productivity per employee. In fact, no such evidence could be found from the sample of nurseries examined, indicating that IRCA was not an impediment for the industry during this period.

Finally, results of the labor study completed in March 1993 and compared to a similar study completed in 1988, gave no indication of labor supply restrictions. In fact, perhaps the most conclusive evidence was in the assertions of respondents: (1) the number who claimed that supplies were adequate or abundant increased from 82 percent in 1988 to 89 percent in 1993, and likewise, (2) the number who stated that immigration laws *did not* adversely affect labor supplies grew from 47 to 60 percent, a substantial increase.

NOTES

1. The Florida Agricultural Statistics Service collects limited information for foliage, cut greens, potted flowering plants, bedding plants, cut flowers, and flowering hanging baskets. Of these, the foliage sector comprises the most complete data set. Information for woody ornamentals is excluded. However, from a study conducted by the authors in 1989, sales from a sample of 104 wholesale woody ornamental nurseries had an estimated value of $125 million (Hodges and Haydu 1990).

Part IV. IRCA and Agriculture: The View from 2000

The gap between the goals and outcomes of United States immigration reforms has been greater in agriculture than in other sectors. Since the unanticipated consequences of IRCA seem in the early 1990s to be more important than its anticipated effects, what are the effects likely to be in 2000 of the mid-1980s efforts to legalize the farm work force and to stop the entry of unauthorized workers?

Sumner (Chapter 21) asserts that the supply of labor to agriculture is quite elastic, and that, while IRCA did not alter the nature of this supply, it may have shifted the supply outward. Sumner believes that, because of IRCA, farmers want to pay a lower wage to farm workers, apparently to pass onto them any costs associated with compliance. He believes that lower farm wages were a predictable effect of IRCA, largely because IRCA should not have been expected to reduce illegal immigration.

Is there a surplus of farm workers? Even though IRCA may have increased the supply of farm workers, Sumner argues that, because average hourly farm earnings continue to be above the minimum hourly wage, there is no farm labor surplus. Furthermore, more farm labor regulations, he believes, would only further depress farm wages.

Sumner concludes that more data and data analysis are necessary to understand what is happening in the farm labor market. In order to provide more useful analysis, Sumner notes that it is important to have clear objectives and the data to test hypotheses that flow from them.

Huffman (Chapter 22) also begins with an I-told-you-so perspective: With Mexicans able to earn in one hour in the United States the equivalent of a day's wages at home, and with technology making it easier to counterfeit documents, he argues that it was naive to expect IRCA to stop illegal immigration. Huffman notes four major effects of IRCA on agriculture:

- Legalization increased the supply of unskilled labor in the U.S. in two ways. Fraud in the SAW program made workers not previously living in the U.S. authorized as U.S. workers, and the families of SAW workers began to arrive illegally.

- Low levels of schooling and inadequate English "trapped" many of these SAWs and their families in seasonal farm labor markets.

- The consequent oversupply of farm workers reduced real farm wages, which began falling in 1981, rose slightly between 1985 and 1988, and then fell.

- Readily available workers and lower real wages increased the competitive advantage of U.S. farmers in supplying the increased quantities of fruits and vegetables that U.S. consumers are demanding.

Huffman examines trends in fruit and vegetable consumption and notes that, even though U.S. demand is shifting toward relatively easy-to-mechanize vegetables, falling real wages may retard mechanization in the 1990s.

Heppel (Chapter 23) notes that the farm labor problem — defined by Heppel as low wages and inadequate working and living conditions for farm workers — has been bemoaned in federal government reports for most of the twentieth century, but farm workers remain "the poorest of the working poor" in the 1990s. IRCA failed to unleash the farm labor market reforms that some hoped for: higher wages for legal workers in an agriculture that treats seasonal labor as a scarce commodity.

What is to be done? Heppel mentions farm worker problems that range from low earnings to inadequate housing, and calls for more labor law protections and better enforcement of them to begin the transformation of the farm labor market that IRCA failed to set in motion. She notes that it may be possible for the United States and Mexico to cooperate on a binational labor policy that recognizes current labor market links, as was done under the *bracero* program. She also calls for a debate over whether cheap farm work is really needed to provide Americans with inexpensive food.

Villarejo (Chapter 24) notes that IRCA reversed the trend that began in the 1960s to eliminate special consideration of agriculture under immigration and labor laws. However, IRCA so clearly failed to have its intended effect of slowing illegal immigration that Villarejo believes IRCA may mark the end of agricultural exceptionalism in labor and immigration matters. Villarejo ticks off the reasons why IRCA may represent a watershed in agricultural exceptionalism:

- Agriculture is losing its political clout, while politicians sensitive to farm worker concerns are gaining power.

- The hired farm work force in California is more foreign than ever before, and there are few prospects for inducing Americans back into the fields.

Labor-intensive agriculture in the twenty-first century should, according to Villarejo, be larger and more export oriented, but just as dependent on immigrant workers. Villarejo thinks it unlikely that the United States will be able to control the entry of unskilled immigrant workers; instead, he calls for recognition of the likely reality of continued immigration and the establishment and enforcement of minimal international labor standards, reforms in farm worker service programs, and government support for self-help farm worker groups.

21

Immigration Reform:
Lessons for Future Research and Policy?

DANIEL A. SUMNER

This chapter summarizes impressions from the research and presents a perspective about what we may have learned that could be of use for the next round of debate and policy action on seasonal farm labor. Concerns related to seasonal farm labor, whether tied to immigration or poverty or other issues, have been with us for many years and are not likely to be resolved soon. Researchers, policy officials, and those in between have a responsibility to try to build on the IRCA policy experiment and on the studies that it generated.

Farm Labor Supply and Demand

In analyzing the quantity of labor services supplied, the cost of labor to employers, and the net wage earned by employees, it is useful to ask what IRCA and related policy events have meant for labor supply and demand functions. Little of the data analysis presented has been in the form of rigorous hypothesis testing (that phase of the research is ongoing). However, from the variety of commodity and location case studies, one may reasonably gather some tentative notions about changes in the supply-and-demand conditions for seasonal labor in agriculture.

The supply of seasonal labor to U.S. agriculture remains quite elastic. Available evidence suggests that IRCA did not change the relative ease of border crossings or limit the access of potential workers in other ways. IRCA did not cause the high elasticity of supply of seasonal labor, but IRCA did not seal the border and thus did not change a long-standing feature of the market. Because of the high supply elasticity, a small increase in the demand for labor (say, because of some reduction in the contribution of family labor or some expansion in the importance of those agricultural industries that use seasonal labor), would cause only a very small wage-rate increase.

Moreover, efforts to improve wages or working conditions, by restricting the availability of seasonal labor in a particular market or nationally, through unionization or other means, are unlikely to prove successful unless stronger border barriers are erected. So long as a large reserve of workers continues to be available at the same or slightly higher wages, the real wage for seasonal farm labor is not likely to increase. Even if demand were quite inelastic, which is unlikely for more than a few seasons, with an almost perfectly elastic supply, very large labor-quantity reductions would be required to move the wage noticeably.

But, IRCA did not just leave supply conditions unchanged. There is substantial evidence from the research that the supply of labor has shifted due to IRCA. That is, workers seem willing to work in seasonal agriculture at somewhat lower real wages than before IRCA was passed. The supply has shifted for three main reasons. First, by providing documentation for SAWs and other immigrants, IRCA provided a more secure opportunity to work in the United States for several hundred thousand experienced seasonal farm workers. The added costs of employment these workers had to bear previously, because of their status as undocumented workers, was removed by the law. Second, this large pool of now legal, documented workers has naturally made the migration costs of new immigrants lower. The importance of a migration network facilitated by successful past migrants has been well documented. Now the pool of recent migrants from Mexico living securely in the United States is larger. Finally, changes in the way immigration laws are enforced have lowered the likelihood of undocumented workers being picked up. Therefore, lost work time caused by immigration enforcement is lower, which allows workers to be willing to work at somewhat lower wages.

The demand curve for seasonal labor has probably declined in response to IRCA. IRCA probably reduced the wage employers are willing to pay. However, there are forces in IRCA that could shift the demand for labor in the opposite direction, so the effects may well be ambiguous. The evidence is not clear.

The major factors in IRCA affecting employers directly are the additions to administrative costs of complying with the new regulations. Any factor that raises even the perception of higher nonwage labor costs, due to higher transaction or administrative costs of employing seasonal workers or due to significant potential losses from employer sanctions, reduces wage offers from employers. There is evidence in the studies that, at least in the early years, growers perceived significant costs associated with complying with IRCA rules.

However, with immigration enforcement relying less on surprise in field raids, employers are experiencing less lost work time and the smaller costs associated with finding replacement workers at short notice. This tends to increase the wage offer, but since this factor also lowers the costs borne by workers, the effect on the observed wage is ambiguous.

Nonwage employment costs can also have significant distributional effects within farming that have not been studied. Growers with large operations have an advantage in learning about and complying with complex government regulations. Also, costs of learning about and complying with new regulations are largely independent of the number of employees. The lower cost of employing seasonal workers in a more variable and complex environment probably provides size economies affecting the costs of agricultural production.

IRCA's Effects

The net result of these changes in the supply-and-demand conditions for seasonal labor is lower wages than would have occurred without IRCA. Is that what should have been expected? Since I felt that employer sanctions were going to be enforceable, and given that sealing the border was not practically or politically feasible, it would have been very unlikely that IRCA could have raised wages for seasonal farm labor in the United States. I do not agree with the assertion of some that the government was unusually lax or that there was a conscious policy to maintain low farm wages. Rather, given the basic economics of the situation, it is hard to envision a different result from IRCA.

There has been considerable discussion of a labor surplus in agriculture, which seems counter to the facts of the case. It seems clear that the real or relative wage in seasonal agriculture has not risen, and indeed has fallen in many areas. However, there is little evidence that *unemployment* has increased for seasonal farm workers more than in other occupations. I do not take the word surplus simply to mean low wages; rather, it means, in labor markets as in other markets, that there are more qualified workers willing to work at the prevailing wages than can get jobs at those wages. In fact, downward wage flexibility is exactly what assures market clearing. The fact that in most markets wages are above the legal minimum is evidence that there is no surplus. This situation does not mean that wages are high or fair. It only means that if there were excess supply, that is, if additional supplies were available at the current wage, then that wage would be likely to fall even further.

In future years, a substantial portion of the supply of labor for seasonal agriculture in the United States will continue to respond to earning opportunities in Mexico, as well as to employment opportunities outside of agriculture in this country. If that remains true, a key factor affecting the seasonal agriculture wage in the United States will continue to be economic conditions in Mexico. Substantial migration will continue in any case, but economic growth in Mexico, from NAFTA or from other sources, should tend to reduce the flow over time.

To stem the migration in the short run would require actions that could be possible but do not seem likely. Some have suggested additional regulations on seasonal farm employment. But to the extent that more regulation raises the cost of employing seasonal farm workers, it increases the costs of labor services to producers while actually lowering the effective net wage to workers. This regulatory approach may achieve some other social goals, such as driving jobs out of agriculture or out of the United States. However, it is unlikely to cause the farm labor market to provide more income to workers.

Research Implications

There is more information about seasonal labor markets now than a few years ago, but it will take additional study to understand more fully the information available. For data analysts, this suggests more effort to deal with statistical inference and estimation issues. More careful data analysis will allow us to measure the trends that seem evident and may allow some understanding of factors behind the observations. Of course, more structured statistical analysis will also help us to discard some of the mistaken notions that seem to arise from a more casual approach to evidence.

Data issues capture the attention of researchers and have long been a source of frustration for those looking at seasonal farm labor markets. It should be acknowledged that different approaches to data collection lead naturally to different definitions of what constitutes the seasonal farm labor force and contributes to different objectives. The surveys that focus on farm employers and ask about labor services employed during a survey week are likely to show substantial

variability in hours and wages that are due to survey-week characteristics and weather variations rather than labor market changes. In
addition, these data cannot be particularly useful in measuring worker
characteristics, such as the amount of farm work they do. These data
do tell us how many workers are employed, at what earning levels,
at which locations, on which tasks, related to which commodity, and
during which specific weeks of the year. All of these bits of information have their uses.

Random national surveys of households are bound to generate
small samples of those that do seasonal farm work and undercount
populations outside the mainstream because of low income, language,
or culture. They also miss people that may work in the United States
but reside for parts of the year outside the survey area. These data
may tell us characteristics of people in the survey population who
have worked in seasonal farm labor during the year, and it can be
useful to compare those workers and their experiences with others in
the sample who have not worked in seasonal agriculture, but they
cannot be expected to tell us much about the seasonal farm labor force
itself. Surveys that attempt to gather information from seasonal farm
workers are not providing some information about the workers and
jobs. These data seem to be confirming much of what those familiar
with the market have known from casual observation. However, the
systematic documentation of these facts has been useful, as is the development of a system to track changes in this population. Researchers often request additional data to expand the samples and to fill the
gaps of the types of survey just mentioned. However, it may also be
useful to make renewed attempts to make the information that we do
have more available to a broad range of analysts. That is, the excuse
that the data do not exist is not enough without more effort to make
the data that do exist more available and to exploit that information
more thoroughly.

Conclusions

This chapter began with the observation that seasonal farm labor
issues will continue to be with us for many years. For the research
community the continuing challenge is to better understand the effects of the recent round of policy changes and to build our basic
knowledge about the seasonal farm labor market. When we enter the
next round of intense policy activity, we should be better prepared to
anticipate the effects of policy changes and more secure and convinc-

ing in our expectations and explanations. In making better policy, the challenge is to recognize that consequences of policy choices relate to the reality of farm labor markets, rather than an idealized market with limited responses to incentives created. It is a boring truism that it is important to clarify policy goals and to face the conflicts of those goals with each other and with the markets themselves.

22

Immigration and Agriculture in the 1990s

WALLACE E. HUFFMAN

The Immigration Reform and Control Act of 1986 (IRCA) contained provisions having the intent of changing the supply and demand for labor in the United States and alleviating many social problems associated with earlier illegal immigration. Many of the proponents of IRCA were optimistic in the mid-1980s about how IRCA would slow illegal immigration. Issues of fairness to workers had been addressed by legalizing individuals who had a history of working as undocumented workers in U.S. agriculture and the rest of the economy; needs for a short-term guaranteed supply of seasonal agricultural service workers to agriculture had been taken care of in new SAW (Special Agricultural Worker) and RAW (Replenishment Agricultural Worker) programs; penalties for U.S. employers who hired undocumented workers were imposed; and the INS was supposed to tighten border controls and greatly reduce illegal immigration rates. In the mid-1980s, some believed that IRCA would permanently solve the problems associated with illegal immigration to the United States.

The Report of the Commission on Agricultural Workers (USCAW 1993) summarized the reasons why IRCA failed to meet its objectives. Some had anticipated that the technology of document counterfeiting might be advancing more rapidly than the technology of counterfeit-document detection and that these problems could be sufficiently great to subvert the intentions of IRCA. A more basic problem, however, seemed to underlie IRCA's poor performance. This problem was the huge wage (and standard-of-living) differential between Mexico and the United States that existed during the 1980s. Wage differences have continued to differ by factors on the order of 8 to 15 times, and standard-of-living differences seem similar. IRCA did not directly address the large incentives that Mexicans, especially low-wage ones, have for meeting the conditions for SAW status and for continuing to migrate illegally for work to the United States. These incentives have

[425]

not changed appreciably since the mid-1980s, and a North American Free Trade area is one of the few prospects on the horizon that might eventually reduce intercountry wage and income differences.

The economic forces behind international trade in labor and agricultural commodities were addressed by Torok and Huffman (1986). The economic foundations of this chapter continue to be applicable, and establishment of a North American Free Trade Area might reduce barriers to trade in final commodities, although it does not currently address international migration.

The objective of this chapter is to summarize a few key effects of IRCA on U.S. agriculture and the U.S. economy, and to examine likely major developments in immigration issues facing agriculture during the 1990s.

A Few Key Effects of IRCA

A Large Increase in Labor Supply

IRCA appears to have had a much larger short-term and long-term impact on the U.S. labor supply of low-skilled labor than expected. First, the applications/approvals and the regular amnesty and SAW programs were much larger than pre-IRCA forecasts. Over 1,759,000 individuals who were illegal aliens and claimed continuous residency in the United States between January 1, 1982 and May 1, 1986 could apply for immediate U.S. resident status. About 1,655,000 individuals were approved and given rights to work anywhere in the United States. Only a very small share of these individuals (maybe 4%) appear to have been or to be employed in U.S. agriculture. Other undocumented workers, who had worked at least 90 days in U.S. seasonal agricultural services (SAS) field work during April 30, 1985 to May 1, 1986, could apply for SAW status (USCAW 1993).

About 1,272,000 individuals applied for SAW status and about 1,037,000 previously illegal workers were granted SAW status, which gave them immediate temporary U.S. resident status and authorization to work in any occupation. The number of SAWs approved was much larger than expected (Mines, Gabbard, and Samardick 1993). Both programs covered individual workers, but did not authorize legal status for family members who did not also meet the special provisions. SAWs could later apply for permanent resident status.

Second, the new workers legalized under IRCA were largely Spanish speaking and had attained a low level of schooling. English-speaking ability was generally poor, and most had strong ties to families and friends back home in Mexico. A large share of these individuals chose to be part of a transnational labor force, with strong local ties to a geographical area (and friends) in both the United States and in a foreign country (largely Mexico).

Given the new legal status and improved U.S. earning potential of the legalized workers, they became an international transfer agent for bringing more illegal undocumented workers to the United States. This occurred in the following way: These legalized workers were able to purchase better cars and trucks (transportation services) and had more resources to spend on family living, so they frequently brought undocumented family members and other relatives and friends to the United States when they returned from their homeland. Thus, workers legalized under IRCA seem to have had a multiplier effect on available low-skilled labor. This was totally unanticipated, and nobody really knows exactly what the multiplier has been. It was undoubtedly reduced somewhat by the economic recession that started in 1990 and for which the recovery in employment has been very slow (USPres. 1993). Also, U.S. real wage rates for high school dropouts declined by 18.2 percent for the decade 1979–89 (Mishel and Bernstein 1993).

Third, because of the low average schooling attainment and poor English-speaking ability of new IRCA workers, there was a large degree of occupational immobility. Also, the SAWs are seasonal workers and have a high frequency of returning to their home country. These are all factors that slowed their integration into mainstream U.S. culture and society, slowed their learning of English, and greatly lowered their exit rate from SAS work (Mines, Gabbard, and Samardick 1993). Thus, both the short-term and long-term effects of the SAW program on availability of SAS (and probably other low-skilled) labor has been much greater than anticipated.

Furthermore, this long-term persistence for work in the same location has created incentives for permanent settlements in rural areas, particularly in California, Washington, Oregon, Texas, and Florida (USCAW 1993). These settlements have greatly increased the demand for low-priced housing services and for social services for largely a Spanish-speaking culture.

A Modest Decline in Real Wage Rates

The conclusions in the overall *Report of the Commission on Agricultural Workers* and in the research reports (USCAW Staff 1993, Duffield and Vrooman 1993) were that IRCA did not have any significant effect on U.S. wage rates, either the farm wage rate or the farm wage relative to the nonfarm wage rate. Furthermore, the empirical evidence presented in these chapters largely supports this conclusion.

However IRCA may have caused a decline in the real earnings of farm and nonfarm workers. My evidence supports the conclusion that IRCA did not change the wages of farm labor relative to the wages of nonfarm labor. The data that I used were taken from CAW's report (USCAW Staff 1993, p. 792; see also Duffield and Vrooman 1993).

What are the reasons for my new interpretation of roughly similar economic information? First, the argument developed in the previous section of this chapter was that IRCA had a much larger impact on the available supply of low-skilled labor (for example, high school dropouts) in the United States than expected. This is an outcome that implies a larger impact on U.S. wage rates than what would have been anticipated in pre-IRCA years. Second, considerable evidence exists that all U.S. (and Mexican) labor markets, differentiated by skill type and geographical location, are interconnected through labor mobility (migrating and training). Economic incentives in the form of a positive compensating differential are generally required to induce individuals to change their human capital stocks, including geographical migration for work. See the research by Rosen (1986), Torok and Huffman (1986), Tokle and Huffman (1991), and Huffman (1993b). Furthermore, Huffman (1980), Tokle and Huffman (1991), and Huffman (1993b) show that U.S. farm and nonfarm labor markets have become highly integrated during the past 25 years, and Huffman (1993b) shows that compensating differentials, after an allowance for the cost of living and the likelihood of being employed, are now very small across the United States. Thus, these are arguments that it would be *unusual* to find an effect of IRCA on the ratio of farm-to-nonfarm wage rates. These arguments are leading up to the hypothesis that we would expect IRCA to have caused a more rapid rate of decline of the real farm wage rate and real nonfarm wage rate than what would have occurred without IRCA.

Figures 22.1 and 22.2 present plots of the real manufacturing wage and real farm wage during 1975–91. In creating these plots, I have chosen to deflate nominal wage rates using the deflator for personal consumption expenditures from the Bureau of Economic Analysis

(National Income and Product Accounts) rather than the Consumer Price Index (CPI). The reason for the switch in deflators is that during 1978–85 the CPI contained a major flaw. It used the nominal interest rate on new home mortgages as a key indicator of the price of services of owner-occupied housing. When U.S. nominal interest rates were deregulated in 1978–79, interest rates on new home mortgages rose abruptly; and given that housing has a weight of about 30 percent in the CPI, the CPI rose "too" rapidly during 1978–85. The rise of the CPI during 1978–85 was 5.2 percent larger than for the deflator of personal consumption expenditures. Thus, real wage rates, obtained using the CPI as the deflator, fell too rapidly or rose too slowly during 1978–85. Thus, the pattern of real wage rates reported in Figures 22.1 and 22.2 for the period 1978–85 are quite different from those reported in the *Report of the Commission on Agricultural Workers* (USCAW 1993).

Figure 22.1. Real Nonfarm Wage in the United States and Selected Regions, 1975–91

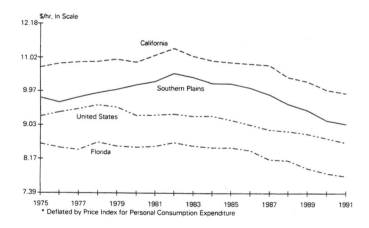

* Deflated by Price Index for Personal Consumption Expenditure

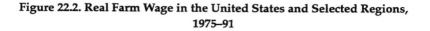

Figure 22.2. Real Farm Wage in the United States and Selected Regions, 1975–91

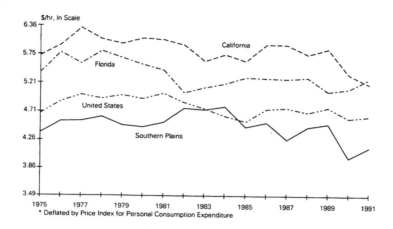

In judging the performance of wage rates, I have chosen to focus on the U.S. average and on averages for California, Florida, and Texas–Oklahoma. The latter three areas were chosen because of the large contributions they make to SAS labor use. In particular, the real manufacturing wage in the United States (and Florida) peaked in 1978 (real compensation peaked later in the mid-1980s because of the value of benefits rising more rapidly than wage rates) and then declined slowly. The real manufacturing wage for California and the southern plains states peaked in 1982. Real manufacturing wage rates for California, Florida, and Texas–Oklahoma declined after 1982, but the rate of decline accelerated after the passage of IRCA in 1986. Recall that the real gross national product (GNP) grew through the second quarter of 1990 (USPres. 1993).

The pattern of the real farm wage during 1975–91 is more irregular than for the real manufacturing wage. The U.S. real wage for farm labor peaked in 1981, fell steadily during 1981–85, and recovered somewhat after 1985. The real farm wage in California and Florida peaked during the late 1970s and then declined until 1983. The real California farm wage recovered some during 1983–89 and then declined sharply. In Florida, the real wage rose during 1982–88, then declined a little. The pattern of the real Texas–Oklahoma farm wage is somewhat different. It has a positive trend from 1975-84, and then a strong negative trend emerges. The break in 1989 is quite sharp and may be partially related to a hard winter freeze in the lower Rio Grande area.

Thus, the patterns that emerge in Figures 22.1 and 22.2 are somewhat different than those presented by the CAW (USCAW Staff 1993). The next step was to see if these differences are statistically significant. Table 22.1 presents econometric results from regressing the natural logarithm of real wage rates to a general time trend and to a post-IRCA trend. The results show an estimated average annual rate of change in real wage rates during 1975–88 and an average annual change in real wage rates during 1989–91.

Why did I choose 1989 as the first post-IRCA year? Although IRCA was passed late in 1986, its legal interpretation and exact procedures for the general worker amnesty and the SAW programs were still to be defined. Furthermore, processing of worker applications under IRCA was slowed because of the much larger than anticipated application rates. Furthermore, even after a majority of the applications was processed, additional time was required for the newly legalized workers to realize additional earnings and modify behavior in accordance. Also, IRCA had attempted to impose stiff employer penalties for hiring undocumented workers, and some time was required before the Immigration and Naturalization Service and the U.S. Department of Justice could work out satisfactory procedures. In fact, the INS chose largely to ignore farm employer sanctions because of the large penalties that might be imposed on employers if they discriminated against a legal worker in the process of checking documents carefully for legal status to work in the United States. High-quality counterfeit documents were frequently easier to obtain than real documents. Thus, I have chosen to include the wage rates for the transitional period of 1986–88 in the pre-IRCA years. (An alternative would have been to exclude wage rates for these years from the statistical analysis.)

In Table 22.1, section III, the fitted equations explain the farm-wage manufacturing wage ratio. They do not show any pre- or post-IRCA trend that is statistically significant. This conclusion supports the the CAW's report. Also, the farm manufacturing average wage differential over 1975–91 is largest for the U.S. average (70%). It is smallest for Florida; but the farm manufacturing wage rates are not significantly different. In California and Texas–Oklahoma, the farm

Table 22.1. Regression Equation Retention of Real Farm and Nonfarm Wage Rates and Relative Wage Rates to a General Time Trend and Post-IRCA Trend, 1975–91

ℓn Wage =	β_1 +	$\beta_2 T$ +	$\beta_3 T \cdot D1 + \mu$;	DW	R^2
I. Real Hourly Farm Wage					
1. A 3-State SUR System					
California	2.065	-.003	-.0005	1.54	
	(10.3)	(1.4)	(1.5)		
Southern Plains (TX, OK)	1.568	0.0006	-.0008	1.84	.47 (system)
	(7.5)	(0.3)	(2.1)		
Florida	2.218	-.006	.0001	1.28	
	(11.8)	(2.8)	(0.2)		
2. United States	1.88	-.004	.00002	1.03	.37
	(13.4)	(2.2)	(0.1)		
II. Real Hourly Manufacturing Wage-Production Workers					
1. A 3-State SUR System					
California	2.467	-.001	-.0008	1.39	
	(24.0)	(0.8)	(4.3)		
Southern Plains (TX, OK)	2.199	.001	-.001	1.00	.88 (system)
	(15.1)	(0.7)	(4.2)		
Florida	2.343	-.003	-.0005	1.30	
	(36.1)	(3.2)	(4.8)		
2. United States	2.584	-.004	-.0003	0.85	.85
	(35.9)	(5.0)	(2.3)		
III. Hourly Farm Wage ÷ Nonfarm Wage					
1. A 3-State SUR System					
California	-.402	-.003	.0002	1.07	
	(2.0)	(1.1)	(0.7)		
Southern Plains (TX, OK)	-.631	-.002	.0003	2.16	.09 (system)
	(3.7)	(1.0)	(1.0)		
Florida	-.124	-.004	.0006	1.17	
	(0.6)	(1.4)	(1.5)		
2. United States	-.703	.0006	.0003	1.12	.25
	(5.3)	(0.4)	(1.3)		

Definitions: T = Time (75 – 91)
D1 = 1 if T ≥ 89 and 0 otherwise
μ = a random disturbance term
Real wage rates are created by using implicated deflator for personal consumption expenditures from Bureau of Economic Analysis, 1987 = 1.00.

wage is on average 40 percent and 63 percent lower than the manufacturing wage, respectively.

In Table 22.1, section II, the results show that the real U.S. manufacturing wage declined at an average rate of 0.40 percent per year during 1975–88 and the average rate of decline accelerated to 0.43 percent per year during 1989–91. In particular, the change in the rate of decline of the U.S. real manufacturing wage during 1989–91 is significantly different from zero. The behavior of the real manufacturing wage in California, Florida, and Texas–Oklahoma is somewhat different. In particular, the real manufacturing wage in Texas-Oklahoma does not have any significant trend during 1975–88. The post-IRCA behavior of the real manufacturing wage in these states is similar. The average rate of decline of the real manufacturing wage is larger during 1989–91 than during 1975–88. Furthermore, a joint test that the coefficient for these three areas are jointly equal to zero is soundly rejected at the 5 percent significance level. Thus, we have evidence that IRCA increased the rate of decline of the real wage in manufacturing in major SAS labor-using states and in the United States overall. This negative post-1988 effect of IRCA on the real wage might be associated with other non-IRCA-induced economic effects, but I remain skeptical.

In Table 22.1, section I, the results show an average annual rate of decline of the U.S. real farm wage of 0.4 percent during 1975–88. The average rate of decline of this wage during 1989–91 is not significantly different than during 1975–89. In California, the average rate of decline of the real farm wage during 1975–88 was 0.3 percent and for Florida, 0.6 percent; but for Texas–Oklahoma, the real farm wage did not have a trend that was significantly different from zero (1975–88). Among these regions, however, Texas–Oklahoma is the only one that shows a more rapid average rate of decline of the real farm wage during 1989–91 than during 1975–88. However, the average rate of decline of .08 percent in Texas–Oklahoma during 1989–91 is smaller than for California and Florida for 1975–91.

My tentative conclusion from this preliminary econometric evidence is that IRCA did have a statistically significant effect on U.S. wage rates. It appears to have increased the rate of decline of the nonfarm real wage rate from what it would otherwise have been. IRCA, however, did not have a statistically significant effect on the ratio of farm-to-nonfarm wage rates because the farm and nonfarm highly interlinked within each state and across states. Low-skilled Mexican workers do earn less than more highly skilled U.S. citizen workers, but the low-skilled and the higher-skilled labor markets are also interconnected and compensating differentials appear to have been unaffected by IRCA. Thus, IRCA may have contributed to the very slow apparent recovery of jobs for U.S. citizens from the business-cycle downturn starting in 1990 by providing newly legalized and new illegal workers with employment.

Also, because of the low average schooling and poor English of new IRCA workers, much greater occupational immobility appears to have occurred than was anticipated. The SAWs in particular are seasonal workers and have a high return frequency to their home country, Mexico. These are all factors that have slowed their integration into the broader U.S. culture and society and their exit from SAS labor. Thus, the long-term effect of the SAW program on SAS crops is greater than could have been anticipated. Furthermore, this long-term persistence has created incentives for permanent settlements in rural locations, such as in rural California, Oregon, Washington, and Florida, that was not anticipated. This has greatly increased the demand for low-priced housing services and social services for a largely Spanish-speaking culture.

Changes in Demand for Labor and Trade

The changes in demand for SAS labor are derived largely from changes in demand for U.S. horticultural and nursery crops and technical change in the production of these crops. Table 22.2 presents information on the quantity of major horticultural crops harvested in 1987, and provides an indication of the amount of work to be done in harvesting these crops. Oranges, apples, grapes, and tomatoes stand out for their large quantities. Table 22.3 ties these crops into 12 SAS or farm-labor regions. California (Region I) and Florida (Region II) stand out for having unusually large areas and quantities of SAS horticultural crops relative to all other regions.

In Table 22.4, some selective trends in the U.S. consumption of fruits and vegetables are summarized. This table shows that over the period 1985–86 to 1989–90 there was a 5.6 percent reduction in per capita total consumption of fruits and berries. The consumption of fresh fruit and berries decreased by a larger rate of 6.5 percent. Per capita fresh citrus was, however, unchanged over this period.

The U.S. consumption of vegetables, melons, and mushrooms increased significantly during 1985–86 to 1989–90. The rate was more rapid for fresh the largest increase in this class of horticultural crops, 17.4 percent for fresh and 13.6 percent for total green vegetables. The

Table 22.2. United States: Number of Farms, Area in Orchards, Vines, Vegetables, and Nursery Crops, by Major Crops, 1987

Crops	Predominant Regions[1]	Number of Farms	Total area in trees, vines, etc. (Acres)	(1,000 sq.ft.) enclosed)	Quantity Harvested (mil lb.)	(mil dol.)
Tree Crops						
Oranges	II, I, & VIII	15,800	918,700		14,776 lb.	
Apples	IV, V, VI, VIII, IX, & XII	36,700	601,000		9,291	
Grapefruit	II, I, & III	5,000	189,400		4,592	
Peaches	VII, I, V, & III	21,000	239,700		2,145	
Plums	I & IV	6,800	151,200		1,884	
Lemons	I, II, & VIII	1,915	68,837		1,795	
Pears	IV, I, & V	10,100	84,247		1,741	
Cherries	VI, IV, I, & V	10,800	131,064		632	
Almonds	I	6,700	427,700		611	
Avocados	I & II	6,902	87,700		540	
English Walnuts	I	8,154	213,628		438	
Pecans	III, VII, VIII, & II	21,431	453,243		174	
Subtotal			(3,158,500)		(38,619)	($5,501)
Ground Crops						
Grapes	I, V, IV, & VI	23,200	833,300		9,911	
Berries						
Strawberries	I, IV, & II	9,398	53,085		959	
Blueberries (all)	VI & V	4,412	59,216		142	
Raspberries		4,297	15,484		66	
Blackberries		2,086	6,679		33	
Other berries		NA	28,960		NA	
Subtotal - Grapes & Berries			(996,724)		(11,111)	($1,583)
Vegetables, Sweet Corn, Melons						
Sweet corn		26,004	671,200			
Tomatoes		14,542	375,500			
Green peas, excluding cowpeas		7,394	290,800			
Snap beans		9,640	289,200			
Lettuce and Romain		2,200	250,200			
Cantaloups		7,322	129,800			
Cucumbers		7,820	129,900			
Dry onions		3,250	115,600			
Broccoli		2,821	100,100			
Asparagus		3,033	97,335			
Sweet peppers		7,021	70,697			
Subtotal			(2,520,332)			($4,698)
Nursery and Greenhouse crops, incl. mushrooms, and sod farms		37,298	578,955	761,996		$5,774

[1]See Table 4 for definition of regions.
Source: See Huffman 1992, p. 758, Table 4.

Table 22.3. Area and Quantity Harvested of Tree Crops, Grapes and Berries, Vegetables, and Nursery and Greenhouse Crops, by Proposed SAS Farm Labor Regions, 1987

Regions	Tree Crops		Grapes & Berries		Vegetables		Nursery and Greenhouse			Total
	Acres (1,000s)	Quantity Harvested (mil lb.)	Acres (1,000s)	Quantity Harvested (mil lb.)	Acres (1,000s)	Quantity Sold (mil lb.)	Open Acres (1,000s)	Enclosed Area (mil sq ft.)	Quantity Sold (mil dol.)	Total Acres (1,000s)
I (CA)	1,261	11,763	726	9,453	775	1,851	64	164	1,412	2,829
II (FL)	742	15,415	6	53	312	698	86	149	823	1,145
III (TX)	204	369	4	5	202	263	30	33	239	440
IV (OR,WA)	270	5,704	56	654	286	256	57	27	325	676
V (NY,PA,NJ,MD,DE)	169	1,751	65	521	353	250	67	106	774	656
VI (MI)	149	1,268	29	167	139	116	22	34	216	347
VII (AL,GA,KY,NC,SC,TN, VA,WV)	319	1,400	14	38	235	270	82	60	587	303
VIII (AZ,NM)	84	957	11	74	126	63	7	5	82	228
IX (MN,WI,IA,MO,IL,IN,OH)	60	462	11	49	726	368	95	108	652	895
XII (ME,VT,NH,CT,RI,MA)	30	282	6	14	42	50	19	35	145	98

Source: Huffman 1992, Table 6, pp. 763.

Table 22.4. Rate and Growth of Fruit and Vegetable Consumption in the
United States, Selected Years, 1979–90

Type Produce		Average per capita consumption (lb.)			Percentage Change	
		1979-80	1985-86	1989-90	1979-80 to 1989-90	1985-86 to 1989-90
Fruits and Berries:						
Citrus:	Total	36.5	36.5	32.7	-11.0	-10.9
	Fresh	26.9	24.1	23.9	-11.4	-0.1
Noncitrus:	Total	80.1	96.9	93.0	14.8	-4.1
	Fresh	59.8	78.0	72.0	18.6	-8.0
Total Fruits & Berries:	Total	113.5	123.6	125.7	7.5	-5.6
	Fresh	86.6	102.3	95.9	10.2	-6.5
Vegetables, Melons & Mushrooms:						
Green vegetables:	Total	52.5	50.0	57.3	8.7	13.6
	Fresh	34.3	33.0	39.3	13.5	17.4
Tomatoes:	Total	76.6	78.8	85.8	11.4	8.5
	Fresh	12.5	15.4	16.0	24.7	3.8
Potatoes, total all types		115.9	124.1	128.4	10.2	3.4
Other vegetables, melons & mushrooms:	Total	102.2	116.6	119.7	15.8	2.6
	Fresh	51.2	58.9	63.5	21.5	7.5
Total Vegetables, Melons & Mushrooms:	Total	347.1	369.4	391.1	11.9	5.7
	Fresh	98.0	107.3	118.8	19.2	10.2

Source: Huffman 1993, p759, Table 5.

growth in per capita total consumption of tomatoes was 8.5 percent and of fresh was 3.8 percent. The consumption of fresh, nongreen (or yellow) vegetables also grew at the relatively rapid rate of 7.5 percent.

Thus, the growth in domestic demand for these horticultural products shifted toward green and yellow vegetables and away from fruits and berries. This trend seems likely to continue into the mid-90s and suggests greater growth for SAS labor in vegetable, melon, and mushroom production than in fruits and berries. Technical labor-saving advances might, however, modify this trend.

The potential for international trade in horticultural crops seems likely to change during the 1990s. First, IRCA has reduced the real wage in the United States, and this seems likely to have reduced the quantity demanded of horticultural products imported from Mexico and other countries. This is based largely on the implications from the multiequation agricultural trade and immigration model by Torok and Huffman (1986) and Huffman (1986). Furthermore, because a large share of the SAWs are relatively young, they can be expected to have impacts on U.S. labor availability for all of the decade of the 1990s.

Second, lowering the wage for SAS labor has undoubtedly increased U.S. exports of labor-intensive horticultural products — ones in which the United States has a comparative advantage, for example, citrus, grapes, apples, peaches, and lettuce exports to Canada. These conclusions are based on a selective summary of U.S.–Canadian and U.S.–Mexican trading patterns in horticultural products (see Table 22.5), and an econometric analysis of U.S. exports of fresh lettuce to Canada (Huffman 1986). This latter study showed a large impact of a lower California farm wage rate on the quality of U.S. fresh lettuce exported to Canada during 1956–83. These results seem likely to apply broadly to exports of other labor-intensive horticultural crops.

The North American Free Trade Area (NAFTA) seems likely to also change trade patterns in horticultural products between the United States and its neighbors, Mexico and Canada. My forecasts are based on an examination of current trade patterns and geographical locations of product, especially relative to the borders, as follows. Products that are currently produced in the United States but near the Mexican border will most likely move to Mexico in time (Huffman 1993). These are crops that are produced in southern California, Arizona, New Mexico, and Texas. Cuban producers are also likely to become strong competitors in the future with those in south Florida, Louisiana, California, and Hawaii. These are largely crops for the winter and early spring U.S. fresh fruit and vegetable markets: tomatoes, cucumbers, peppers, cantaloupes, broccoli, melons, cauliflower, cabbage, onions, sugar, and pineapple. The United States might also lose a little market to Canada, particularly in tuber potatoes, carrots, and apples.

The United States can, however, expect to experience some increases in demand for exports of horticultural products from its neighbors to the north and south. During the 1990s the market potential is much larger in Canada than Mexico. Increases appear likely for citrus, grapes, peaches, pears, and plum exports to Canada. For Mexico, additional exports seem most likely for pears and plums.

Conclusions

IRCA has had some expected but also many unexpected effects in the United States. It increased the quantity supplied of low-skilled labor, as expected, but it had the unexpected effect of contributing to the immigration of additional undocumented individuals. Although the initial estimates of the depressing effects of IRCA on real U.S. wage rates were less than expected, the analysis presented here brings with it the conclusion that IRCA did in fact reduce U.S. real wage rates — both farm and nonfarm — and that these effects were largely diffused throughout the United States. When data for the 1990s become available, more research to quantify these effects is needed.

During the 1990s, the growth in demand for green and yellow vegetables appears likely to be larger than for fruits and berries. Because vegetables are grown largely on or close to the ground, technical advances in labor-saving devices can occur more rapidly than for tree crops. The decade of the 1990s appears, however, likely to have unusually low average rates of growth of aggregate labor demand. This means that higher real wage rates are unlikely to provide much incentive to undertake costly mechanization. Furthermore, if modest limits are placed on welfare tenure, many welfare recipients may need to take work. Their re-entry into the U.S. labor force during the decade of the 1990s could put further downward pressure on the real U.S. wage rates of largely low-skilled workers.

Comparisons for economic conditions of transnational population groups that spent part of each year in the United States and part in a low- or moderate-income country are difficult. Because housing, health, and environmental goods have large income elasticities, U.S. standards for these goods generally exceed by a large margin the standards in the originating countries of immigrants. Thus, substandard

U.S. conditions, which immigrants frequently face in agriculture and some other sectors, may equal or exceed standard conditions for these goods in their originating countries. One can see that comparisons of economic conditions of immigrants most likely should be made relative to the originating country and to the United States and an average computed.

It is easy to see that the economic incentives for further immigration of undocumented persons can be expanded beyond those included in real wage differences when illegal immigrants are given access to most social services of U.S. citizens. To help control access to jobs and social services, the United States needs a genuine personal identification system; otherwise, it will ultimately extend work opportunities and social program coverage to a large share of the population of Mexico and Central America. Some people might find this to be fair, but the financial reality is that the burden of paying for extending social program benefits will fall on U.S. citizens; there is no "deep pocket" elsewhere to pay, and outsiders can obtain greater benefits only if U.S. citizens on average receive fewer benefits than otherwise would be the case. U.S. citizens might be prepared to greatly extend access to jobs and social services to a much larger population, but these are choices that they and politicians should confront directly.

For IRCA-legalized individuals and their families, the issue of who pays for their social services seems likely to become a growing concern during the decade of the 1990s. Should growers' associations or consumers pay? Should local governments, state governments, or the federal government pay? Although some people believe that most of the benefits from cheap(er) SAS labor (and other technologies) go into larger profits of growers and processors of agricultural products, this is not the case. Except for a few specialty crops, there are a relatively large number of U.S. producers of agricultural products compared to U.S. manufacturers of cars, televisions, and washing machines, etc. Competition among these producers results in the benefits of cheap(er) labor and improved technologies being transferred rather quickly to consumers in lower (than otherwise would be the case) prices of agricultural products (Huffman and Evenson 1993). For example, over the period 1978–92, the prices received by U.S. farmers decreased by 3.9 percent per year relative to the prices of all goods purchased by U.S. consumers (measured with the implicit price deflator for personal consumption expenditures). Prices paid by U.S. farmers for inputs relative to this same consumer price index declined by only 1.1 percent per year over the same period. Consumers could decide that their food is cheap enough and not want lower food costs because of the use of cheap labor or new technologies.

Because consumers largely benefit and they are spread throughout the United States, the local and state governments are inefficient agencies for providing, collecting revenues for, and paying for these social services. Furthermore, local differences in programs can have major effects on the choice of the exact location of U.S. residence. Perhaps the federal government should bear a major share of the cost and use excise or income tax receipts to pay for them.

Although some of the problems of agricultural labor are attributed to the farm labor contractors (FLCs) who serve as intermediaries, this institution exists and has expanded largely because of economic incentives in federal and state legislation. The FLCs, as other businesses that provide employers with temporary employees, became profitable when the economic incentives for inter- versus intrafirm transactions changed. This was one of the important issues considered by Coase (1952) in his noted article, first printed in 1937, on the nature of the firm and the importance of transaction costs, which was expanded on by Stigler (1951) and Rosen (1983). IRCA and some preceding federal legislation raised the relative cost of intrafirm labor transactions relative to the use of outside farm labor contractors. Hence, it is not surprising that FLCs grew rapidly as a source of SAS labor during 1986–92 (Mines, Gabbard, and Samardick 1993). This is just another example of how new legislation changes economic incentives and the extent of use of institutions in unanticipated ways, as economic agents attempt to minimize the impacts of legislative and other changes on their expected well-being.

Much greater effort is needed to examine the likely implications of a North American Free Trade Area on labor and product markets. The advances over the past 15 years in computer technologies and communications have facilitated the breaking of previous long manufacturing processes into a sequence of smaller processes that can be contracted out to the lowest cost producer in the world economy. Final assembly of the components may occur in a moderately advanced country or near the sight of ultimate use of the final good, depending on the nature of trade policies. Modern manufacturers have, however, made domestic content policies a farce, as they find ways to get around them. These changes have increased greatly the competition that high school dropouts and other low-skilled U.S. people face in the U.S. labor market. It is competition that will not go away, and agriculture will feel the pressure, too. Much research is needed that combines theories from international trade, labor economics, and public choice economics with good econometric expertise into an empirical examination of trade and immigration issues facing North America.

In conclusion, IRCA has revealed many surprises. More undoubtedly will occur. Furthermore, the conclusions presented by Torok and Huffman (1986) for dealing with illegal immigration still stand. It is a complex problem involving factor and product markets, trade and immigration policies of both countries, and general policies affecting economic growth and social policies. No easy fixes exist!

23

Immigration and Agriculture: Policy Issues

MONICA L. HEPPEL

Immigration and Agricultural Labor

For years, seasonal farm labor in the United States has been an occupation performed by members of minority groups and characterized by low wages and poor working conditions. Mexican-born workers supplied western agriculture, while African Americans were more prevalent on eastern farms. Today, Mexican-born farm workers are employed throughout the country, as more and more farmers turn to workers born in third-world countries to fill their labor needs. Federal and state commissions have long noted this dependence on foreign workers and made recommendations to alleviate the poverty that characterizes the occupation, but farm workers today remain the poorest of the working poor.

Some hoped that IRCA would break this cycle by ending the use of undocumented workers in agriculture. This would have limited the supply of workers, eventually leading to improved farm wages and working conditions. However, with the continued employment of unauthorized workers, "business-as-usual" continues in the post-IRCA era. We thus see a continuation of both a "farm labor problem" and an "illegal immigration problem."

The Farm Labor Problem

Theodore Roosevelt's 1909 Commission on Country Life set the tone for a host of commissions[1] throughout the century that were to investigate the living and working conditions of hired agricultural workers:

> The farm labor problem is complicated by the fact that the need for labor is not continuous, there are no

conveniences of living for the laborer, the long hours,
the wanting of companionship, and in some places,
apparently low wages (USPres. 1909, p. 42).

Reporting on agricultural labor problems in California, the
LaFollette Commission in 1942 noted:

There is an inadequate income for many workers and
their families. Supplementary funds from public-re-
lief sources, dispensed on a broad scale, are necessary to maintain even a subsistence level for a great
majority of them. However, many are not eligible and
relief funds are not always adequate (USSenate 1942,
p. 156).

Harry Truman's 1951 Commission on Migratory Labor detailed
a series of problems faced by migrant farm workers in the United
States:

Farm workers earn a little over one-third, but well
below one-half the average wage paid in manufacturing. They receive some perquisites, such as housing, which increase their real wages. They don't have
other perquisites that industrial employees do, such
as sickness benefits, pensions, welfare plans, paid
vacations, and paid holidays (USPres. 1951, p. 17).
. . . Much, if not most, of the on-job housing of migratory farm labor in the United States is below minimum standards of decency (p. 138). . . . The "permanent" housing in which migrants live for the six to
eight months of the year when they are not working
on the crops is among the most deplorable in the
nation (USPres. 1951, p. 144).

In 1992, the Commission on Agricultural Workers (USCAW 1993)
found that legislation introduced in response to these reports produced a set of regulations that, when enforced, improved conditions
for many farm workers.[2] It also found, however, that little had changed
in the overall living and working conditions for hired agricultural
workers in the United States, despite the accumulation of reports calling attention to the problems. Farm wages remained at approximately
one-half the levels of nonfarm wages and seasonal employment continued to result in low annual earnings.

The Illegal Immigration Problem

Mexican-born workers have long been a mainstay of U.S. agriculture, originally in the West and more recently throughout the country. Many have been authorized to work in this industry, both as legal immigrants and through a nonimmigrant contract labor program. Many others, however, have found it easier to enter and find jobs in the United States without work authorization.

Illegal immigration was a major problem that the enactment of IRCA was to alleviate. Sanctions against employers who hired unauthorized workers were supposed to reduce the magnet of employment that prompted workers to enter the United States illegally. When IRCA was being debated, the agricultural industry stepped forward and acknowledged its dependence on an illegal work force. Based on this historic dependence, the industry sought, and was granted, special treatment under the act. After a generous amnesty program and a delay in enforcement of sanctions, agricultural employers, like all other employers in the United States, were no longer allowed to hire unauthorized workers.

Despite IRCA's mandate to control illegal immigration, the Commission on Agricultural Workers found that

> Since the enactment of IRCA, unauthorized immigration into the U.S. agricultural labor force has continued. This is a function of the availability of agricultural jobs and the ineffectiveness of both employer sanctions and the border control effort. There is no indication that this will change in the near future (USCAW 1993, p. 92).

Immigration and Farm Labor

Labor-intensive agriculture has relied on an immigrant work force since its inception. African slaves in the South, European immigrants in the Midwest, and Chinese and Mexican workers in California allowed for agricultural development to continue apace. The availability of immigrant workers who were either willing or forced to labor for low wages and under relatively poor working conditions has left its mark on the farm labor market. This relationship was highlighted in the 1951 president's report on migratory labor:

> Statements made by nonfarm employers in the early
> 20s regarding the need for immigrant labor are the
> same as those heard now for agriculture. When im-
> migrant labor was denied these other industries, they
> were compelled to develop working conditions and
> job standards compatible with the expectations of
> American workers. Farm employers of migratory
> labor on the other hand continue to offer jobs and
> working conditions that are not better and in many
> respects are worse than those offered three and four
> decades ago (USPres., p. 22).

The historic availability of immigrant workers for agriculture has
not only kept wages low, but has worked against the widespread
adoption of labor-saving technologies to increase worker productivity or a reorganization that would allow the industry to rely on a
much smaller work force. When 120 tomato pickers are cheaper than
60 pickers and a harvest-assist machine, there is no economic incentive to adopt new strategies that would benefit workers. Labor-intensive agriculture has come to be characterized by a self-reinforcing
cycle based on the availability of cheap workers, which leads to depressed wages and poor working conditions, which then leads to a
lack of available domestic workers.

Unresolved Policy Issues

Despite the intentions of policy makers, IRCA has neither prevented the employment of unauthorized workers in agriculture nor
deterred continuing illegal immigration. Although many farm workers obtained legal status through IRCA's Special Agricultural Worker
(SAW) program, by 1992 the farm labor force once again included a
significant, and increasing, proportion of unauthorized workers. One
easy explanation is the availability of fraudulent work authorization
documents, allowing both workers and employers to circumvent the
law. Other relevant factors include the lack of enforcement of employer sanctions in agriculture, despite the industry's historic use of
undocumented workers; a low priority given to farm labor issues by
the U.S. Department of Agriculture in its dealings with farmers; and
a generally scattered approach by the federal government toward
agricultural labor, with specific programs and policies emanating from
a number of sources and generally defined by the same categories
and assumptions that were established more than 50 years ago.

With the failure of employer sanctions, continued illegal immigration has contributed to a national agricultural labor surplus, with the predictable stagnation of wages, little progress toward improved working conditions, and the lack of any widespread adoption of enlightened labor management techniques. Farm worker annual earnings remain low and periods of unemployment and underemployment remain frequent.

Since IRCA, labor-intensive fruits, vegetables, and horticultural specialty production is expanding, as measured by both cash receipts and harvested production.[3] Farms are becoming larger and more lucrative, with these larger operations employing a growing proportion of hired farm workers. Individual producers are making such expansion decisions with little or no thought as to whether seasonal workers will be available to harvest their investment. Such availability has become a given.

IRCA has accelerated certain trends in the structure of the agricultural labor market, as well. With increased competition for agricultural jobs, particularly in the immigrant-receiving states of California, Texas, and Florida, farm workers are moving further afield in search of jobs. This increased internal migration has led to a far-reaching "Latinization" of the hired farm work force and of many rural communities.

IRCA also appears to have accelerated the trend toward a reliance on farm labor contractors. More workers, particularly new immigrants, are finding employment through farm labor contractors, while more employers are turning to such contractors to provide recruitment and direct supervision of workers, as well as to alleviate many of the responsibilities inherent in an employer/employee relationship.

Farm Labor Policy Issues

In a very real sense, farm labor in the United States is not a problem. A surplus of workers allows farmers to produce enough food so that U.S. consumers enjoy fresh produce the year round while spending a smaller proportion of their household income on food than in any other nation. This implicit "cheap food policy," a result of short-sighted political compromises rather than conscious policy formulation, comes at an enormous cost to farm workers. To date, policy makers have attempted to tinker with the system to improve the lives of workers, but have not addressed the basic question of what a "good" system of labor-intensive agriculture should be, how it could be achieved, and what the fallout would be from implementing such a system. Until a long-overdue national debate about an overall restructuring of labor-intensive agriculture occurs, policies that affect seasonal farm workers will remain piecemeal.

Given the farm labor system that is in place, there are still legislative changes that could help alleviate some of the problems. Farm worker earnings could be raised by expanding labor legislation to afford protection to more farm workers. This would involve removing exclusions for agriculture under certain legislation, such as the National Labor Relations Act. It would also entail a restructuring of other legislation, such as the Fair Labor Standards Act, to take into account the current seasonality of much agricultural employment.

The role of farm labor contractors has, to date, eluded effective policy intervention. The use of a non-English-speaking work force is one reason used to explain the ubiquity of the system. Another is the desire by fixed-site employers or growers to shift the regulatory responsibilities and day-to-day headaches of dealing with a large, seasonal work force and a high rate of worker turnover. Such a shift in responsibility, however, is not easy to achieve. Courts have consistently maintained the joint employer status of growers faced with penalties for violations incurred by labor contractors. Yet the continuing legal possibility of not being held to be a joint employer, as defined under the Fair Labor Standards Act, for violations of the Migrant and Seasonal Agricultural Worker Protection Act, has led to lengthy and expensive court battles over this issue. Clear, legal specification of liability would assist both parties. It is particularly important that responsibility for the actions of farm labor contractors be maintained, given the historic abuse suffered by workers under this system.

Services to farm workers is another area that must be reevaluated. Federal programs for migrant health, education, and job training were instituted under a certain set of assumptions and with a certain view of the migrant labor system. Their basic underpinnings must be examined in light of the changing nature of the occupation and the demographic characteristics of "new" immigrants: young men from Mexico.

The effectiveness of any changes to the farm labor system, however, depends on the willingness of federal and state governments to undertake the enforcement of labor standards in agriculture. Such enforcement has been inadequate. Agencies responsible for enforcement are underfunded for enforcement tasks. Further, enforcement efforts among those agencies have been poorly coordinated. The signal that those in power lack the interest and the will to ensure that farm workers are at least minimally protected has far-reaching implications.

Immigration Policy Issues

Not surprisingly, the Immigration Reform and Control Act has served neither to reform or control unauthorized immigration. Social scientists have long recognized a basic fact that policy makers have attempted to avoid — that immigration is a social process unlikely to respond quickly, if at all, to legislative dictates. The United States may be able to do a better job of managing its immigration, though control will remain elusive.[4] The fallout from a continuing inability to "control" immigration from Mexico is likely to be frustration directed at the immigrants themselves, rather than at the system that has encouraged and sustained large immigrant flows.

IRCA continued the importation of temporary workers to agriculture by making only minor changes in the nonimmigrant guest worker program — the H-2A program. With scattered exceptions (such as in North Carolina), few employers who were not previously H-2 users have adopted this program. This is not surprising, given the current oversupply of agricultural workers. Nevertheless, by limiting debate on this issue, Congress simply postponed an inevitable discussion of a program that institutionalizes a category of workers who are tied to a particular employer, are prevented from negotiating their terms and conditions of employment with that employer, and are excluded politically from the society in which they are working. Such a program raises the basic concern of how such policies coincide with the principles upon which this country was founded.[5] Nonimmigrant worker programs raise other policy issues, including the tendency of such programs to depress wages and working conditions in areas and industries where they are widely utilized.

Other policy questions were raised by the agricultural provisions of IRCA. One was the inherent shift in the balance away from the principle of family unification, which had guided U.S. immigration policy since 1965, and toward an emphasis on employment-based immigration. By legalizing workers because of their participation in a specific occupation, the recipients of SAW amnesty were predominantly male. At its inception, no provisions were specifically included to allow the wives and children of the new permanent resident aliens to join their husbands and fathers in the United States.[6] In addition, the Replenishment Agricultural Worker (RAW) provisions, which were part of IRCA, held out the reward of permanent resident status contingent upon an immigrant's willingness to work in agriculture for a specified period of time. Employers utilizing RAW workers were to have had no requirements, however, to attempt to recruit and retain domestic workers, including those granted legal status through the SAW program. The balance between the often competing priorities of family reunification and employment-based immigration that the SAW and RAW programs represent should be one that is subject to public deliberation and debate.

The Need for a Binational Policy

The long-standing relationship between the United States and Mexico should not be ignored in policy debate. The development of many sectors of the U.S. economy has relied on the "flexibility" Mexican labor has provided. Socially, culturally, and economically, the 2,000-mile political boundary between the countries has proven extremely porous.

The *bracero* program was one example of a binational agreement that recognized the economic contribution of Mexican workers and stipulated protections for these workers when they were in the United States. The abuses that *bracero* workers suffered, largely as a result of noncompliance by U.S. employers, should not negate this precedent.

The scope of Mexican migration to the United States, both legal and illegal, requires that policy debate regarding this issue take place in both countries. As the two countries become increasingly integrated economically and enter into a more comprehensive trading relationship, the opportunity to negotiate labor standards and the movement of people between the two countries within this binational economy should be seized. Mexican immigration to the United States has provided social and economic benefits as well as costs to both countries. The reality, however, is that migration streams that began in response to economic needs — on both sides of the border — have become self-perpetuating social processes that cannot be "controlled" through regulation. Policy debate must start from that point, recognize the range of costs and benefits that immigration has and continues to provide, and develop realistic alternatives designed to protect individuals caught up in the historical tides that have determined the present relationship between the two countries.

Conclusion

IRCA has not only failed to curtail the employment of unauthorized Mexican workers in U.S. agriculture, which would have reduced the farm labor supply, but, through a variety of mechanisms outlined in the 1992 *Report of the Commission on Agricultural Workers* (USCAW 1993), has actually increased the supply of farm labor. The trends that have characterized labor-intensive agriculture in the United States for many years have continued and actually appear to have been accelerated by IRCA. There are larger farms that hire more Mexican-born workers for seasonal jobs under the authority of a farm labor contractor. The need for a comprehensive reexamination of both farm labor policy and immigration policy has become critical. The gap between legal, political, and social realities in both these realms has never been greater.

One underlying question that should provide the basis for a discussion of policy options regarding immigration and agricultural labor is the extent and manner in which the United States should subsidize the agricultural industry by ensuring the availability of low-cost workers. The questions of appropriate wage levels and types of organizational adjustments that agriculture should be required to make before additional foreign workers are to be admitted to this country need to be examined, determined, and then enforced. If market adjustments are allowed to determine the price of labor, the country may arrive at the point at which it will have to consciously decide whether it should import fruits and vegetables or "import" workers willing to labor under conditions that will keep such commodities competitive. Such a decision should not be stumbled into as a result of a lack of coherent agricultural policy.

NOTES

1. In addition to the 1909 Country Life Commission (USPres. 1909), the Senate Committee on Education and Labor (the LaFollette Commission) submitted a comprehensive analysis of agricultural labor problems and policies in California agriculture (USSenate 1942). In 1951, during the height of the *bracero* program, the President's Commission on Migratory Labor issued a report, with recommendations, on the social, economic, health, and educational conditions among migrant farmworkers, both alien and domestic (USPres. 1951). This was followed by a Senate Subcommittee on Migratory Labor report in

1961. In 1967, the National Advisory Commission on Food and Fiber included a section on hired farm labor in its report (NACFF 1967). Finally, the federal Commission on Agricultural Workers issued its report, with recommendations, in 1992 (USCAW 1993). Each of these commissions acknowledged the continuation of a farm labor problem and made policy recommendations to improve the conditions of hired farm labor. Some of these recommendations have been enacted, although a great many have been rediscovered and re-proposed in each report.

2. For example, OSHA regulations, while not uniformly enforced, require that farm workers at large farms be provided access to drinking water and sanitary facilities in the fields, while several states mandate that unemployment insurance be provided to a majority of farm workers.

3. Between 1980 and 1985, cash receipts for fruit, vegetable, and horticultural specialty products increased by 21 percent, while between 1985 and 1990 the increase was 38 percent. Vegetable production increased 13 percent during that first five-year period, while fruit production decreased by 23 percent. However, in the latter five-year period, 1985 to 1990, vegetable production increased by 17 percent and fruit production by 7 percent (USCAW 1993).

4. This could entail, for example, a higher cap for legal immigrants from Mexico or a system of border registration for a set number of temporary migrants that would allow these workers access to the U.S. job market without tying them to a particular employer or type of employment.

5. For a thoughtful discussion of this contradiction, see Calavita (1989).

6. This was partially remedied by a family fairness policy, whereby the Immigration and Naturalization Service did not deport some family members of IRCA-legalized residents, and by a specific number of visas for such family members under the 1990 Immigration Act. Many family members, however, must either enter the country illegally or wait for several years before obtaining visas that will allow them to live together as a family in the United States.

24

Agricultural and Immigration Issues in the 1990s

DON VILLAREJO

Thirty years ago U.S. agriculture enjoyed special exemptions from many labor laws that secured basic rights for workers in all other industries. Farm workers were excluded from the protection of the National Labor Relations Act, minimum wage standards, universal workers compensation insurance requirements, and both social security and unemployment insurance programs. These exemptions were thought to be necessary because, at the time these laws were enacted, a great many farmers were among the ranks of the lower-income segment of the U.S. population and were thought to need special protection. During the 1930s, when much of this legislation was enacted, family farmers had considerable influence in Congress and state legislatures.

Among the special privileges enjoyed by U.S. agriculture until the early 1960s was the *bracero* program, which provided for the special admission of foreign nationals to work in the fields. Though the program ended some three decades ago, many farmers and some farm workers still view that program as an ideal solution to the needs of farm employers for gaining access to a capable labor force.

As a result of the farm workers' rights movement led by César Chavez, basic rights won by industrial workers in the 1930s have now been extended to agricultural employees in a number of states, most notably California, and, for federal minimum wage protection and certain other rights, to all hired farm workers in the nation. Other states, such as Oregon and Washington, are facing increased pressure to include agricultural employees in programs nominally intended to serve all workers but from which farm workers have been excluded.

Step by step, farm workers have begun to enjoy the same bundle of rights that other workers regard as fundamental: a minimum wage, social security, unemployment insurance, workers compensation insurance protection, disability insurance, and collective bargaining.

[453]

The special exemptions enjoyed by agriculture are slowly but consistently eroding.

With respect to immigration, the Immigration Reform and Control Act of 1986 (IRCA) represents what I believe will be the last time that U.S. agriculture will be able to successfully obtain special privileges not shared by other segments of the business community. Clearly, the failure of IRCA either to stop the flow of immigrants or to improve wages and working conditions by restricting the labor supply shows that, under present conditions, it will take much more than an act of Congress to achieve either objective. As far into the future as anyone can imagine, there will be a large and continuing flow of Mexican and Central American nationals into the United States. It is naive to believe that this flow of humanity will be stopped by pieces of paper signed in our nation's capital.

The failure of IRCA also means that the agricultural industry has lost much, if not all, of its credibility to claim a need for special immigration programs. Amidst all the cries of "labor shortage," the evidence is incontrovertible that a substantial surplus of agricultural labor characterizes this market. This surplus has meant that employers have also been able to press real wages even further down.

Among the reasons why the agricultural industry will not be able to successfully push through new special immigration programs is that the composition of the Congress has changed considerably as a result of the 1990 census. Agricultural interests are now even less important and will probably be concentrating their energies on retaining USDA price and income support programs or on cushioning the impact of environmental legislation. Finally, the Latino/Latina representation in Congress is likely to continue to increase into the future.

IRCA, in my view, was a minor blip in a much larger and more important phenomenon. *The dominant social dynamic of the farm labor market for the past generation has been unprecedented levels of immigration from Mexico and Central America.* Driven by economic decline and civil war, millions of political and economic refugees have fled north seeking simply to survive. Many have entered the United States without proper authorization to work in this country. The Statue of Liberty might well be moved to Tijuana, Juárez, or Mexicali in response to the great migration of the last quarter century.

This great south-north flow of humanity has meant that agricultural investors have been able to greatly expand development of orchards, vineyards, nurseries, and vegetable plantings. Labor-intensive agriculture has expanded greatly over the past 20 years fueled by a 50 percent increase in per capita consumption of fresh fruits and vegetables and by greatly expanded exports to Asia and Europe.

Less obvious is the decline of resident family farming and its replacement by large-scale agricultural businesses. Hired workers have largely supplanted farmers and unpaid family labor in many parts of the country. The effect of this trend is increased reliance on hired workers.

As is becoming evident from the results of the National Agricultural Workers Survey (NAWS) and the Commission on Agricultural Workers (CAW) case studies, the hired farm work force in the United States is primarily composed of foreign-born workers. In California, at least 92 percent of hired farm workers are foreign born, mostly Mexicans and Central Americans. Thousands more enter the United States each year seeking work and the possibility of legalization through future Special Agricultural Worker (SAW) programs. *We are more dependent on foreign-born hired farm workers today than at any other time in this century.*

An aspect that is not widely known or understood is that U.S.-born children of foreign-born farm workers usually choose not to seek work in agriculture. A careful random survey of minor children of Mexican farm workers in California's San Joaquin Valley showed that less than 1 in 30 would even consider doing hired farm work. Not surprisingly, they want a better life. *As far into the future as anyone can envision, the California farm labor force will be almost entirely foreign born.*

It is one of the ironies of our age that our nation's leaders seek to resolve the dilemma of the increasing reliance on a foreign-born labor force by ignoring the reality of how the process of immigration actually works. Immigrants do not randomly enter a country hoping to survive. Most often, people leave their homes under duress, either political or economic, just seeking to survive. Many know that native-born U.S. citizens will not take certain types of jobs: farm work, live-in nannies, manual labor of various types, and lower-level service jobs. So those who are willing to do this work head for places where jobs like these are held by people like themselves. Most often they seek places where kinfolk, however distantly related, are already established, or places where others from their village or region are living. This makes it likely that the new immigrants will find places to live and assistance in locating jobs.

These informal immigrant networks provide the vital link between those who have established a beachhead and those who come afterward. In many cases those who follow later are spouses, children, or parents of those who came before. Most often, the new arrivals are a critical part of the earning strategy of the household unit. To seek to cut off one member of the family through immigration policy not only aims to cut familial ties but also strikes at the heart of household survival strategies. Thus, selective policies of admitting immigrants without respecting familial rights will, as in the case of IRCA, ultimately fail because they do not take into account centuries of family survival strategies and traditions. The only way to successfully counter the flow of immigration from Central and South America is to invest in the development of the sending countries. Without new investment and development strategies, we can expect an endless cycle of new migrants in California.

Future Labor Demand

Expansion of the U.S. production of many fruit, nut, vegetable, and horticultural crops appears likely to continue, driven both by increases in the domestic consumption of certain products and by increased exports of many of these commodities. The most rapid increases are likely to be in the export portion.

It is surprising to learn of the sharp increases in the production of fruits and vegetables in the United States over the past 20 years. However, the evidence is compelling: A 70 percent growth in the physical volume of production was recorded in that 20-year period. California has dominated this growth. These data probably understate the increase in the volume of production of labor-intensive commodities since nursery crops are not included in the data.

Total U.S. agricultural exports have suffered in the past several years as a result of greater competition from European and other producers, as well as from the worldwide recession of the early 1990s. The sector most directly affected has been field crops, especially grains. However, U.S. fruit, nut, berry, vegetable, and horticultural product exports have continued to expand, despite the expected adverse impact of the recession on high-priced commodities. In nominal dollars, the combined total of U.S. exports of fruit, nut, vegetable, seed, berry, and nursery products increased by 82 percent in just the past six years. California alone accounts for three-quarters of all U.S. exports of these commodities and, as of 1991, approximately 40 percent of the state's fruit and nut crop production was exported.

A little recognized factor in the importance of international trade for U.S. agricultural producers is that, generally, annual per capita consumption of fresh fruits and vegetables is much greater in other nations than it is in the United States. In both Japan and France, for example, annual per capita fresh vegetable consumption is twice that of the United States. In Turkey, it is four times that of the United States.

This increase in production and exports suggests that we need to examine changes in labor demand that may accompany changes in production volume. We have used the Mamer and Wilkie (1990) labor coefficients for California crops to compare 1989 overall temporary labor demand with similar data computed by Runsten and LaVeen (1981) 15 years earlier. We find an overall 21 percentage *increase* in labor demand implicit in this comparison.

Globalization of production as well as distribution will probably continue to increase under the new rules for trade. Indeed, the point of the new rules is to facilitate, if not directly stimulate, additional international commerce. The opening of the borders of the European Community and the inclusion of Spain and Portugal have already resulted in greater agricultural trade in the EC. Despite the continuing impasse in the Uruguay Round of the General Agreement on Tariffs and Trade (GATT), there will ultimately be a greater international flow of goods, including agricultural commodities. It is worth noting that disagreements about the terms of GATT with respect to agricultural trade appear to be the main stumbling block to the successful conclusion of an agreement. The U.S. seeks terms more favorable to our agricultural exports, while the Europeans, especially the French, want terms more favorable to their producers.

The largest single world market, at this writing, is the European Community. Therefore, much of the world's agricultural production intended for export is targeted to the EC. Those nations with geographic areas suitable for counter-seasonal production enjoy a particular advantage in the competition for these markets during the "off-season." Regions with this capacity have sought to develop their productive resources with the intention of serving this market. It is now possible for European residents to enjoy fresh melons and grapes during the long winter, if one is able and willing to pay the price. Dole Fresh Food Company has recognized this point and has become the main exporter of table grapes from both Chile and the United States, taking advantage of "two" summer seasons each year.

In recent years, California agricultural exporters have geared much of their efforts to serving the two potentially most lucrative markets: the EC and Japan. Altogether, Pacific Rim nations account for 47 percent of California's agricultural exports, including 19 percent to Japan. About 20 percent of the state's agricultural exports are now shipped to the EC.

Well-organized industry leaders, such as Dole and Sun World International, Inc., have implemented strategies designed to reach these markets. According to Doug Barker, executive vice-president of Sun World, "If you're not shipping 30 percent of your product overseas, you're depending too heavily on the domestic market." Sun World reports exporting 85 percent of its Valencia oranges, 65 percent of its grapefruit, 50 percent of its lemons, 40 percent of its grapes, and 45 percent of its tomatoes.

As an illustration of the impact of exports on production trends, consider table grapes. California table grape exports, in pounds, increased by 67 percent in the ten-year period of 1981–90. Exports today represent 21 percent of total fresh grape shipments. Hong Kong is now the third most important destination for California table grapes, ranking behind Los Angeles and New York, but well ahead of all other U.S. cities.

Dave Runsten (this volume) has pointed out that increased production will also result in heightened competition; some firms and regions may be forced out of producing certain crops. Much of the concern in the agricultural community about the North American Free Trade Agreement centers on the impact of heightened competition under the new rules, particularly fresh tomato production in Florida versus Mexico.

The trend to increased dependence on immigrant manual farm labor supplied from less developed nations will be likely to continue in northern-hemisphere production areas. Within the EC it appears that a large share of manual farm labor in Spain is now supplied by immigrants from North Africa, whereas in Italy manual farm labor is being supplied in large part by sub-Saharan Africans. Estimates from Spanish government sources place the number of undocumented North African immigrants working in the Spanish farm sector at more than 1 million.

Within the United States it appears likely that Mexicans and Central Americans will continue to assume a larger and larger share of the hired farm labor force. Already, there are reports of Mexicans supplanting Puerto Ricans in New Jersey. The displacement of African Americans by Mexicans in the hired farm labor force in the South appears to be nearly complete. One surprising finding of the NAWS is the relatively small share of the U.S. farm labor force represented by African Americans.

There is a great reluctance on the part of the EC and U.S. nationals to enter the hired farm work force, obviously associated with the low levels of wages and benefits available and the difficulty of the manual labor required. Therefore, the displacement of the EC and

U.S. nationals will proceed as rapidly as immigrants are able to build the necessary networks to gain access to these jobs.

The continuing hardships experienced by many Mexicans and Central Americans, as well as the displacement of hundreds of thousands of peasants from *ejidos,* which are expected in the next period of time, suggests that "push" factors will play an important role in making large immigrant populations available within the United States.

Zabin et al. (1993) and Runsten and Kearney (forthcoming) have suggested that the impoverishment of rural areas of Oaxaca has facilitated the movement of large numbers of indigenous immigrants into Sinaloa, Baja California, and, eventually, the United States. The appearance of the Mixtec farm workers in California and as far east as Pennsylvania and Maine reflects the broadening and deepening of the immigration of farm workers into the United States.

Gabbard and Mines (this volume) have pointed out the importance of self-generated immigrant "networks," as well as the crucial role of language and cultural intermediaries such as labor contractors and crew leaders in facilitating the entry of new immigrants into the farm labor force. They have suggested, correctly, I believe, that the recent dramatic increase in the utilization of labor contractors in U.S. agriculture is largely associated with the large pool of new immigrant labor being absorbed into the agricultural sector.

Policy Implications

First, as its title suggests, the Immigration Reform and Control Act of 1986 has two components: immigration reform and immigration control. With respect to the former, IRCA accomplished the critical task of providing an opportunity for undocumented agricultural workers in the United States to regularize their status. The large number of undocumented workers employed in U.S. agriculture in the mid-1980s was a serious problem throughout the Sun Belt and certain other regions, and was becoming more serious with each passing year. IRCA dealt with that problem, although some argue that it did so imperfectly. The Immigration and Naturalization Service was ill prepared for the magnitude of its task, largely as a result of a striking underestimate of the number of persons likely to be eligible for the SAW program.

With respect to immigration control, IRCA was a catastrophe. Not only did the mechanism for control — employer sanctions — have important loopholes (such as the infamous provision regarding "knowingly" hiring unauthorized workers), the basic concepts of the reform provisions were in fundamental conflict with the control provisions. That is, IRCA essentially says that if you want a chance to be legally authorized to work in the United States, the place to be working as an undocumented worker is the United States, not Mexico. The 1986 "IRCA surge" in immigration from Mexico is associated with the widespread understanding of the implied meaning of the reform provision.

In the epoch of globalization of the economies of the nations of the world, immigrant farm laborers are probably among the most advanced segments of our labor force. As the work of Palerm (1988, 1991) and his students has demonstrated, we are increasingly seeing transnational families working both as farm laborers and as farmers, with residences established in both the United States and Mexico. In that sense, farm workers, like soccer-style place kickers, may be harbingers of the globalized labor force of the future.

The most important policy implication of the experience of the post-IRCA period to date is that large-scale immigration of farm workers to the United States continues, largely for reasons beyond IRCA's ability to address. The evidence suggests that today, as in the early 1980s, we have a farm labor force in the United States that includes a very large number of undocumented workers and that their number will continue to grow with each passing year.

As a nation we have yet to come to grips in a serious way with the fact that we seem to need, and therefore have, a large, and increasing, number of undocumented persons in our labor force to do the jobs (farm labor and domestic work) that U.S.-born persons will not do. In an open and free society, the border is not like a gate that can be opened and closed. Rather, the issue today is whether we are willing to cooperate with other nations to seek to establish minimum standards for labor across all borders. Can we develop rules for labor standards that are analogous to the rules we seek to establish for moving commodities across borders?

Second, many, if not most, of our farm worker service agencies, both governmental as well as private or nonprofit, have been very slow to adjust to the rapid recent changes in the U.S. farm labor force. From obsolete and bureaucratic definitions of farm workers to the desire of agencies to build empires or protect their turf, the real needs of workers have been sadly neglected in all too many instances. Rare is the agency in which current farm workers provide real policy direction to guide its program.

Third, recent efforts to assess the status of farm workers and suggest approaches to improve their situation have essentially ignored the central importance of strengthening organizations that directly represent workers themselves. Both the Commission of Agricultural Workers and the Farm Workers Services Coordinating Council (California) have avoided the issue of helping to rebuild farm labor or mutual benefit self-help associations. As has been found, farm workers represented by labor unions in the work place are better paid, have more benefits, and experience better working conditions than farm workers not represented by unions. Support for labor organization in agriculture is badly needed. It is to be hoped that the new administration in Washington will provide a more sympathetic environment for workers than did the two previous ones.

Fourth, we need to strengthen the efforts of the National Agricultural Workers Survey (NAWS) (USDOL 1991) to provide accurate, detailed information about the farm labor force. Other data collection efforts, for example the surveys conducted by the USDA (formerly known as QALS) (USDA 1975ff), although now more frequent, provide little, if any, information about employment patterns, demographics, economic status, utilization of government programs, and other vital data. NAWS has helped to humanize our understanding of the farm labor force in a way that reports of employment numbers can never accomplish. If there are objections to the NAWS results, then we should insist that the USDA cooperate with the U.S. Department of Labor to strengthen the NAWS methodology and sharpen its questions.

Appendix A. CAW Executive Summary

Introduction

The Commission on Agricultural Workers was authorized on November 6, 1986 by section 304 of Public Law 99–603, the Immigration Reform and Control Act (IRCA). Its purpose was to study the effects of the act on the agricultural industry, with special emphasis on perishable crop production.

IRCA was primarily a measure to control illegal immigration through a system of penalties against employers who hired unauthorized workers. The act also included special provisions to legalize some aliens present in the United States prior to 1982. The act gave agriculture additional time to adjust to a fully legal work force. These provisions included a program to legalize a large portion of the existing agricultural work force — the Special Agricultural Worker (SAW) program, and included the Replenishment Agricultural Worker (RAW) program that could admit additional agricultural workers during a four-year period beginning in 1989 if farm labor shortages developed. The Commission on Agricultural Workers (CAW) was asked to conduct an overall evaluation of the SAW provisions and review specific questions regarding the legislation's impact on agriculture and the overall agricultural labor market.

Had employer sanctions succeeded in preventing the employment of unauthorized workers in agriculture, a more stable work force could have been expected. This, in turn, could have led to several possibilities: higher real wages and improved overall working conditions for domestic farm workers (including SAWs); improvements in personnel practices; increased mechanization; decreased fruit, vegetable, and horticultural specialty (FVH) production; higher food prices for consumers; an efficient job-matching system; increased use of the nonimmigrant H-2A program; and implementation of the probationary immigrant RAW program. On a national level, none of these has occurred, due primarily to an oversupply of labor resulting from

[463]

continuing illegal immigration and a stagnant economy with little job growth in competing sectors.

The SAW program legalized many more farm workers than expected, providing employers with access to a large, legal work force. No labor shortages that would have triggered implementation of the RAW program have occurred. Those who had hoped that IRCA would lead to a tighter labor market, with resulting improvements in wages and working conditions for farm workers, have been disappointed. Despite this lack of direct impact on the agricultural system, however, IRCA has had important unanticipated and indirect effects on the U.S. farm labor market.

The commission concluded that what perishable crop agriculture needs most is a more stable labor market. Many of its recommendations are intended to bring the sector closer to such stability. Stabilization of the labor force and the adoption of more effective labor management techniques will allow the agricultural industry to best compete in the global marketplace.

Findings

Six years after IRCA was signed into law, the problems within the system of agricultural labor continue to exist. Since the mid-1980s, the living and working conditions of hired farm workers have changed but seldom improved. Nominal wages have increased, partially as a result of mandated changes in the federal, and in some cases state, minimum wage, but these increases have seldom kept pace with inflation, causing real wages to fall. In some areas piece rates have risen, while in others they have remained the same or have been replaced by hourly wages. In most areas an increasing number of newly arriving, unauthorized workers compete for available jobs, reducing the number of work hours available to all harvest workers and contributing to lower annual earnings. Increasing numbers of workers are covered by state-mandated unemployment insurance, but employers are less likely to provide such nonmandated benefits as housing, meals, and transportation.

Despite the fact that some FVH producers face economic losses, there has been a significant expansion of fruit, vegetable, and horticultural specialty production in the United States, an increase in cash receipts from that production, and a rise in farmland prices.

Rather than a stabilization of the labor supply, there is a general oversupply of farm labor nationwide. Unauthorized immigrants continue to cross the southern border in large numbers. While the majority find employment in industries other than agriculture, a significant number join the farm labor force. With fraudulent documents easily available, employer sanctions have been largely ineffective. As such, IRCA has not provided an economic impetus for the changes in the personnel practices of agricultural employers that could lead to improvements in the lives of seasonal farm workers. After a slight improvement in 1985, the decline in real wages and annual earnings for farm workers that began in 1978 has continued.

The continuing influx of unauthorized workers causes problems for both farm employers and farm workers. On the one hand, the surplus of labor in most areas militates against improvements in wages and working conditions for seasonal agricultural employees. On the other hand, unauthorized workers are continuing to find employment, aided by the widespread availability, sophistication, and low cost of fraudulent documents. While it is unclear whether stemming the supply of new unauthorized immigrants, or even removing unauthorized workers from the current labor supply, would lead to labor shortages, illegal immigration has a negative effect on workers who are faced with increasing job competition and employers who are concerned about their continuing access to a legal labor supply.

An Overall Evaluation of the SAW Provisions of IRCA

Notwithstanding fraud in the SAW program and the paucity of reliable data on the size of the perishable crop work force, it appears that the number of undocumented workers who had worked in seasonal agricultural services prior to IRCA was generally underestimated. However, the majority of SAW-eligible workers have indeed gained legal status. This has added substantially to the number of legally available agricultural workers. Nevertheless, due to the continuing participation of unauthorized workers in the farm labor market, the objective of a legal work force has not been achieved and agricultural employers still face the possibility of having needed workers removed from their fields at critical times. Likewise, the labor surpluses, due in large part to the presence of unauthorized workers, have frustrated the development of a stable farm labor market with improved wages and working conditions.

In retrospect, the concept of a worker-specific and industry-specific legalization program was fundamentally flawed. It invited fraud, posed difficult definitional problems regarding who should or should not be eligible, and ignored the long-standing priority of U.S. immigration policy favoring the unification of families.

The statutory provisions in the RAW program for determining the existence of a national labor shortage were compromised by the inclusion of hypothetical questions. The findings of "no labor shortages" in each of the four program years, however, comport with all other evidence. Furthermore, although it was never implemented, the RAW program would not have been an efficient vehicle for dealing with local or regional agricultural labor shortages.

The Extent to Which the Agricultural Industry Relies on the Employment of a Temporary Work Force

Due to its seasonal nature and its reliance on a biological production process, much of labor-intensive agriculture consists of a succession of short-term tasks. With the extent of specialization that characterizes the industry of fruit, vegetable, and horticultural specialty production, the practice is that many workers move not only from task to task but from employer to employer to make a living.

The Impact of the SAW Provisions on the Adequacy of the Supply of Agricultural Labor

Through IRCA, nearly 1.3 million foreign workers applied for legal resident status under the SAW program, a number that far exceeded expectations. To date, over one million of these applications have been approved. Most of these applicants were young Mexican men. Although the SAW program's eligibility criteria required a history of field work in perishable commodities, there is no way to know what proportion of applicants were actively engaged in such work at the time their applications were approved. The U.S. Department of Labor's National Agricultural Workers Survey (NAWS) estimates that in 1990, 29 percent of the perishable crop work force and 39 percent of the harvest work force was composed of SAWs (USDOL 1991). Commission-sponsored case studies conducted in 1990 and 1991 found that between 28 and 77 percent of the harvest labor work force in different regions and commodities was made up of SAWs (USCAW, App. I, 1993). While the majority of SAWs have made application in California and other home-base states, newly legalized workers are increasingly dispersing into farm labor markets throughout the Northeast and Midwest.

The Impact of the General Legalization Program on the Supply of Agricultural Labor

There were nearly 70,000 applicants under the general legalization program of IRCA who listed their occupation at the time of filing as "farming, forestry, and fishing." We do not know how many of those were employed in agriculture or how many other legalization applicants had worked in farming at some other time of the year. The NAWS indicates that about 3 percent of the 1990 seasonal agricultural services (SAS) work force was made up of general legalization recipients (USDOL 1991).

One of the impacts of both the SAW and general legalization programs has been to facilitate the legalized farm workers' continued participation in the farm labor force. Legalization has led to increased settlement in the United States, thus improving the year-round availability of farm workers. Legalization has also facilitated the process of cyclical migration, whereby workers return each year to their homes in other countries.

The impact of legalization on the supply of labor cannot be isolated from that of the presence of large numbers of post-IRCA unauthorized workers. Settlement in the United States has increased the number of "anchor households" whose presence stimulates both the entry and employment of future authorized and unauthorized immigrants. At the same time, as in the past, increased cyclical migration has served to encourage additional unauthorized migrants to travel north in search of work.

The Impact of Employer Sanctions on the Supply of Agricultural Labor

Employer sanctions have been ineffective at preventing, and have not significantly curtailed, the employment of unauthorized workers in agriculture. The number of unauthorized workers, most of whom are young men traveling alone, is significant and growing. Equipped with the easily available fraudulent work-eligibility documents, they compete for entry-level jobs in the harvest labor work force.

A consequence of the failure of employer sanctions to significantly reduce the supply of unauthorized immigrants in U.S. agriculture is the diffusion of unauthorized farm workers from such traditional immigrant settlement areas as California and Florida to the Midwest and eastern seaboard in search of jobs. Farm labor contractors often serve to recruit and transport workers from the traditional settlement areas into new parts of the country.

The Extent to Which SAWs Have Continued to Work in Seasonal Agriculture

A large proportion of SAWs who were identified as working within agriculture after obtaining legal status appear to continue to work in seasonal agricultural services.

These SAWs' continuing participation in the farm labor force is the result of limited English-speaking ability, low educational levels, personal preferences, and relatively low wages and high unemployment rates in other industries. In the six years since the enactment of IRCA, the poor state of the U.S. economy has curtailed SAWs' movement out of agriculture.

Generally, worker decisions regarding post-IRCA work patterns and employment options are strongly affected by such factors as family considerations, length of experience in the U.S. labor market, and occupational history. As a result, there is no typical post-IRCA work adaptation, but instead a broad spectrum of working and living patterns. However, as the work force ages and gains experience in the United States, gradual attrition of SAWs from agricultural work can be expected.

The Adequacy of the Supply of Agricultural Labor in the United States and the Need to Further Supplement this Supply with Foreign Labor

Despite an expanding perishable-crop industry, the national supply of agricultural labor has been more than adequate for the past several years. In many parts of the country, however, a significant portion of the harvest work force is unauthorized. The majority in this latter group have entered the agricultural labor force since 1986.

With the existing agricultural labor supply, there is currently no need to supplement the farm work force with additional foreign workers beyond those authorized through the existing H-2A program. However, questions regarding the long-term adequacy of the agricultural labor supply persist. For example, more effective enforcement of employer sanctions would affect the supply of farm labor and could necessitate access to additional legal foreign workers.

The Impact of the SAW Provisions on the Wages of Domestic Farm Workers

In general, agricultural wages have stagnated in the post-IRCA period while average nominal wage rates have risen. Most farm workers' real (adjusted for inflation) hourly, weekly, and annual earnings have fallen. The discrepancy in agricultural wage rates between western growers, who prior to 1986 paid higher wages than eastern growers, has been reduced. This is the result of western growers having held wages steady while nominal wages paid by eastern growers have risen — in part due to increases in the federal minimum wage.

Earnings for western seasonal harvest laborers, particularly those hired through farm labor contractors, have seen the sharpest decline. Any possibility for a general increase in farm labor earnings has been offset by the presence of large numbers of unauthorized workers and the general economic downturn affecting all industries. It is thus impossible to determine the precise effect of the SAW program on wages.

The Impact of the SAW Provisions on the Working Conditions of Domestic Farm Workers

The SAW program has had little direct impact on working conditions for farm workers. Since IRCA, labor surpluses have contributed to stagnating and sometimes deteriorating conditions, particularly for harvest workers. There has also been increased crowding in farm worker settlement areas. Because most employers have had no difficulty attracting and retaining workers, there has been little incentive for them to increase benefits or generally improve working conditions for farm workers. The abundance of labor, the increased internal migration of SAWs, and the increase in men traveling alone, combined with higher building costs, increased regulations, and community resistance, have also reduced incentives for agricultural employers to upgrade and expand farm worker housing.

The Impact of the SAW Provisions on the Ability of Farm Workers to Organize

There are fewer agricultural workers under union contract now than when IRCA became law. While there is evidence that legal status has allowed some workers to become more vocal in attempts to improve their job conditions, labor surpluses have generally interfered with workers' ability to organize. Even prior to IRCA, this ability had been curtailed by changes in the general economy, the political climate, and management decisions affecting the agricultural work place — such as the increasing reliance on farm labor contractors in California.

The Extent to Which Citizen and Permanent Resident Farm Workers are Unemployed and Underemployed

Virtually all groups of legally authorized farm workers (citizens, pre-IRCA permanent resident aliens. and IRCA-legalized resident aliens) who hold seasonal jobs are unemployed at some point during the course of the year. The extent of unemployment and underemployment is influenced strongly by the supply of labor, crop size and prices, and the availability and quality of mechanisms for matching workers with jobs.

Since IRCA, largely because of labor surpluses, unemployment rates among farm workers have remained high, particularly during the off-peak season. Furthermore, there is no effective job placement system. Therefore, there is a great deal of variance in terms of an individual's access to work. Such access ultimately depends on the strength of individual farm worker connections. Farm workers' various economic strategies often include periods of nonfarm work, reliance on unemployment insurance, and time spent in their home countries. Workers, however, would prefer to have access to more work.

The Extent to Which Agricultural Employers' Problems in Securing Labor Are Related to the Lack of Modern Labor Management Techniques

The abundance of labor has meant that most agriculture employers are not experiencing problems in securing adequate supplies of labor. As a result, there have been few incentives for producers to adopt modern labor management practices.

Modern labor management techniques in agriculture would include the adoption of benefit packages. seniority systems, mechanisms for dispute settlement, and techniques to reduce the size of the work force and increase the efficiency and amount of work available per worker. Adoption of such practices would allow employers to better utilize the local domestic work force. Instead of adopting such techniques to stabilize the labor force, many agricultural producers are turning to labor market intermediaries. Quality labor management in recruitment, compensation, and retention, which would have become necessary in tighter labor supply situations, would result directly in increased employment stability and lower rates of turnover.

Whether Certain Geographic Regions Need Special Programs or Provisions to Meet Unique Needs for Agricultural Labor

There is a great deal of variation among regions in the need for labor and the reliance on workers who migrate. Climate, cropping patterns, the size and structure of farms, and proximity to labor sources are attributes that distinguish one region from another.

Since IRCA, growers generally have been able to make planting decisions and assume that workers would be available. However, some regions or, more accurately, individual producers within a region, may experience spot labor shortages. To respond to these situations, better systems for recruiting and training workers, transporting them to their jobs, and providing housing are necessary. When these efforts are ineffective, utilization of the H-2A program can be relied upon to prevent economic loss.

*The Impact of the SAW Provisions on the Ability
of Crops Harvested in the United States to Compete
in International Markets*

The SAW provisions have had no direct impact on the ability of crops harvested in the United States to compete in international markets. With the current oversupply of labor, producers have experienced no labor-induced increases in production costs. The export of labor-intensive, perishable commodities reached an all-time high in 1991.

In the long run, however. the short-term positive impacts of a labor surplus on the industry's export performance may be offset by a tendency to dampen the industry's commitment to technological and organizational improvements that could increase production efficiency. The ability of the U.S. fruit, vegetable, and horticultural specialty industry to compete with other nations, particularly Mexico, stems from its advantages of a highly developed infrastructure, high-quality produce, and the ability to offset higher labor costs with greater productivity.

Recommendations

Many of the problems in the farm labor market can be alleviated by a more structured system. Such a system would be characterized by more effective recruiting and job matching, reduced worker turnover and higher retention rates, a more dependable labor supply, institutionalized opportunities for training and advancement, and a better balance between labor supply and labor demand. Such a system would further address the needs of seasonal farm workers through higher earnings and the needs of agricultural employers through increased productivity and decreased uncertainty over labor supply. Market mechanisms would provide the incentives that would ultimately lead to and maintain this stabilization.

A stable and reliable work force is critical to the long-term health of the industry and would thus provide clear benefits to both workers and employers. One element necessary for achieving this goal is curtailing the employment of unauthorized workers in agriculture through the effective enforcement of immigration controls. A second is an effective system for recruiting agricultural workers. Farm labor contractors, who currently serve this matching function, must be more effectively regulated. In addition, alternatives to the use of contractors must be developed. These alternatives could be implemented by the government or through the private sector, with either worker or grower associations taking over the function of matching workers with jobs.

The agricultural worker provisions of IRCA were supported by both industry and labor groups. This indicates that common interests between these two traditional adversaries can be found. It is in the spirit of finding such common ground for the benefit of labor, the industry, and the public that the commission has reviewed its findings and developed the following recommendations.

Any changes in various federal programs called for in this report should be achieved by the reallocation of existing resources and not from increased federal fees or taxes. The commission's recommendations can and should be realized through a reprioritization and re-examination of the efficiency of current programs.

SAW/RAW as Immigration Policy

In crafting the agricultural compromise that allowed IRCA to pass, Congress created an industry- and worker-specific legalization program of unprecedented scale. It proved to be extremely difficult to implement the SAW program and to control for fraud within it. By legalizing workers but not their families, the Special Agricultural Worker (SAW) program deviated from the long-standing immigration policy of promoting family unity and contributed to the growing pool of immigrants who illegally enter the United States. Furthermore, special treatment for an industry requires difficult definitional distinctions to determine who should and who should not be eligible for the benefits of legalization.

The worker-specific or industry-specific legalization programs contained in IRCA should not be the basis of future immigration policy.

The Implementation of IRCA

The commission's evaluation of the SAW program leads it to conclude that in its implementation, some of the program parameters regarding worker eligibility for legalization were unduly restrictive.

Specifically, the commission believes that the exclusion of sugarcane and sod workers from SAW program eligibility was unwarranted. Field work and workers in these two commodities are close enough to other qualifying work and workers to have warranted inclusion within the SAW legalization provisions of IRCA. The exclusion of sugarcane and sod workers is relatively easy to rectify inasmuch as these workers filed applications for SAW status during the course of litigation. Notwithstanding a recent appellate court decision, the legal status of sod workers should not be rescinded and the applications of sugarcane workers now in the government's possession should be adjudicated.

Congress should amend the Immigration Reform and Control Act to specifically include field work in sugarcane and sod within the definition of seasonal agricultural services.

The Effects of IRCA on the Supply of Hired Agricultural Workers

Attacking the pull factors. A major factor contributing to the current oversupply of agricultural workers in the United States is continuing illegal immigration. This migration consists of family members and friends of those legalized under IRCA and others who are seeking work or who have been recruited. Increased monitoring of the southern border has not stopped the well-established immigration pathways between Mexico and the United States. With employer sanctions failing to curtail the employment of large numbers of these unauthorized immigrants, unauthorized entrants have continued to supplement the pool of available workers in the United States.

As long as the employment of unauthorized workers continues, it will be difficult to have a structured and stable agricultural labor market. Employers will continue to be threatened with the potential loss of a portion of their work force, while workers will suffer the consequences of stagnating wages and deteriorating working conditions.

Illegal immigration must be curtailed. This should be accomplished with more effective border controls, better internal apprehension mechanisms, and enhanced enforcement of employer sanctions. The U.S. government should also develop a better employment eligibility and identification system, including a fraud-proof work authorization document for all persons legally authorized to work in the

United States, so that employer sanctions can more effectively deter the employment of unauthorized workers.

If these immigration control mechanisms ultimately prove ineffective, after being given ample time for implementation, alternatives to employer sanctions should be explored.

Attacking the push factors. In addition to the Commission on Agricultural Workers, IRCA established a Commission for the Study of International Migration and Cooperative Economic Development. This commission was asked to explore measures that might mitigate migration pressures within the sending countries. Focusing on the principal reason people migrate — the lack of adequate economic opportunities at home — it recommended various ways to stimulate economic growth in developing countries.

The Commission on Agricultural Workers agrees with its sister commission on the need for cooperative efforts that encourage economic development.

Supplementing the labor force. An important part of the mandate of the Commission on Agricultural Workers is the question of whether the domestic agricultural labor force should be supplemented with legal foreign workers during periods of labor shortages. The authors of IRCA sought to create statutory mechanisms that would respond to a shrinking agricultural labor supply. The RAW program was one of those mechanisms. Streamlining and continuing the H-2A program was the other.

The RAW program was designed specifically to supplement the farm labor force in the event of a worker shortage as U.S. perishable crop agriculture was adjusting to the new circumstances IRCA was expected to bring about. However, with the failure of immigration control efforts, the projected tightening of the labor supply, with concomitant changes in labor-intensive agriculture, has not materialized. Nor does the commission believe that this situation will change in the near future.

Neither an extension of the RAW program nor any new supplementary foreign worker programs are warranted at this time. However, adequate labor supply-and-demand statistical programs must be continued and an ongoing analysis performed to provide a basis for immediate action by Congress or the Administration, should future agricultural labor shortages develop.

By "immediate action" the commission means whatever measures may be necessary to avert crop losses due to a shortage of labor, without weakening protections for domestic workers. Such measures could include the effective recruitment of domestic, including Puerto Rican, workers through cooperative efforts between private- and public-sector agencies (including unions) and the importation of legal foreign workers.

Accurate determinations of the adequacy of the agricultural labor supply are a critical prerequisite to successful responses to crises.

The U.S. Departments of Labor and Agriculture should continue to coordinate efforts to determine the adequacy of the agricultural labor supply, including the proportion of legally authorized workers. This information should identify possible agricultural labor shortages and the need for any new supplemental foreign agricultural worker program. These findings should be reported to Congress annually.

The Department of Labor should continue to sponsor research on farm workers and their evolving migration patterns, both within the United States and between the United States and immigrant-sending countries. The department should issue an annual report on the status of U.S. farm workers.

The commission recognizes that there is an oversupply of workers in most agricultural labor markets. Yet there continue to be instances of crop-specific local labor shortages. This indicates that the supply of workers is not coordinated well enough with the demand for workers. This mismatch encourages some employers in some areas to rely on the U.S. Department of Labor-administered H-2A program.

The H-2A program has been the target of both criticism and litigation. Many growers feel that the H-2A procedures are too cumbersome to respond to emergency situations in a timely fashion. In testimony before the commission, growers argued that the complicated procedures and costs involved, particularly the need to provide worker housing, make the program prohibitively expensive. In view of the criticism of and the litigation surrounding the current H-2A program, its provisions should be reviewed.

Congress should reexamine the H-2 provisions of the Immigration and Nationality Act, with the goal of having the same program available to agricultural and nonagricultural employers, both in law and regulation, for the importation of temporary nonimmigrant workers.

Matching Agricultural Workers with Agricultural Jobs

The fruit, vegetable, and horticultural specialty industry will continue to rely on a seasonal work force for many tasks. The seasonal nature of the demand means that many farm workers are unable to find enough work during the course of the year to be economically self-sufficient. Resolving this dilemma is a precondition not only for the survival of the economically at-risk population, but also for the long-term viability of the industry.

The U.S. Employment Service. Individual agricultural employers will continue to need seasonal workers for relatively short periods of time. The coordination between available jobs and available workers, however, can be improved. The U.S. Department of Labor's Employment Service currently plays a negligible role in farm placement — despite a statutory mandate to do so. The commission is sensitive to the relatively high cost of offering placement services for short-term jobs, and is cognizant of the mission conflict inherent in the Department of Labor's dual mandate of enforcing work-place and housing standards while soliciting jobs for workers. Nonetheless, the commission urges the department to consider redirecting its efforts to utilize the resources devoted to the agricultural sector more effectively.

The Department of Labor's Employment Service should develop a new or alternative system for recruiting qualified farm labor to meet agriculture's constantly changing labor needs. The focus of this initiative should be to facilitate the efficient movement of farm workers from one crop task or job to the next, including the use of job itineraries or annual worker plans. Special funding provisions should be made for the development of this program.

The U.S. Employment Service should expand efforts to distribute information to farm workers regarding labor needs by crop and area, including updates reflecting rapidly changing conditions. Outreach efforts should include information on the availability of housing for arriving workers.

The Department of Labor should consider separating the job-matching function of the U.S. Employment Service from its nonlabor exchange functions.

Congress should establish a task force to oversee implementation of the above recommendations.

Farm labor contractors. In recent years, farm labor contractors (FLCs) have increasingly filled the role of matching seasonal workers with jobs. Concurrently, the extent to which the U.S. Employment Service performs this function has declined steadily. Workers employed by farm labor contractors generally receive lower wages and are employed under working conditions inferior to those offered to farm workers hired by other agricultural employers.

Farmers who hire labor contractors can be held jointly liable for these contractors' violations of labor laws. Many times, however, there are special problems in collecting judgments against farm labor contractors. FLCs, even when fully responsible for the actions resulting in judgments, are often unable to satisfy them. In cases of joint liability, the payment then falls on the fixed-site employer, or farmer. Furthermore, unlicensed farm labor contractors are thought to be more likely to violate protective labor statutes than licensed ones. The commission believes that the regulatory system that controls farm labor contractors must undergo significant reforms.

Licensing for farm labor contractors should be conditioned on adequate bonding for the services provided.

Participation in and adequate completion of educational and training programs, and a testing procedure, should be a prerequisite to an individual being certified as a farm labor contractor.

Persons who knowingly use the services of an unlicensed farm labor contractor should be solely liable for any claim arising from any violation of the applicable federal protective labor statutes or regulations.

Despite specific legislation designed to regulate farm labor contractors, the laws and regulations that govern farm labor contracting have failed to adequately protect many farm workers. At the same time, they have provoked much litigation. In the absence of collective bargaining agreements, there has been an increase in the use of farm labor contractors, particularly in California. In light of this trend, steps must be taken to better protect employers and employees. Regulation of farm labor contractors is another area in which the utility and efficacy of exceptional treatment for agriculture should be reconsidered.

The body of the federal statutes and regulations that govern contracting activities in the farm and nonfarm labor markets should be reexamined to bring all contracting activities under a uniform federal system.

Managing Agricultural Labor

In addition to the farm labor market's organizational problems, many individual labor-intensive farm operations continue to be characterized by organizational inefficiency. Worker dissatisfaction and turnover increase employer costs and reduce farm worker earnings. Mutual employer-worker benefits are possible through the adoption of modern labor management techniques. Yet both managers and workers receive little training on how these might be accomplished.

The U.S. Department of Agriculture's Extension Service, the U.S. Department of Labor's Employment Service, and state agencies and universities should undertake a major effort to educate growers, packing house operators, farm labor contractors, workers, and worker organizations in the need for and benefits of improving labor management practices in agriculture. This effort should include encouraging appropriate state agencies and universities in those states with a high incidence of hired farm labor to employ advisors, teachers, and researchers with expertise in farm labor management. Research on the need for and benefits of good labor management in agriculture, and the development of training and teaching programs, including outreach programs, in modern labor management must be supported.

Differential Treatment of Hired Agricultural Workers

There are several areas of labor law in which exceptional treatment for agriculture is no longer warranted and puts agricultural workers at a disadvantage compared to other workers. This dichotomy can also be confusing and costly to agricultural employers.

Unemployment insurance. Seasonal agricultural workers are by definition employed for less than a full year. As a result, significant unemployment and underemployment among farm workers is common — particularly within the fruit, vegetable, and horticultural specialty sector. Continuing labor surpluses have contributed to even lower average annual earnings. It is thus essential that FVH farm workers be placed under the full protection of unemployment insurance programs and that the differential coverage of farm workers across the unemployment insurance programs of various states be eliminated.

Agricultural employees should be provided with federal/state unemployment insurance coverage that provides them with protection against unemployment comparable to that of other workers in the United States.

The commission recognizes the difficulty of administering unemployment insurance programs when multiple employers are involved in a relatively short period of time, and believes that close monitoring would be necessary to ensure program integrity.

Workers compensation. An additional area requiring policy attention concerns access to workers compensation benefits for agricultural workers. Agricultural employment remains one of the most hazardous occupations in the country. Yet many farm employees are poorly served by the uneven coverage of hired agricultural workers under state workers compensation statutes. This must be remedied.

Congress should encourage all states to provide workers compensation insurance coverage to agricultural employees comparable to that of other workers in the United States. Workers compensation should be the exclusive remedy for all U.S. employees.

A recent judicial decision in *Adams Fruit Company v. Barrett* may have a negative effect on the likelihood of states to include farm workers within workers compensation legislation. The court held that a state workers compensation program does not serve as the exclusive remedy for injuries resulting from violations of the Migrant and Seasonal Agricultural Worker Protection Act. Ongoing negotiations in Congress are attempting to address this possibility.

The commission recognizes the difficulty of administering workers compensation programs when multiple employers are involved in a relatively short period of time, and believes that close monitoring would be necessary to ensure program integrity.

The right to organize and bargain collectively. Farm workers face special problems if they attempt to organize and bargain collectively in order to improve their working conditions. Effective organizing is made more difficult by the fact that farm workers are essentially powerless, both in objective terms and relative to the agricultural employers who oppose organizing. This powerlessness is compounded by the explicit exclusion of agricultural employees from legislation designed to afford U.S. workers this basic right.

Farm workers should be given the right to organize and bargain collectively, with appropriate protections provided to all parties.

Living Conditions for Hired Agricultural Workers

Housing. The scarcity of low-cost housing continues to disrupt the lives of many seasonal agricultural workers. Thus, a serious examination of the overall farm labor system cannot be complete without paying proper attention to the housing needs of farm workers. Yet, those who try to address this problem, in addition to obtaining only seasonal returns on their investment, also face administrative problems and community resistance in their attempts to provide adequate low-cost housing for farm workers. Few states have been eager to finance, much less operate, housing projects for seasonal farm workers. The result is that workers live where they can — many in overcrowded, substandard units or temporary camps, or with no shelter whatsoever. The commission believes that housing seasonal farm workers is a critical step to developing a stable and committed labor force. Housing must thus be addressed more realistically and effectively. A necessary first step in this process is a reassessment of the federal government's role in supporting farm worker housing.

The Farmers' Home Administration, as well as any other agencies involved in providing funds for farm worker housing, should review their current housing programs to facilitate and expand funding to provide publicly and privately constructed or rehabilitated housing for agricultural workers.

Federal and state standards for farm worker housing should be reviewed to allow more flexibility in the design and construction of conceptually different seasonal farm worker housing that responds to the needs of workers and is economically viable. This could take the form of in-transit camp sites, trailer camps, or direct housing subsidies to farm workers. Federal and state funds should be provided for such construction and assistance.

Services to farm worker children. The conditions in which many children of seasonal farm workers live is of critical concern to the commission. An unstable child-care environment, and, at best, intermittent access to health care and educational opportunities place an unjust burden on these children. Between fiscal years 1988 and 1992, federal funding for migrant health, education, and day-care programs increased 28 percent, yet many children remain unserved and most children are at risk.

Providing adequate services for farm worker children must take into account the mobility of the parents, language problems, and other elements that are unique to the farm worker population.

Federal, state, and local government programs and policies should ensure that the children of migrant and seasonal farm workers are provided the same opportunities for health care, education, day care, and child development as are provided to the children of other workers in the United States.

Coordinating farm worker service programs. A 1992 report by the Administrative Conference of the United States, prepared at the request of the National Commission on Migrant Education, addressed the problem of regulatory barriers to coordinating federal programs that provide benefits to migrant and seasonal farm workers.

The Commission on Agricultural Workers endorses the recommendations of the Administrative Conference to ensure the efficient and effective delivery of services to farm workers and their dependents by improving the coordination of farm worker programs at the federal, state, and local levels. A better data collection system could help to achieve these goals.

An interagency coordinating council on migrant and seasonal farm worker programs should be established to strengthen national coordination of these service programs. The council would be charged with identifying specific coordination tasks to be accomplished, in most cases under the primary responsibility of a designated lead agency.

A farm worker database designed to meet the needs of federal agencies involved in designing and allocating resources at the local level for programs for migrant and seasonal farm workers should be created. This database should be developed by a multi-agency federal group representing the various farm worker programs.

Within the President's budget there should be an ongoing review of all programs that expend funds for delivery of services and enforcement of protective statues for farm workers and their children to determine whether these programs are effectively meeting farm worker needs.

Enforcement of Farm Worker Protective Statutes

Enactment of the above recommendations would help stabilize the farm labor market. However, it is clear that a sustained commitment to enforce protective legislation for farm workers is also essential. The commission was made aware of numerous violations of worker protective statutes in both its research and hearings. Clearly this must stop.

The enforcement of protective statutes for farm workers should be made more effective, and should be coordinated within the federal enforcement agencies and between responsible federal and state enforcement agencies. When it is apparent that an agency has failed to properly interpret, enforce, and apply laws and regulations assigned to it for enforcement, and an agricultural agency has relied upon that agency's interpretation of those laws and regulations in the conduct of its business, then the agency and not the agricultural employer should be held legally responsible for any monetary damages, fines, and penalties arising therefrom. All laws related to farm labor should be uniformly enforced by the agencies concerned, so that employers not in compliance do not gain an unfair competitive advantage over those employers in full compliance with the various laws and regulations.

The U.S. Department of Labor's Farm Labor Coordinated Enforcement Committee should play a leading role in the allocation and coordination of resources to effectively enforce agricultural labor standards. To this end, it should convene a special task force of industry, labor, and public representatives to conduct a comprehensive review of present U.S. Department of Labor enforcement policies and practices that affect farm labor contractors, farm workers, and farmers, including enforcement strategies. This task force should direct enforcement toward those violations that endanger the health and safety of workers, while encouraging agencies to work with employers to correct the problems arising from other violations.

Conclusion

Congress charged the commission with addressing issues specific to the effects of IRCA, but also to certain long-term concerns that surround agricultural labor in the United States. In addressing its mandate, the commission identified certain significant trends that are changing the structure of agriculture. Increasing economies of scale in production and distribution have led to more levels of supervision, changing the relationship between employers and employees. The globalization of agriculture has allowed labor-intensive fruits, vegetables. and horticultural specialty products to be grown and distributed by countries whose wages are only a fraction of comparable U.S. wage costs. How FVH agriculture in the United States can remain competitive in the world market while affording a decent life for its workers was central to the theme of the commission's discussions.

The commission examined the farm labor market in the aftermath of a policy designed to reduce the number of people who worked and lived illegally in the United States, curb illegal immigration, and provide an adequate legal agricultural work force. While the SAW program was successful in legalizing large numbers of workers, ineffective enforcement of employer sanctions and inadequate border controls have curbed neither illegal immigration nor the employment of unauthorized workers in agriculture.

The 1986 law and its effects cannot be analyzed as if they occurred in a vacuum. The overall U.S. economy, consumer expectations of inexpensive yet high-quality fresh produce, changing patterns of agricultural production, the economic and political situation in immigrant-sending countries, and changing trade opportunities and challenges must all be taken into account. Furthermore, international migration must be understood as the social process it is. The strength of networks in continuing migration patterns and as a time-tested and preferred method of worker recruitment has proven to be unyielding, despite expectations that changes in the legal framework would alter long-standing labor market dynamics.

There is currently an oversupply of agricultural workers at the national level. The attrition of newly legalized workers from agriculture has been much less rapid than expected. At the same time, illegal immigration has continued and the work force in many parts of the country now includes proportions of unauthorized workers that rival those of a decade ago. This is particularly clear in the case of the harvest work force. That work force has become increasingly

"Latinized" and more likely to be employed through farm labor contractors than in the pre-IRCA period. Because of the labor surplus, wages and working conditions for most hired farm workers have remained relatively stagnant since the passage of IRCA. For many seasonal (particularly short-term) workers, real earnings and working conditions have deteriorated.

The response of the United States to competition from countries that pay even lower wages should be the development of a more structured and stable domestic agricultural labor market with increasingly productive workers. Industries must modernize to remain successful in the increasingly competitive international market place. Agriculture is no exception.

To assure its long-term competitive position, agriculture must improve its labor management practices. This report suggests ways to stabilize the labor force, improve productivity, and increase earnings for farm workers through longer periods of employment.

In a fundamental way, IRCA's goal of controlling illegal immigration may be better served by such reforms. The prospect of employment is a magnet for unauthorized workers. To the extent that job opportunities are secured by legal workers in a more stable labor market, the pull factor for illegal immigration is reduced.

Appendix B. Agricultural Labor:
A Review of the Data

VICTOR J. OLIVEIRA AND LESLIE A. WHITENER

Introduction

Efforts to estimate the impact of the Immigration Reform and Control Act (IRCA) of 1985 on the U.S. agricultural industry and its workers have been hampered by the lack of detailed and reliable data on the Nation's hired farm work force. While several national-level data sets containing information on hired farm workers are available, there is currently no one comprehensive data set that provides the necessary detailed information to help us understand IRCA-related changes in the supply, demand, wages, earnings, benefits, and characteristics of hired farm workers at both the local and national level. Furthermore, because actual data estimates vary widely among these sources of data, conclusions about the effects of IRCA may vary depending on the data source used for analysis.

In this chapter, we review the major data sources which are most useful for examining agricultural labor issues, including those related to IRCA. To be included in our review, the data source must

- measure some aspect of hired agricultural employment, wages, earnings, or labor expenditures;

- provide data at the national level;

- use a statistical survey or formal estimating procedure to generate data;

- be collected on a periodic or regular basis; and

- be available to the public.

The eight sources of hired farm worker data that met these criteria differ in terms of survey methodology, definitions and concepts, frequency of data collection, survey reference period, geographic detail and types of data collected (Table C.1). Each data set has advantages and disadvantages; some are more appropriate for IRCA-related research than others. This chapter provides general background on each data source, discusses the most relevant uses and limitations of the data, and identifies several major points that users should consider when selecting a data set for analysis or interpreting study findings based on these data. A set of tables presenting geographic distributions of farm labor data from these different data sources is also included.

Agricultural labor data originate from three major sources: establishments that employ workers, households which supply workers, and agencies that administer employment-related programs. The data sources reviewed here are classified according to these three categories.

Establishment Surveys

The major sources of establishment data are the Census of Agriculture, the Farm Labor Survey, and the Farm Costs and Returns Survey. Establishment surveys collect data directly from the farm employer and generally provide information on the characteristics of the job or the farm, but not on the demographic characteristics of the workers. Since data are collected from the employer and not the worker, establishment surveys are more likely to include unauthorized alien workers who may avoid survey enumerators because of their illegal status or who may be missed in household samples.

A growing concern regarding establishment surveys is the increased incidence of farms producing crops, livestock, and poultry under contract where the contractor and not the farmer makes most of the production decisions. These "nontraditional" forms of production arrangements make data collection more difficult when farm operators are asked to estimate labor use or costs.

Census of Agriculture

Responsible agency .... Bureau of the Census,
                                      U.S. Department of Commerce
Date of origin 1840
Frequency of data collection .... Every 5 years
Reference period .... Year
Degree of coverage .... Sample of farms
Geographic detail .... U.S., state, county

The Census of Agriculture has been conducted by the Bureau of the Census periodically since 1840. It is the leading source of statistics about the Nation's agricultural production and is the most comprehensive source of agricultural data available at the county level. The Census is now conducted every 5 years, for years ending in 2 and 7. Data are available in published volumes for each state and the United States (U.S. Department of Commerce 1989) and in public-use summary computer tapes. The Bureau of the Census will provide special tabulations for users at cost (see Oliveira 1991 and Runyan and Oliveira 1992 for examples of expense data obtained from special tabulations). Summary data for the 1992 Census of Agriculture are scheduled to be released in the fall of 1994.

Methodology. The Census of Agriculture is a mail survey of U.S. farms and ranches. The mailing list for the 1987 Census generated 1.8 million useable questionnaires. After adjusting for nonrespondents, survey data were expanded to the estimated 2.1 million farms in the United States. To reduce respondent burden, some questions were asked only of a sample of farms; data on hired and contract labor expenditures were collected from this sample of about 616,000 farm operators in 1987.

Types of data available. The Census of Agriculture provides separate estimates of expenses for hired workers, contract labor, and customwork and machine hire at the national, state, and county level. Expenditure data can be examined by 3-digit Standard Industrial Classification (SIC) of farms, value of agricultural products sold, size of farm in acres, type of organization, and selected operator characteristics. Data on the number of persons working fewer than 150 days or 150 days or more are collected periodically and will be available in the 1992 Census.

Data uses and limitations.

- The Census of Agriculture offers the most complete geographic coverage of hired and contract farm labor use as measured by labor expenditures. Expenditure data can be used to indicate the relative magnitude of labor use and to estimate the share of total production expenses attributed to labor by size and type of farm.

- The Census of Agriculture, along with the Farm Costs and Returns Survey, are the only data sources that collect information on customwork (activities such as spraying or threshing where a person is paid a combined rate for use of equipment and labor). However, expenses for labor involved in customwork are combined with expenses for machine hire and cannot be separated out.

- Data on the number of hired workers working less than 150 days and 150 days or more are collected only periodically. Days worked are not reported cumulatively for all jobs held by a farm worker, but refer only to work on a particular farm. Thus, worker data are subject to double-counting (especially for workers in the less than 150 days worked category) since workers are reported by each of their employers during the year. Data on the number of contract workers are not collected.

- The Census does not collect information on the demographic and job characteristics of hired and contract workers.

- Data on hired workers refer to all hired workers on the farm, including bookkeepers, secretaries, and mechanics who are generally not considered to be hired farm workers. Expenditure data do not include payment in-kind, such as lodging or meals provided to workers.

- Census data are collected only once every 5 years and may not reflect the most recent changes in the farm labor situation.

Farm Costs and Returns Survey

Responsible agency .... Economic Research Service and National Agricultural Statistics Service, U.S. Department of Agriculture
Date of origin 1984
Frequency of data collection ...... Annual
Reference period .... Year
Degree of coverage .... Sample of farms
Geographic detail .... U.S., region, state[1]

The Farm Costs and Returns Survey (FCRS) is conducted annually by the U.S. Department of Agriculture's National Agricultural Statistical Service (NASS) and the Economic Research Service (ERS). The FCRS is used to determine production input costs for various commodities (as mandated by Congress), estimate farmers' net farm income, and determine the characteristics and financial situation of farm operators and their households. Information is collected for a variety of farm expenditures, including farm labor. These data are published annually by NASS in the Farm Production Expenditures series and by ERS in the Economic Indicators of the Farm Sector series (U.S. Department of Agriculture, 1992, 1994a, 1994b).

Methodology. The FCRS is a probability sample featuring multiple frame sampling from a list frame (consisting of mostly larger, more specialized operations) and an area frame of small land areas to account for farms not on the list. The FCRS is conducted in all states except Alaska and Hawaii. Data are collected for the entire year. Each year the useable sample size is about 12,000 farms.

In the past, the FCRS undercounted the number of farms due to problems with both undercoverage and nonrespondents. Beginning with the survey of the 1992 calendar year, NASS and ERS implemented adjustment procedures for expanding FCRS data to provide more complete coverage of all U.S. farms. Data for 1991 were re-summarized using the new estimation procedures. Data prior to 1991 will not be revised and will not be directly comparable to more recent FCRS data. See USDA (1993a) for information on the new estimation procedures associated with the FCRS.

Types of data available. ERS and NASS publish separate estimates of farm labor expenses based on the FCRS. ERS's expense items are consistent but not identical with those published by NASS. NASS's estimates are derived solely from the survey based on 48 states. In the ERS series, data for Alaska and Hawaii are estimated separately and added to the 48-state statistics. Some ERS components are also adjusted for conceptual differences between the ERS and NASS publications.

NASS publishes annual estimates of labor expenditures, number of farms reporting labor expenditures, and average labor expenditures per farm for the United States, 10 farm production regions, 5 gross sales classes, and crop and livestock farms.

ERS publishes separate estimates of expenses for cash wages, employers' contribution to Social Security, perquisites, contract labor, and machine hire and customwork at the national level. The FCRS itself yields reliable statistics only at the national and regional level. However, ERS publishes estimates of combined hired and contract labor expenses for states. These estimates are derived from regional totals allocated to states based on distribution patterns constructed by ERS staff using alternative data sources such as the Census of Agriculture.

Data uses and limitations.

- The FCRS provides annual, detailed data on labor expenses, such as value of perquisites and employer's contribution to Social Security, that are generally unavailable from other sources. Expenditure data can be used to indicate the relative magnitude of labor use by size and type of farm.

- The FCRS, along with Census of Agriculture, are the only data sources that collect information on customwork. However, expenses for labor involved in customwork are combined with expenses for the use of equipment and cannot be separated out.

- The survey does not collect information on the number of hired or contract workers or on their demographic characteristics.

- Data on hired workers refer to all hired workers on the farm, including bookkeepers, secretaries, and mechanics who are generally not considered to be hired farm workers.

- Estimates of labor expenses provided by NASS and ERS differ. Users should be aware of the differences when selecting the estimates most appropriate for their research.

- Due to the new estimating procedures associated with the FCRS, data prior to 1991 are not strictly comparable with more recent data, thereby limiting historical analyses.

Farm Labor Survey

Responsible agency .... National Agricultural Statistics
Service, U.S. Department of
Agriculture
Date of origin 1910
Frequency of data collection ...... Quarterly
Reference period .... Week
Degree of coverage .... Sample of farms
Geographic detail .... U.S., region, selected states

The Farm Labor Survey (FLS), conducted by the U.S. Department of Agriculture's National Agricultural Statistics Service (NASS) provides seasonal estimates of farm employment. Information on wages of hired farm workers collected in this survey is used in Federal determinations of the Adverse Effect Wage Rates for H-2A workers and is a component of USDA's Parity Index. The survey has been conducted since 1910, although numerous changes in the frequency, detail, and methodology have occurred since that time. Data are reported quarterly in the NASS publication, Farm Labor (see U.S. Department of Agriculture, 1993b).

Methodology. The survey is conducted 4 times per year (in January, April, July, and October) in all states except Alaska. During each quarterly survey, data are collected for a one-week period that includes the 12th of the month. The FLS is a probability survey using a list of farm operators supplemented by an area frame of respondents not on the list. The list sample for each survey period contains about 11,000 farms and the area frame contains another 3,500 farms. Useable reports are obtained from about 80–90 percent of the sample. To collect information on the number of contract workers, a sample of farm labor contractors are interviewed in California and Florida. All other states collect information on contract labor directly from the farm operators in the survey. The survey is collected by mail, telephone, and personal interview.

Types of data available. The FLS provides quarterly estimates of the number of hired workers (including those expected to be employed 150 days or more and 149 days or less), average weekly hours worked, and average wage rates for hired workers for 16 separate states and 15 regions. Hourly wage rates for hired workers are provided by type of worker (field, livestock, supervisor, and other) and method of pay. Information on number of agricultural service workers, their hours worked, and their wage rates are provided for the United States, California, and Florida. In addition, annual average wage rates of hired workers are published for all states.

Data uses and limitations.

- The FLS offers the most detailed and timely information available on hired farm labor wage rates by state. Data are collected 4 times per year and released one month after collection.

- Data do not measure total hired labor use during the year since data collected represent a total of only 4 weeks out of the year. Because of the seasonality of hired farm work, estimates for 4 one-week periods may not accurately reflect labor use during the rest of the year.

- The survey does not collect information on the demographic characteristics of workers.

- Farm operators who hire a crew leader or agricultural service firm on a contract basis may be unable to accurately estimate the number of contract workers working on their farms because they do not hire the workers directly. Labor contractors are interviewed directly only in California and Florida.

- Data on hired workers refer to all hired workers on the farm, including bookkeepers, secretaries, and mechanics who are generally not considered to be hired farm workers.

- Duplication in numbers of workers may occur when an individual works on more than one farm during the survey week; the number of hours worked and wages earned are not duplicated.

- Data on the number of hired workers are limited to 16 states and 15 regions; data on the number of contract workers are reported for California, Florida, and the U.S. total.

- Hourly wage rates are derived by dividing gross weekly wages by hours worked per week, and are therefore influenced more by workers who work longer hours.

- Persons involved in the operation of a farm, but who pay themselves a regular salary, such as a shareholder in a corporation or partnership, are counted as hired workers.

Household Surveys

Three major sources of household data on farm labor are the Decennial Census of Population, the Current Population Survey, and the National Agricultural Worker Survey. Household surveys generally provide more detailed economic and demographic information on hired and contract workers and less information on the characteristics of the job or farm.

Decennial Census of Population

Responsible agency .... Bureau of the Census,
U.S. Department of Commerce
Date of origin 1790
Frequency of data collection .... Every 10 years
Reference period .... Week
Degree of coverage .... Sample of households
Geographic detail .... U.S., state, county

The Decennial Census of Population was first conducted by the Bureau of the Census in 1790 and has been taken every 10 years since. The Census counts persons at their usual residence—the place where the person lives and sleeps most of the time—to determine how many representatives each state will have in Congress. The Census also collects household data on the demographic and economic characteristics of the U.S. population which are used for a variety of statistical purposes by government agencies, businesses, and public and private groups. Data are available in published volumes and in public-use microdata computer tape files (see U.S. Department of Commerce 1992a, 1992b).

Methodology. Information from the 1990 Census was collected primarily through mail surveys—95 percent of the population was enumerated by the mailback procedure. Some data, including number of people, race, Spanish origin, gender, age, and household relationship, were based on a complete count of households. Other information such as employment status, occupation, industry of employment, and previous year's income was based on a 17-percent sample of households. All employment data were based on the respondent's chief job activity or business during the reference week, generally the last week of March or the first week of April. Total wage and salary income before deductions is combined and reported for all jobs held in the previous year.

Types of data available. The Census reports a variety of demographic and employment information on U.S. individuals and their households. Employment data are reported by 2-digit SIC code (livestock production, crop production, agricultural services, and horticultural services) and for various occupational categories of hired farm workers, such as farm managers, horticultural specialty managers, supervisors of farm workers, farm workers, and nursery workers. Demographic data as well as information on number of weeks worked, usual hours worked per week, and wages earned during the previous year are also available by occupation and industry.

Data uses and limitations.

- The Decennial Census provides demographic and employment information on all U.S. residents at the county, state, and national level and allows comparisons to be made between hired farm workers and other occupation and industry groups. Since data are collected from all persons in a household, information can be obtained on the other members of the farm worker's household.

- Census data are collected only once in every 10 years and may not reflect recent changes in agricultural employment or earnings.

- Census data undercount the number of hired farm workers since the employment data are based on primary activity during the reference week, generally the last week of March or the first week of April. Because of the seasonality of agricultural work, many hired farm workers would not be counted in this occupational category since they did not work on farms during these time periods.

- Census data are more likely to exclude seasonal workers and characterize year-round hired farm workers—a group more likely to be white, older, engaged in livestock production, and dependent on hired farm work as a primary activity (Whitener 1984).

- Foreign nationals who had not established a residence in the United States as well as migrant farm workers who live most of the year outside the United States would not be included in the data. Census data are also more likely to miss households that have non-traditional or complex housing arrangements, move frequently, speak a non-English language, and have members with low educational levels. These key factors suggest an undercount of racial/ethnic minorities and probably farm workers.

- Wage and salary income is reported collectively for all jobs worked during the previous year and cannot be tied to the specific occupation or industry reported by respondents as their major activity during the census year.

- The Census does not report information on the characteristics of the farm or the farm job beyond industry or occupational classification.

Current Population Survey

Responsible agency .... Bureau of Labor Statistics,
                                                  U.S. Department of Commerce
Date of origin 1940
Frequency of data collection ...... Monthly
Reference period .... Week
Degree of coverage .... Sample of households
Geographic detail .... U.S., state[2]

The Current Population Survey (CPS) is conducted each month by the Bureau of the Census for the Bureau of Labor Statistics. The survey, conducted since 1940, is the source of the official Federal statistics on employment and unemployment in the United States and is also used to collect demographic information on the population. Results from the survey are published in BLS's monthly Employment and Earnings (U.S. Department of Labor 1994), in periodic Bureau of Census reports, and USDA's biennial Profile of Hired Farm workers (Runyan 1994). Data are also available in public-use computer tapes from the Bureau of the Census.

Methodology. The CPS is based on a probability sample of households, designed to represent the U.S. civilian noninstitutional population. Each month, the CPS collects information from a sample of households in all 50 states. About 57,000 households are interviewed in the CPS each month; about two-thirds are interviewed by telephone and the remainder by personal interview. Employment data are collected for individuals 15 years of age and older but published for those 16 years and older. The data collected refer to the activity or status reported for the week containing the 12th of the month.

Each month, workers in about one-quarter of the interviewed households are asked additional questions on weekly hours worked and earnings. These data are combined into the annual earnings microdata file which includes all records from the 12 monthly surveys during the year that were subject to having these questions on hours worked and earnings asked. Average weekly earnings are derived from this file.

Each March, supplemental questions are added to the CPS to obtain information on family characteristics, income during the previous year, weeks worked, and occupation and industry of longest job during the previous year. Data from this March Annual Demographic File are based on the full CPS sample plus an additional 2,500 Hispanic households.

A redesign of the CPS, affecting the questionnaire, data collection methods, and the data processing system, began in January 1994 to help improve information on the labor force. Some of these changes, such as those designed to clarify the concept of unpaid family worker and the addition of questions determining the second occupation of multiple jobholders, may lead to more comprehensive data on farm employment. See U.S. Department of Labor (1993b) for more information on the redesign of the CPS.

Types of data available. The CPS provides a variety of demographic and employment information on the U.S. work force. Employment information is provided by 2-digit SIC codes and for various occupational categories of hired farm workers, including farm managers, horticultural specialty managers, supervisors of farm workers, farm workers, and nursery workers. Demographic data as well as information on hours worked and weekly earnings of hired farm workers are available.

Data uses and limitations.

- The CPS provides employment and demographic information on the entire U.S. work force, allowing comparative analyses between farm workers and other occupation and industry groups. The CPS is conducted on a monthly basis and provides the most timely data of all the data sources reviewed here. Since data are collected from all persons in a household, information can be obtained on other members of farm workers' households.

- The CPS is based on a sample of households and is likely to undercount hired farm workers. Individuals living in more unconventional living quarters such as trailers or labor camps are more likely to be missed in the CPS, and studies suggest that these persons are more likely to be Hispanic, a group that comprises a large portion of the hired farm work force in the United States. In addition, unauthorized foreign nationals doing farm work in this country may not be counted because they tend to avoid survey enumerators due to their illegal status.

- The CPS classifies employed persons according to the job at which they worked the greatest number of hours during the survey week. As a result, farm workers who spent more time during the survey week at a nonfarm rather than a farm job would not be included as farm workers.

- The small sample size raises questions about generalizing from sample data to describe all U.S. farm workers. On average, about 500 hired farm workers are interviewed each month. The annual earnings file is comprised of records from fewer than 1,400 farm workers. Also, the CPS sample is too small to allow publication of reliable estimates for states.

The National Agricultural Workers Survey

Responsible agency .... Office of the Assistant Secretary for Policy, U.S. Department of Labor
Date of origin 1989
Frequency of data collection .... 3 times per year
Reference period .... Week
Degree of coverage .... Sample of households
Geographic detail .... U.S., 6 regions

The National Agricultural Workers Survey (NAWS) is a national survey of perishable crop field workers commissioned by the U.S. Department of Labor (DOL). The NAWS was first conducted in 1989 to meet DOL's Federally mandated responsibility to estimate the supply of agricultural workers performing Seasonal Agricultural Services (SAS) as required by IRCA. Although DOL is no longer required to estimate annual changes in the labor supply, the NAWS continues to collect a variety of demographic, employment, and earnings data on SAS workers each year. Results from the survey are published annually in DOL reports (see Mines et al. 1991) and data are available on public-use computer files.

Methodology. Each fiscal year, the NAWS collects information from personal interviews with 2,000 to 2,700 randomly selected workers performing Seasonal Agricultural Services. The sample was selected using probabilities proportional to farm labor expenditure data from the 1987 Census of Agriculture. Using this method, a sample of 73 counties covering 25 states was selected to represent 12 distinct agricultural regions. No fewer than 4 counties were chosen from each region. Interview cycles lasting six to ten weeks are conducted three times a year beginning in January, May, and September to reflect the seasonality of agricultural work. For each cycle, approximately 30 of the 73 counties are selected randomly as interviewing sites. A random sample of SAS employers is generated from a county list of employer names obtained from Federal agencies and other sources. Employers are contacted and a random sample of workers are selected for interview. Site selection and interview allocations are proportional to seasonal payroll size.

Types of data available. NAWS provides detailed information on the demographic and employment characteristics of SAS workers, including their legal status; literacy and education; family composition; income, assets, and use of government programs; and employment history, earnings, and job characteristics. NAWS also collects information on migrant farm workers defined as SAS workers that traveled 75 miles or more from home while looking for work or from job to job during the year.

Data uses and limitations.

- NAWS provides detailed social and economic characteristics on SAS farm workers—the group of workers most likely to be affected by immigration reform. NAWS is currently the only national level data source that provides information on the characteristics of migrant farm workers.

- NAWS data include only workers performing Seasonal Agricultural Services—in effect, perishable crop field workers. Livestock workers are excluded from the survey.

- Published data do not provide estimates of the number of SAS workers.

- NAWS data have been collected only since 1989 and do not allow long-term historical comparisons of patterns and trends.

- The small sample size and the use of simple random sampling of employers and workers raises questions about the ability to generalize from the data. Only 2 percent of all counties and half of the states in the United States were sampled.

The geographic detail in published reports is limited to six regions; Alaska and Hawaii are not included in the data.

Administrative Records

The administrative records of Government agencies which operate employment-related programs provide employment, earnings, or expenditure data typically obtained from employers. This information is usually less detailed than other types of data and the universe of the administrative data may be limited by program definitions. The major sources of farm labor data from administrative records are the Bureau of Labor Statistics Unemployment Insurance data and the Bureau of Economic Analysis Employment and Income data.

Bureau of Labor Statistics Unemployment Insurance Data

Responsible agency .... Bureau of Labor Statistics,
U.S. Department of Labor
Date of origin 1978
Frequency of data collection ...... Quarterly (published on annual
basis)
Reference period .... Week
Degree of coverage .... Census of Unemployment Insurance covered farm employers
Geographic detail .... U.S., state, county

The Unemployment Insurance (UI) Program was enacted in 1938 to provide unemployed workers with partial income in a temporary period of involuntary unemployment. In 1978, agricultural labor was first covered by unemployment insurance under the Federal Unemployment Tax Act. The program is state-administered with Federal participation. Employer taxes are the major source of financing for the compensation program. Employers of agricultural labor are required to pay UI coverage for their employees if they: (1) paid wages of $20,000 or more for agricultural labor during any calendar quarter in the current or preceding calendar year or; (2) employed at least 10 persons in agricultural labor for some portion of a day in each of the 20 different weeks during the current or preceding calendar year with each day being in a different calendar week. Each state establishes the tax structure and requirements for qualification, and some states have stricter requirements than others. These size-of-firm coverage provisions apply to crew leaders as well as farm operators. State and national totals are published in the Bureau of Labor Statistics (BLS) annual report, Employment and Wages (see U.S. Department of Labor 1993a). Unpublished county level data are available on a cost-reimbursable basis from BLS.

Methodology. The UI data file is a byproduct of the administration of the Unemployment Insurance program. Qualifying employers are legally required to file quarterly tax reports with their respective state employment security agency which compiles data from these reports to send to BLS. Employers report their employment for the payroll period including the 12th of each month comprising the quarter and total wages paid during the quarter.

Types of data available. UI data provide a complete sum of wages paid to workers on farms and in agricultural service establishments that are covered by the Unemployment Insurance program. Summary statistics are available on annual average employment, total wages paid, annual wages per employee, and average weekly wages. Information is reported at the county, state, and national level by industry. Data are reported at the 4-digit SIC code.

Data uses and limitations.

- UI data report the numbers and earnings of hired farm-workers at the 4-digit SIC code—the finest level of detail among the data sources reviewed here—and provide information at the county level.

- Generally, only hired farm workers employed on the larger farms are covered under the UI program. BLS estimates that about 44 percent of all workers in agricultural industries are covered (U.S. Department of Labor 1992).

- State comparisons of UI data must consider the differences in laws and coverage among states. For example, California, Florida, Minnesota, Rhode Island, and Texas have laws that are more inclusive than others.

- UI data measure jobs not persons, and individuals working for more than one covered firm during the quarter are reported more than once; wages are not duplicated.

- Data are reported by industry not occupation. Therefore, data on agricultural industries will include all wage and salary workers on the farms, such as bookkeepers and mechanics who are not generally considered to be hired farm workers.

- Since employment is measured for only one week during each month, but wages refer to wages paid in all 52 weeks, the average weekly wages reported by BLS do not take into account variations in employment levels during the other 40 weeks of the year.

- Average weekly wages do not control for hours worked and are therefore affected by the ratio of full-time to part-time workers. Computation of average annual wages per employee are computed by dividing total annual wages by annual average employment. (This estimate of average annual wages per employee fails to account for seasonality of employment and turnover among workers.) Average annual wage per employee is then divided by 52 to derive average weekly wages per employee. (This accepts the unlikely premise that workers worked all 52 weeks at farm work during the year.)

Bureau of Economic Analysis Employment and Income Data

Responsible agency .... Bureau of Economic Analysis,
U.S. Department of Commerce
Date of origin 1967
Frequency of data collection ...... Annually
Reference period .... Year
Geographic detail .... U.S., state, county

The Bureau of Economic Analysis (BEA) has provided annual estimates of personal income by industry for states and local areas since 1929. Since 1967, BEA has also prepared annual estimates of employment by industry, as a supplement to its personal income series. Together, the data provide estimates of the number of farm wage and salary workers and their income at the state and county level. Employment and income data at the state and county level are not currently published but are available to the public on computer printouts and tapes from BEA.

Methodology. BEA uses information collected by others to prepare its employment and income estimates. For most industries, earnings and employment estimates are compiled from administrative records, such as the state unemployment insurance (UI) programs. However, because of the large numbers of farm wage and salary workers not covered under the various administrative programs, BEA uses different estimating procedures to estimate their numbers.

BEA estimates the annual average wage and salary employment on farms by state based on data from USDA's Farm Labor Survey.[3] An annual average number of wage and salary workers by region is estimated by averaging the 4 survey weeks of regional data from the Farm Labor Survey. The regional data are distributed into individual states in proportion to cash farm wages as estimated by the USDA's Farm Costs and Returns Survey. State data are then distributed into counties using data on the number of workers working 150 days or more from the Census of Agriculture. In nine states (Arizona, California, Rhode Island, Connecticut, Delaware, Florida, Hawaii, Massachusetts, New Jersey, and Washington) UI data are considered to have adequate coverage and are used to allocate numbers of workers to counties. County level data on the number of corporate farms (used to represent the number of corporate officers) reported in the Census of Agriculture are added to these figures to generate total farm wage and salary employment.

BEA estimates farm wages and salary earnings by taking estimates of cash wages by state from the Farm Costs and Returns Survey and distributing it to the county level using data on hired labor expenditures from the Census of Agriculture. In nine states (Arizona, California, Rhode Island, Connecticut, Delaware, Florida, Hawaii, Massachusetts, New Jersey, and Washington) UI data are used to allocate county level income. To this figure is added the value of pay-in-kind such as food and lodging as reported in the Farm Costs and Returns Survey which is then distributed into counties using data on the number of workers working 150 days or more from the Census of Agriculture. County level estimates of the salaries of corporate officers included in the data on hired workers are based on amounts reported in the Census of Agriculture. BEA makes further adjustments to account for sources of other labor income, such as employer contributions to private pension and welfare funds.

Types of data provided. BEA employment and income files provide information on the annual average number of farm wage and salary workers and their wage and salary earnings at the U.S., state, and county level.

Data uses and limitations.

- BEA provides the only annual estimates of the employment and income of farm wage and salary workers at the county level.

Table B.1. General Characteristics of the Major Data Series

Data series	Frequency of data collection	Survey reference period	Level of greatest geographic detail	Number of workers	Labor expenditures	Demographic characteristics of workers	Data provided on: Wages or earnings per worker		
							Hourly	Weekly	Annual
Establishment Surveys:									
Census of Agriculture	Every 5 years	Year	County	Yes[1]	Yes	No	No	No	No
Farm Costs and Returns	Annually	Year	Region	No	Yes	No	No	No	No
Farm Labor	Quarterly	Week	Region[2]	Yes	No	No	Yes	Yes[3]	No
Household Surveys:									
Decennial Census of Population	Every 10 years	Week[4]	County	Yes	No	Yes	No	No	Yes
Current Population Survey	Monthly	Week	State[5]	Yes	No	Yes	Yes[3]	Yes	No
National Agricultural Worker Survey	3 times per year[6]	Week[4]	Region	No	No	Yes	Yes	No	Yes
Administrative Records:									
Unemployment Insurance Program	Quarterly[6]	Week[7]	County	Yes	Yes	No	No	Yes	Yes
BEA Data	Annually	Year	County	Yes	No	No	No	No	Yes

[1]Data on the number of workers who worked less than 150 days and those who worked 150 days or more are collected periodically.

[2]Annual average wage rates are reported at the State level, while quarterly data on number of workers and wage rates are reported at the regional level and for selected States.

[3]Can be estimated from data provided.

[4]Some information is collected for the entire year.

[5]State data available but not published, due to small sample size and low reliability of the estimates.

[6]Data are published on an annual basis.

[7]Data on employment are based on a one-week period for each month comprising the quarter, while data on wages are based on the entire quarter.

- Calculations used in deriving estimates of employment and personal income come from different data sets based on different years which may not be strictly comparable.

- Data include all workers on the farm, even bookkeepers and secretaries not generally considered as hired farm workers.

- BEA data do not collect information on the demographic and employment characteristics of hired and contract workers.

- BEA reports separate estimates for employment and earnings in agricultural services. However, data on contract labor cannot be separated out of the broader group of agricultural services, which includes non-farm work such as lawn and garden services, landscape planning, veterinary services for pets, boarding kennels, and others.

Points to Consider

Agricultural employment data vary widely among different secondary sources and users should understand the methodological and conceptual differences that underlie the data sets they choose for their research and program evaluations. While data shown in Table B.2 are not strictly comparable, they provide a general indication of the magnitude of variation in estimates among different data sources. For example, the annual average number of hired farm workers (excluding contract or agricultural service workers) ranged from 698,000 reported by the 1991 Unemployment Insurance Program to 904,000 reported by the 1990 BEA employment and income data—a difference of over 200,000 workers. However, the 1992 Census of Agriculture reports 3.8 million hired workers employed by farm operators during the year. Estimates of the number of contract or agricultural service workers range from an annual average of 54,000 reported by the 1992 CPS Earnings File to an annual average of 256,000 reported by the 1993 Farm Labor Survey—a difference of 200,000. Labor expenditure data also differed, ranging from $12.5 million reported in the 1991 Unemployment Insurance data to $15,300 reported by the 1992 Census of Agriculture.

A detailed analysis of variations in employment and labor expense estimates is beyond the scope of this chapter (but see Daberkow and Whitener 1986; Martin and Martin 1994). However, in general, these data variations are due to differences in the population universe examined, concepts and definitions used, age criteria, time of data collection and employment reference period. For example, some data series limit the population universe for which data are collected. The Unemployment Insurance data are collected only for those employers covered by the UI program; the National Agricultural Workers Survey includes only those workers who performed Seasonal Agricultural Services. Also, agricultural labor data are affected by the unknown number of unauthorized aliens who work in agriculture. These workers are more likely to be included in establishment surveys such as the Farm Labor Survey or in the NAWs which makes initial survey contact at the place of work. Unauthorized aliens are more likely to be missed in household surveys because many live in non-standard housing units or because they wish to avoid detection. Several data sets do not include contract or agricultural service workers or do not distinguish between these and other hired workers when reporting data.

Definitions of farm workers and labor expenditures differ among the data sets and data collection. Establishment and administrative data, for example, are more likely to include bookkeepers, accountants, and other professional staff people who work on farms; household data do not. Some expenditure data include payments-in-kind and fringe benefits; while others report only cash wages. Age criteria also differ among the data sources. The Current Population Survey, for example, collects employment information only on persons 15 years of age and over while the Census of Agriculture and Farm Labor Survey have no age criteria.

Because of the seasonality of agriculture, worker estimates differ depending on the employment reference period. The Census of Population, for example, collects employment information based on respondent's chief activity during one reference week during the year, generally the last week of March or first week of April. The NAWS conducts interviews during a 6–10 week period at three times during the fiscal year. The CPS is conducted monthly and the Farm Labor Survey is taken quarterly; annual averages can be computed for both. FCRS and the Census of Agriculture report expenditure data representing the entire year. Also, data users should consider differences in survey coverage, sample size, frequency of data collection, historical availability, and data access when selecting sources for analysis.

Selection of the most appropriate data source to explore farm labor issues depends upon individual research questions and objectives. Data users should also keep in mind that many Federal agencies are currently facing budget reductions and must control costs.

Cost-cutting measures may result in the modification, reduction, or elimination of existing agricultural employment data in the near future, and potential data users should continue to closely monitor changes in federal data sources.

NOTES

1. Not all data are reported at the state level.

2. State data are not published.

3. At this time, BEA has no published documentation explaining in detail the methodology used in estimating farm employment.

Table B.2. Numbers of Workers and Expenditures for Labor as Reported by the
Major Data Series for the Most Current Year for which Data Are Available

Data series	Workers			Expenditures for labor		
	Hired	Contract	Total	Hired	Contract	Total
	Number (in thousands)			*Dollars (in millions)*		
Establishment surveys:						
Census of Agriculture						
(1987)	3,802[1]	--	3,802[1]	12,962[2]	2,324	15,286[2]
Farm Costs and Returns						
ERS (1992)	--	--	--	12,043	2,016	14,060
NASS (1992)	--	--	--	NA	NA	14,500
Farm Labor (1993)	857[3]	256[3]	1,113[3]	--	--	--
Household surveys:						
Census of						
Population (1990)	NA	NA	1,099	--	--	--
Current Population						
Survey Earnings						
File (1992)	749[4]	54[4]	803[4]	--	--	--
National Agricultural						
Worker Survey (1991)	--	--	--	--	--	--
Administrative Records:						
Unemployment						
Insurance (1991)	698[5]	254[5,6]	952[5,6]	9,893[7]	3,356[6,7]	12,482[7]
BEA Data (1990)	904[5]	NA	NA	--	--	--

-- = Data are not collected.

NA = Data not available.

[1] Total hired workers employed by Farm operators during the year; subject to double-counting.

[2] Excludes pay-in-kind.

[3] Annual average computed from survey data, based on 4 weeks of data.

[4] Annual average computed from survey data, based on 12 weeks of data.

[5] Published annual average.

[6] Data reported here do not include veterinary and horticultural services performed on the farm.

[7] Refers to total wages and excludes pay-in-kind and employer's contributions to social security, workers' compensation, health insurance, unemployment insurance, private pensions, and welfare funds.

Table B.3. Labor Expenses Reported in the 1987 Census of Agriculture

State	Hired labor expenses $1000	Hired labor expenses % of U.S.	Contract labor expenses $1000	Contract labor expenses % of U.S.	Total labor expenses $1000	Total labor expenses % of U.S.
Alabama	140,414	1.1	15,095	0.6	155,509	1.0
Alaska	3,928	0.0	176	0.0	4,104	0.0
Arizona	190,442	1.5	72,997	3.1	263,439	1.7
Arkansas	223,124	1.7	25,890	1.1	249,014	1.6
California	2,922,390	22.5	967,377	41.6	3,889,767	25.4
Colorado	209,675	1.6	26,105	1.1	235,780	1.5
Connecticut	77,980	0.6	3,901	0.2	81,881	0.5
Delaware	23,911	0.2	3,674	0.2	27,585	0.2
Florida	937,571	7.2	367,000	15.8	1,304,571	8.5
Georgia	252,721	1.9	35,626	1.5	288,347	1.9
Hawaii	178,788	1.4	6,406	0.3	185,194	1.2
Idaho	245,990	1.9	32,386	1.4	278,376	1.8
Illinois	300,090	2.3	15,302	0.7	315,392	2.1
Indiana	209,089	1.6	16,567	0.7	225,656	1.5
Iowa	259,210	2.0	19,833	0.9	279,043	1.8
Kansas	239,629	1.8	25,166	1.1	264,795	1.7
Kentucky	202,545	1.6	34,570	1.5	237,115	1.6
Louisiana	146,667	1.1	11,560	0.5	158,227	1.0
Maine	61,086	0.5	7,348	0.3	68,434	0.4
Maryland	93,631	0.7	7,688	0.3	101,319	0.7
Massachusetts	77,337	0.6	6,406	0.3	83,743	0.5
Michigan	318,276	2.5	22,488	1.0	340,764	2.2
Minnesota	261,649	2.0	21,620	0.9	283,269	1.9
Mississippi	168,464	1.3	16,705	0.7	185,169	1.2
Missouri	190,051	1.5	20,629	0.9	210,680	1.4
Montana	107,632	0.8	15,244	0.7	122,876	0.8
Nebraska	254,132	2.0	18,344	0.8	272,476	1.8
Nevada	31,652	0.2	3,672	0.2	35,324	0.2
New Hampshire	21,601	0.2	1,140	0.0	22,741	0.1
New Jersey	115,161	0.9	11,665	0.5	126,826	0.8
New Mexico	115,633	0.9	32,608	1.4	148,241	1.0
New York	336,461	2.6	18,485	0.8	354,946	2.3
North Carolina	388,338	3.0	41,893	1.8	430,231	2.8
North Dakota	99,790	0.8	9,370	0.4	109,160	0.7
Ohio	259,501	2.0	16,012	0.7	275,513	1.8
Oklahoma	144,750	1.1	23,629	1.0	168,379	1.1

Table B.3. (con't).

State	Hired labor expenses		Contract labor expenses		Total labor expenses	
	$1000	% of U.S.	$1000	% of U.S.	$1000	% of U.S.
Oregon	367,047	2.8	31,329	1.3	398,376	2.6
Pennsylvania	352,456	2.7	25,590	1.1	378,046	2.5
Rhode Island	9,076	0.1	508	0.0	9,584	0.1
South Carolina	111,836	0.9	15,223	0.7	127,059	0.8
South Dakota	95,956	0.7	10,668	0.5	106,624	0.7
Tennessee	138,434	1.1	24,663	1.1	163,097	1.1
Texas	729,915	5.6	125,550	5.4	855,465	5.6
Utah	72,014	0.6	6,866	0.3	78,880	0.5
Vermont	38,323	0.3	2,359	0.1	40,682	0.3
Virginia	188,671	1.5	17,113	0.7	205,784	1.3
Washington	601,614	4.6	58,402	2.5	660,016	4.3
West Virginia	26,956	0.2	3,520	0.2	30,476	0.2
Wisconsin	362,356	2.8	18,705	0.8	381,061	2.5
Wyoming	57,677	0.4	8,831	0.4	66,508	0.4
United States	12,961,640	100.0	2,323,904	100.0	15,285,544	100.0

Source: U.S. Department of Commerce, Bureau of the Census, 1987
Census of Agriculture.

Table B.4. Number of Workers Reported in the 1992 Census of
Agriculture

State	Working less than 150 days	Working 150 days or more	Total number	% of U.S.
Alabama	31,958	11,918	43,876	1.2
Alaska	940	154	1,094	0.0
Arizona	20,358	14,003	34,361	0.9
Arkansas	35,414	17,172	52,586	1.4
California	495,607	182,287	677,894	17.8
Colorado	32,059	14,365	46,424	1.2
Connecticut	6,848	4,568	11,416	0.3
Delaware	3,226	1,504	4,730	0.1
Florida	93,010	68,037	161,047	4.2
Georgia	46,805	17,779	64,584	1.7
Hawaii	5,737	9,008	14,745	0.4
Idaho	56,778	15,453	72,231	1.9
Illinois	63,213	20,763	83,976	2.2
Indiana	50,425	14,909	65,334	1.7
Iowa	85,399	21,779	107,178	2.8
Kansas	42,735	16,784	59,519	1.6
Kentucky	181,657	18,128	199,785	5.3
Louisana	29,388	12,068	41,456	1.1
Maine	19,566	3,385	22,951	0.6
Maryland	14,067	6,386	20,453	0.5
Mass	8,445	4,024	12,469	0.3
Michigan	83,923	20,501	104,424	2.7
Minnesota	84,671	19,509	104,180	2.7
Mississippi	26,732	14,353	41,085	1.1
Missouri	56,692	17,164	73,856	1.9
Montana	19,168	8,669	27,837	0.7
Nebraska	46,679	16,722	63,401	1.7
Nevada	3,446	2,312	5,758	0.2
New Hampshire	3,183	1,351	4,534	0.1
New Jersey	14,366	8,175	22,541	0.6
New Mexico	19,236	8,506	27,742	0.7
New York	45,858	22,224	68,082	1.8
North Carolina	126,611	27,185	153,796	4.0
North Dakota	29,860	7,115	36,975	1.0
Ohio	63,785	19,238	83,023	2.2

Table B.4. (con't).

State	Number of workers			
	Working less than 150 days	Working 150 days or more	Total number	% of U.S.
Oklahoma	40,582	12,613	53,195	1.4
Oregon	99,646	21,060	120,706	3.2
Pennsylvania	41,987	25,098	67,085	1.8
Rhode Island	770	565	1,335	0.0
South Carolina	29,077	9,049	38,126	1.0
South Dakota	25,409	8,502	33,911	0.9
Tennessee	157,687	14,608	172,295	4.5
Texas	142,915	61,107	204,022	5.4
Utah	20,072	6,050	26,122	0.7
Vermont	5,103	3,398	8,501	0.2
Virginia	54,251	16,046	70,297	1.8
Washington	216,683	34,251	250,934	6.6
West Virginia	14,152	3,322	17,474	0.5
Wisconsin	77,855	31,589	109,444	2.9
Wyoming	8,623	4,438	13,061	0.3
United States	2,882,657	919,194	3,801,851	100.0

Source: U.S. Department of Commerce, Bureau of the Census, 1992
Census of Agriculture

Table B.5. Farm Labor Expenses Based on the 1992 Farm Costs and
Returns Survey

State	Total hired and contract labor expenses	
	$ millions	% of U.S.
Alabama	126.7	0.9
Alaska	3.3	0.0
Arizona	207.4	1.5
Arkansas	266.4	1.9
California	3,809.1	27.1
Colorado	168.8	1.2
Connecticut	78.6	0.6
Delaware	28.1	0.2
Florida	910.9	6.5
Georgia	213.7	1.5
Hawaii	215.7	1.5
Idaho	176.5	1.3
Illinois	316.1	2.2
Indiana	228.7	1.6
Iowa	327.4	2.3
Kansas	193	1.4
Kentucky	202.4	1.4
Louisiana	158.6	1.1
Maine	62.4	0.4
Maryland	97.4	0.7
Massachusetts	81	0.6
Michigan	385.4	2.7
Minnesota	385.5	2.7
Mississippi	187	1.3
Missouri	196.6	1.4
Montana	98.3	0.7
Nebraska	219.6	1.6
Nevada	31.3	0.2
New Hampshire	21.8	0.1
New Jersey	107.8	0.8
New Mexico	99.6	0.7
New York	332.9	2.4
North Carolina	363.1	2.6
North Dakota	78.6	0.6
Ohio	254.9	1.8

Table B.5. (con't).

State	Total hired and contract labor expenses	
	$ millions	% of U.S.
Oklahoma	155.2	1.1
Oregon	357.1	2.5
Pennsylvania	352.9	2.5
Rhode Island	11.4	0.1
South Carolina	95.5	0.7
South Dakota	81	0.6
Tennessee	151.8	1.1
Texas	798.1	5.7
Utah	50.3	0.3
Vermont	37.9	0.3
Virginia	180.5	1.3
Washington	583.4	4.1
West Virginia	30.2	0.2
Wisconsin	489.3	3.5
Wyoming	50.6	0.4
United States	14,059.5	100.0

Source: U.S. Department of Agriculture, Economic Research Service, 1992 Farm Costs and Returns Survey.

Regional estimates of farm labor expenses are derived from the Farm Costs and Returns Survey. The Economic Research Service distributes these regional estimates by State using expenditure data from the Census of Agriculture.

Table B.6. Number of Hired Workers Reported by the 1993 Farm Labor
Survey

State and region	Quarterly estimates				Annual average	
	Jan. Thou.	April Thou.	July Thou.	Oct. Thou.	Thou.	% of U.S.
Northeast I	35	43	53	57	47	5.5
New York	19	23	25	31	24.5	2.6
Other	16	20	28	26	22.5	2.6
Northeast II	28	37	42	32	34.75	4.0
Pennsylvania	20	22	24	19	21.25	2.5
Other	8	15	18	13	13.5	1.6
Appalachian I	25	36	67	52	45	5.2
North Carolina	13	25	49	32	29.75	3.5
Virginia	12	11	18	20	15.25	1.8
Appalachian II	29	30	41	34	33.5	3.9
Southeast	33	39	55	39	41.5	4.8
Florida	54	61	48	63	56.5	6.6
Lake	45	54	86	77	65.5	7.6
Michigan	14	17	29	28	22	2.6
Minnesota	10	14	25	24	18.25	2.1
Wisconsin	21	23	32	25	25.25	2.9
Cornbelt I	39	48	66	57	52.5	6.1
Cornbelt II	20	26	31	31	27	3.1
Delta	24	37	57	41	39.75	4.6
Northern Plains	23	39	43	46	37.75	4.4
Southern Plains	54	64	85	59	65.5	7.6
Oklahoma	9	14	15	13	12.75	1.5
Texas	45	50	70	46	52.75	6.1
Mountain I	17	23	29	26	23.75	2.8
Mountain II	13	14	24	23	18.5	2.1
Mountain III	13	22	19	16	17.5	2.0
Arizona	7	13	11	11	10.5	1.2
New Mexico	6	9	8	5	7	0.8
Pacific	30	59	92	56	59.25	6.9
Oregon	12	23	38	17	22.5	2.6
Washington	18	36	54	39	36.75	4.3
California	120	188	215	211	183.5	21.4
Hawaii	9	8	9	8	8.5	1.0
United States	611	828	1062	928	857.25	100

Source: U.S. Department of Agriculture, National Agricultural Statistics
Service, 1993 Farm Labor Survey.

Table B.7. Average Annual Hourly Wage Rates for Hired Workers
Reported in the 1993 Farm Labor Survey

State	Hourly wage rates (Dol.)
Alabama	5.53
Arizona	6.03
Arkansas	5.84
California	6.56
Colorado	5.84
Delaware	6.56
Florida	6.62
Georgia	5.94
Hawaii	9.47
Idaho	6.05
Illinois	6.27
Indiana	6.71
Iowa	6.39
Kansas	6.33
Kentucky	5.35
Louisiana	5.63
Maryland	6.16
Michigan	6.36
Minnesota	6.78
Mississippi	5.26
Missouri	5.79
Montana	5.56
Nebraska	6.12
Nevada	6.87
New Jersey	7.56
New Mexico	5.95
New York	6.16
North Carolina	5.54
North Dakota	6.49
Ohio	5.96
Oklahoma	5.93
Oregon	6.75
Pennsylvania	6.02
South Carolina	5.64
South Dakota	5.69

Table B.7. (con't).

State	Hourly wage rates (Dol.)
Tennessee	5.81
Texas	5.34
Utah	6.25
Virginia	5.83
Washington	6.98
West Virginia	5.03
Wisconsin	5.29
Wyoming	5.49
Other States (CT, ME, MA, NH, RI, VT)	7.4
United States	6.25

Source: U.S. Department of Agriculture, National Agricultural Statistics Service, 1993 Farm Labor Survey.

Average wage rate is average of the published wage rates for each survey week weighted by number of hours worked during the week. The annual average is based on data collected for January, April, July, and October.

Data exclude Alaska.

Table B.8. Number of Hired Farmworkers Reported in the 1990
Decennial Census of Population

State	Number of workers	% of total
Alabama	13,300	1.1
Alaska	500	0.0
Arizona	21,000	1.8
Arkansas	20,300	1.7
California	237,500	20.4
Colorado	16,600	1.4
Connecticut	3,600	0.3
Delaware	1,900	0.2
Florida	71,100	6.1
Georgia	30,200	2.6
Hawaii	6,900	0.6
Idaho	18,600	1.6
Illinois	31,200	2.7
Indiana	20,500	1.8
Iowa	26,000	2.2
Kansas	16,300	1.4
Kentucky	24,600	2.1
Louisiana	14,200	1.2
Maine	5,800	0.5
Maryland	9,100	0.8
Massachusetts	5,800	0.5
Michigan	23,700	2.0
Minnesota	23,200	2.0
Mississippi	19,400	1.7
Missouri	22,700	2.0
Montana	10,200	0.9
Nebraska	18,900	1.6
Nevada	2,900	0.2
New Hampshire	2,000	0.2
New Jersey	9,400	0.8
New Mexico	11,900	1.0
New York	28,900	2.5
North Carolina	35,100	3.0
North Dakota	9,000	0.8
Ohio	21,300	1.8
Oklahoma	17,500	1.5

Table B.8. (con't).

State	Number of workers	% of total
Oregon	25,200	2.2
Pennsylvania	22,400	1.9
Rhode Island	,600	0.1
South Carolina	13,600	1.2
South Dakota	8,700	0.7
Tennessee	16,700	1.4
Texas	106,200	9.1
Utah	6,000	0.5
Vermont	5,300	0.4
Virginia	18,900	1.6
Washington	37,000	3.2
West Virginia	2,300	0.2
Wisconsin	28,400	2.4
Wyoming	4,700	0.4
NA	14,400	1.2
United States	1,161,500	100.0

Source: U.S. Department of Commerce, Bureau of the Census, 1990
Census of Population and Housing Public Use Microdata (1 Percent)
Sample.

Data exclude Alaska.

NA=State not identified

Table B.9. Number of Workers and Median Weekly Wages Reported in
1992 Current Population Survey Earnings File

Farm production region	Number of workers		Median weekly earnings
	Thou.	*Pct.*	*Dol.*
Northeast	57	6.7	220
Lake States	85	10	200
Corn Belt	79	9.3	200
Northern Plains	37	4.4	--
Appalachia	84	9.9	180
Southeast	109	12.9	212
Delta States	49	5.8	--
Southern Plains	72	8.5	200
Mountain	59	7.0	220
Pacific	216	25.5	220
United States	848	100.0	205

Source: U.S. Department of Labor, Bureau of Labor Statistics, 1992
Current Population Survey Microdata Earnings File.

Regions include: Northeast - CT, ME, MA, NH, RI, VT, NY, NJ, PA, MD,
DE, and D.C.; Lake States - MN, WI, MI; Corn Belt - IA, MO, IL, IN, OH;
Northern Plains - ND, SD, NE, KS; Appalachia - VA, WV, KY, TN, NC;
Southeast - SC, GA, AL, FL; Delta States - MS, LA, AR; Southern Plains -
OK, TX; Mountain - MT, ID, WY, NV, UT, CO, AZ, NM; Pacific - WA, OR,
CA, HI, AK.

Annual averages were computed by summing the weekly estimates for
each month and dividing by 12.

-- = Median earnings not shown where base is less than 50,000.

Table B.10. Agricultural Employment and Wages Covered by
Unemployment Insurance in 1990

State	Employment		Wages	
	Number	% of U.S.	$ million	% of U.S.
Alabama	6,431	0.9	94.6	1.0
Alaska	123	0.0	1.8	0.0
Arizona	14,305	2.0	197.9	2.1
Arkansas	8,173	1.2	121.2	1.3
California	228,965	32.7	3,191.2	33.5
Colorado	8,320	1.2	127.2	1.3
Connecticut	5,330	0.8	93.3	1.0
Delaware	1,508	0.2	28.9	0.3
Florida	71,207	10.2	936.1	9.8
Georgia	11,182	1.6	149	1.6
Hawaii	8,870	1.3	177.1	1.9
Idaho	10,724	1.5	142.3	1.5
Illinois	11,212	1.6	179.5	1.9
Indiana	9,820	1.4	147.3	1.5
Iowa	4,824	0.7	74.9	0.8
Kansas	5,603	0.8	105.7	1.1
Kentucky	3,244	0.5	46.8	0.5
Louisiana	5,358	0.8	65.5	0.7
Maine	2,251	0.3	29.8	0.3
Maryland	3,675	0.5	55.4	0.6
Massachusetts	3,562	0.5	63	0.7
Michigan	13,488	1.9	166.4	1.7
Minnesota	8,269	1.2	122.6	1.3
Mississippi	9,171	1.3	114.3	1.2
Missouri	5,998	0.9	81.3	0.8
Montana	2,095	0.3	28	0.3
Nebraska	5,225	0.7	90.8	0.9
Nevada	1,611	0.2	23.3	0.2
New Hampshire	538	0.1	7.3	0.1
New Jersey	7,490	1.1	126.5	1.3
New Mexico	5,245	0.7	67.4	0.7
New York	13,722	2.0	208.6	2.2
North Carolina	14,979	2.1	214.6	2.2
North Dakota	1,152	0.2	16.4	0.2
Ohio	12,153	1.7	159.8	1.7
Oklahoma	3,908	0.6	57.6	0.6

Table B.10. (con't).

State	Employment		Wages	
	Number	% of U.S.	$ million	% of U.S.
Oregon	20,930	3.0	249	2.6
Pennsylvania	15,594	2.2	232.5	2.4
Rhode Island	746	0.1	12.5	0.1
South Carolina	5,709	0.8	64.7	0.7
South Dakota	914	0.1	16.2	0.2
Tennessee	3,749	0.5	51.1	0.5
Texas	43,333	6.2	546	5.7
Utah	1,975	0.3	26.5	0.3
Vermont	760	0.1	10.9	0.1
Virginia	7,605	1.1	103.6	1.1
Washington	59,181	8.4	535	5.6
West Virginia	852	0.1	9.5	0.1
Wisconsin	7,995	1.1	124.6	1.3
Wyoming	1,402	0.2	19.1	0.2
United States	700,476	100.0	9,536.5	100.0

Source: Data derived from Martin, Philip L. and David A. Martin, The Endless Quest: Helping America's Farm Workers. Boulder, CO: Westview Press, 1994.

Data exclude agricultural service and contract workers.

Table B.11. Number of Farm Wage and Salary Workers Reported by the
Bureau of Economic Analysis, 1990

State	Farm wage and salary workers	% of U.S.
Alabama	13,716	1.5
Alaska	132	0.0
Arizona	12,131	1.3
Arkansas	17,691	2.0
California	168,038	18.6
Colorado	16,327	1.8
Connecticut	5,121	0.6
Delaware	1,931	0.2
Florida	52,363	5.8
Georgia	23,418	2.6
Hawaii	9,886	1.1
Idaho	13,702	1.5
Illinois	22,884	2.5
Indiana	15,831	1.7
Iowa	23,476	2.6
Kansas	12,885	1.4
Kentucky	22,293	2.5
Louisiana	11,368	1.3
Maine	4,918	0.5
Maryland	7,258	0.8
Massachusetts	5,076	0.6
Michigan	21,798	2.4
Minnesota	23,892	2.6
Mississippi	14,290	1.6
Missouri	14,548	1.6
Montana	8,153	0.9
Nebraska	14,449	1.6
Nevada	2,712	0.3
New Hampshire	1,538	0.2
New Jersey	6,893	0.8
New Mexico	5,033	0.5
New York	26,807	3.0
North Carolina	29,771	3.3
North Dakota	6,035	0.7
Ohio	17,170	1.9
Oklahoma	10,261	1.1

Table B.11. (con't).

State	Farm wage and salary workers	% of U.S.
Oregon	25,871	2.9
Pennsylvania	24,130	2.7
Rhode Island	672	0.1
South Carolina	11,971	1.3
South Dakota	5,616	0.6
Tennessee	14,943	1.7
Texas	50,236	5.6
Utah	5,249	0.6
Vermont	3,134	0.3
Virginia	15,429	1.7
Washington	44,411	4.9
West Virginia	2,833	0.3
Wisconsin	31,830	3.5
Wyoming	3,877	0.4
United States	903,997	100.0

Source: U.S. Department of Commerce, Bureau of Economic Analysis, 1990 BEA Employment and Income File.

Glossary

Alien: Any person not a citizen or national of the United States.

Anchor households: Households established by immigrants in the United States that serve to ease the transition into U.S. society and the U.S. labor force for new immigrants — both authorized and unauthorized.

Bracero program: The term commonly used to refer to the set of agreements between the United States and Mexican governments that allowed for the importation of contracted agricultural workers from Mexico into the United States. These agreements began August 4, 1942, and ended in January, 1964, with a several-year interruption, during which time there was no formal agreement. Throughout the course of the program between 4 and 5 million Mexican agricultural workers were admitted to the United States for temporary employment.

British West Indies (BWI) program: This program functioned between 1943 and 1952, at first through an intergovernmental agreement and then under the ninth proviso of the Immigration Act of 1917. Between 1943 and 1952 approximately 90,000 workers from the Caribbean were admitted for temporary employment in U.S. agriculture.

Colonias: Hispanic settlements along the southern border of the United States that generally have few zoning restrictions and lack access to municipal services.

Coyotes: Persons who, for a fee, provide assistance to international migrants in crossing the border illegally.

Cyclical migrants: International migrants who spend a portion of the year working in the United States, but return on a regular basis to their homes in another country.

Employer sanctions: Penalties imposed under the Immigration Reform and Control Act of 1986 for the hiring, recruiting, or referring for a fee of unauthorized aliens or for the failure to verify the employment authorization of each new hire.

ESA-92: Employment Security Administration Form No. 92, on which employers report quarterly the number of SAWs employed and person-hours worked. This reporting requirement was in place for the duration of the RAW program to allow the Bureau of the Census to estimate the number of person-hours worked by SAWs as required for the calculation of the RAW shortage number.

Farm labor contractors (FLCs): Labor market intermediaries who, for a fee, recruit, assemble, and supervise crews of farm workers. FLCs are also referred to as crew leaders, crew bosses, or *contratistas*. Many FLCs also provide transportation, housing, or other services to farm workers.

Farm Labor Supply Study (FLSS): A study conducted for the U.S. Department of Labor. The FLSS is one of several research initiatives undertaken by the department to examine the impacts of the Immigration Reform and Control Act of 1986 as mandated in that act. The study dealt specifically with the effect of improvements in wages, working conditions, and enhanced recruitment efforts on the farm labor supply.

Farm Labor Survey (FLS): A survey conducted by the U.S. Department of Agriculture that estimates farm employment, hours worked, and average agricultural wage rates for the United States. These estimates have been published for 80 years.

Fruit, vegetable, and horticultural specialty (FVH) production: A classification that includes the production of most labor-intensive crops. Fruit includes berries, grapes, citrus fruits, deciduous tree fruits, avocados, bananas, coffee, dates, figs, olives, pineapples, tropical fruit, and tree nuts. Vegetable includes all vegetables and melons grown in the open. Horticultural specialties include bedding plants, bulbs, florists' greens, flower and vegetable seeds, flowers, foliage, fruit stocks, nursery stock, ornamental plants, shrubberies, sod, mushrooms, and vegetables grown under cover.

H-2 program: A program authorized under the Immigration and Nationality Act of 1952 that allowed for the importation of temporary foreign workers, including agricultural workers. Employers who anticipated a shortage of U.S. workers could request nonimmigrant alien workers. This involved a request to the U.S. Department of Labor for certification, with final approval by the U.S. attorney general. To obtain approval, the Department of Labor had to certify that a labor shortage existed and that the wages and working conditions of U.S. workers similarly employed would not be adversely affected.

H-2A program: The H-2 program as amended by the Immigration Reform and Control Act of 1986. As amended, the H-2 program is divided into agricultural (H-2A) and nonagricultural (H-2B) employment categories. The procedure for obtaining workers under the H-2A program is very similar to that specified for obtaining agricultural workers through the former H-2 program.

Home-base states: The states of California, Florida, Texas, and the commonwealth of Puerto Rico where most agricultural workers, many of whom migrate to northern states on a seasonal basis to find agricultural work, live.

I-9 form: A form that employers are required to complete by the Immigration Reform and Control Act to verify that they have seen valid identification from each new hire that proves the employee is authorized to work in the United States.

Immigrant: As defined in the Immigration and Nationality Act, an alien in the United States other than a nonimmigrant.

Latinization: The process whereby a high concentration of Latinos move into a certain area in the United States, bringing with them language and cultural attributes, thereby transforming the area into a culturally Latin area.

Mayordomo: An agricultural foreman or field supervisor, hired by either farm labor contractors or farm employers. Often these individuals meet the criteria of farm labor contractors as defined by the Migrant and Seasonal Agricultural Worker Protection Act (AWPA), yet they are distinguished in common usage from labor contractors.

National Agricultural Workers Survey (NAWS): A survey conducted for the U.S. Department of Labor to monitor the supply of workers performing seasonal agricultural services as mandated by the Immigration Reform and Control Act. The study collected information on the characteristics and work patterns, including job history data, used to estimate fluctuations in the labor supply.

Nonimmigrant: An alien legally admitted to the United States for a temporary period of time for a reason specified in the Immigration and Nationality Act.

Nortenización: The process by which villages in Mexico become increasingly socially and economically oriented toward the United States as a result of high levels of northward, primarily cyclical, migration.

Pioneer migrants: An immigrant who moves to a new part of the country or sector of the economy without the benefit of network contacts with friends or family.

Pre-1982 program: The general legalization program of the Immigration Reform and Control Act that granted legal status to unauthorized immigrants who had lived continuously in the United States since January 1, 1982. There were approximately 1.7 million applicants through this program.

Raiteros: Persons who charge farm workers for rides to work. The term includes both short-distance commuting to the fields and long-distance travel. Generally these individuals are not licensed as farm labor contractors to provide transportation services.

Replenishment Agricultural Worker (RAW) program: A program created by the Immigration Reform and Control Act through which additional foreign farm workers could be admitted to the United States in the event that an agricultural labor shortage was determined to exist. The program expired on September 30, 1993.

Seasonal agricultural services (SAS): Defined by the Immigration and Control Act as the performance of field work related to the planting, cultural practices, cultivating, growing, and harvesting of fruits and vegetables of every kind and other perishable commodities. The term was further defined by the U.S. Department of Agriculture. Such work determined eligibility for the SAW program.

Special Agricultural Worker (SAW) program: A program created by the Immigration Reform and Control Act that allowed farm workers who could prove that they had completed at least 90 days of work in seasonal agricultural services between May 1, 1985 and May 1, 1986 to apply for legal status. There were approximately 1.3 million applicants through this program.

Unauthorized workers: The term used to refer to workers lacking the legal status to work in the United States. This term is more accurate than the previously common term "undocumented worker," as many unauthorized workers now possess counterfeit documents.

Upstream states: States that do not have substantial agricultural labor communities and must rely on migrant farm workers who live in home-base states or other countries.

U.S. Employment Service: Provides federal funds and support to associated state agencies that collect and disseminate job market information, place individuals on jobs after ascertaining that those jobs meet minimal conditions of employment, and provide a range of other services associated with matching prospective employees with employers.

Bibliography

Alchian, Armen A., and Harold Demsetz. 1972. "Production, Information Costs, and Economic Organization." *American Economic Review* 62: 777-795.

Allen, Steven G., and Daniel A. Sumner. 1991. "Immigration and Seasonal Labor Usage by North Carolina Tobacco Growers." *Current Issues in Tobacco Economics* 4.

Alvarado, Andrew J., Gary L. Riley, and Herbert O. Mason. 1990. *Agricultural Workers in Central California, 1989.* Sacramento: California Employment Development Department, California Agricultural Studies, Report No. 90-8.

Alvarado, Andrew J., Gary L. Riley, and Herbert O. Mason. 1992. *Agricultural Workers in Central California, 1990-91.* Sacramento: California Employment Development Department, California Agricultural Studies, Report No. 91-5.

Ansberry, Clare. 1993. "Hired Out: Workers Are Forced to Take More Jobs with Few Benefits." *Wall Street Journal* 78, No. 48 (March 11): 1, A9.

Bach, Robert, and Howard Brill. 1990. *Impact of IRCA on the U.S. Labor Market and Economy.* Binghamton, N.Y.: Institute for Research on Multiculturalism and International Labor.

Baumol, William J. 1990. "Entrepreneurship: Productive, Unproductive, and Destructive." *Journal of Political Economics* 98: 893-921.

Belous, Richard S. 1989. "How Human Resource Systems Adjust to the Shift toward Contingent Workers." *Monthly Labor Review* 112 (March): 7-12.

Bhagwati, Jagdish. 1958. "Immiserizing Growth: A Geometrical Note," *Review of Economic Studies* 25: 201-205.

Bhagwati, Jagdish. 1968. "Distortions and Immiserizing Growth: A Generalization." *Review of Economic Studies* 35: 481-485.

Bradshaw, Victoria. 1993. "TIPP: Targeted Industries Partnership Program." *Coastal Grower* (Spring): 24-25.

Brannon, Jeffrey. 1990. "The Border Colonias." *Revista de Estudios Fronterizos*. Mexicali, Mexico: Universidad Autonoma de Baja California, Instituto de Investigaciones Sociales.

Briody, Elizabeth. 1985. "Patterns of Household Immigration to South Texas." Unpublished Ph.D. dissertation, University of Texas.

Brown, G. K., D. E. Marshall, B. R. Tennes, D. E. Bosster, P. Chen, R. R. Garrett, M. O'Brien, H. W. Studer, R. A. Kepner, S. L. Hedden, E. D. Hood, D. H. Lender, W. F. Miller, G. E. Rehkugler, D. L. Peterson, and L. N. Shaw. 1983. *Status of Harvest Mechanization of Horticultural Crops*. St. Joseph, Mich.: American Society of Agricultural Engineers, Special Publication No. 3-83.

Brown, W. 1987. *Wage and Benefits Study for Florida Foliage Association*. Tangerine, Fla.: W. A. Brown & Associates.

Bruce, Alan. 1948. *Farm Labor Contractors in California*. Draft Report to the Labor Commissioner, Department of Industrial Relations, State of California.

Calavita, Kitty. 1989. "The Immigration Policy Debate: Critical Analysis and Future Options." *Mexican Migration to the United States: Origins, Consequences, and Policy Options*, edited by Wayne Cornelius and Jorge Bustamante. San Diego, Calif.: Center for US.–Mexican Studies.

California Assembly, Committee on Agriculture. 1969. *California Farm Labor Force: A Profile*. Sacramento.

California Employment Development Department (CalEDD). 1987–92. *California Weekly Farm Labor Report*. Report Supplement No. 881.

California Employment Development Department (CalEDD). 1991. *California Agricultural Employment and Earnings Bulletin*. Sacramento, November, Table 1-C.

California Employment Development Department (CalEDD). 1992. *Agricultural Employment, 1989 and 1990*. Sacramento, Report No. 882A.

California Senate, Senate Fact Finding Committee on Labor and Welfare. 1961. *California's Farm Labor Problems: Part 1*. Sacramento, p. 105.

Cargill, B. F., and G. E. Rossmiller, eds. 1970. *Fruit and Vegetable Harvest Mechanization*. East Lansing, Mich.: Rural Manpower Center, p. 22.

Carnevale, A. P., Leila J. Gainer, and A. S. Meltzer. 1989. *Workplace Basics: The Skills Employers Want*. Alexandria, Va.: American Society for Training and Developent.

Castro, Janice. 1993. "Disposable Workers." *Time* (March 29): 43-47.

Cavazos, Raul. 1991. "A View of Farm-Labor Contracting in California: Present and Future." In *Farm Labor Research Symposium* by Special Projects Unit Employment Data Section, EDD-LMID. Napa, Calif., June 5-6.

Coase, R. H. 1952. "The Nature of the Firm." *Economica* 4 (1937): 386-405, reprinted in *Readings in Price Theory*, edited by G. Stigler and K. Boulding, pp. 331-51. Homewood, Ill.: Irwin.

Cook, Roberta, Carlos Benito, James Matson, David Runsten, Kenneth Shwedel, and Timothy Taylor. 1992. *North American Free Trade Agreement: Effects on Agriculture, Vol. IV: Fruits and Vegetables*. Park Ridge, Ill.: American Farm Bureau Research Foundation.

Cornelius, W. A., and J. A. Bustamante, eds. 1989. *Mexican Migration to the United States: Origins, Consequences and Policy Options*. San Diego, Calif.: University of California, Center for U.S.–Mexican Studies.

Coughenour, C. Milton, Grace Zilvergerg, and John Hannum. 1991. *The Production and Marketing of Fruits and Vegetables in Kentucky*. Lexington, Ky.: University of Kentucky, College of Agriculture, Department of Sociology, Report No. RS-77.

Cuthbert, Richard W. 1980. "The Economic Incentives Facing Illegal Mexican Aliens in the U.S.: A Case Study at Hood River, Oregon." Unpublished M.S. thesis, Oregon State University.

Daberkow, Stan G., and Leslie A. Whitener. 1986. *Agricultural Labor Data Sources: An Update*. Washington, D.C.: U.S. Government Printing Office, Agricultural Handbook No. 658.

Dale, D. Michael. 1993. Personal Communication, Director, Farmworker Program, Oregon Legal Services, March.

Darrah, Charles. 1991. "Workplace Skills in Context." San Jose State University, mimeo.

Dean, G. W., G. A. King, H. O. Carter, and C. R. Shumay. 1970. *Projections of California Agriculture to 1980 and 2000*. Berkeley, Calif.: California Agricultural Experiment Station, Bulletin No. 847, p. 52.

Diven, Bill. 1990. "Fed Sweep Slows Chile Harvest." *Albuquerque Journal*, September 23.

Donato, Katherine J., Jorge Durand, and Douglas S. Massey. 1992. "Stemming the Tide? Assessing the Deterrent Effects of the Immigration Reform and Control Act." *Demography* 29: 139-157.

Duffield, James A., and Harry Vroomen. 1993. "The Effect of IRCA on Farm Employment and Wages: Intervention Analysis." In *Appen-*

dix I: Case Studies and Research Reports Prepared for the Commission on Agricultural Workers, 1989-1993, to Accompany the Report of the Commission. Washington, D.C. U.S. Government Printing Office, pp. 701-717.

Eastman, Clyde. 1984. *Participation of Undocumented Workers in New Mexico Agriculture, 1983.* Las Cruces, N.M.: New Mexico State University, Department of Agricultural Economics and Agricultural Business, Staff Report No. 35.

Eastman, Clyde. 1991. "Impacts of the Immigration Reform and Control Act of 1986 on New Mexico Agriculture." *Journal of Borderland Studies* 6: 105-130.

Eastman, Clyde. 1992. *Out of the Shadows: The Status of Legalizing Aliens in New Mexico.* Las Cruces, N.M.: New Mexico Agricultural Experiment Station, Research Report No. 661.

Emerson, Robert, Noy Chunkasut, Sharon Moon, and Leo Polopolus. 1991. *Prevailing Wage and Practices Project: Early and Midseason Oranges Interviewer Manual.* University of Florida: Department of Food and Resource Economics, p. 3.

Emerson, Robert, Noy Chunkasut, and Leo Polopolus. 1991. *Florida Orange Harvesting: Selected Employer and Worker Survey Results.* Gainesville, Fla.: University of Florida, Joint IFAS/DLET Project.

Farm Labor Alliance. 1986. "Working with Immigration Reform." *Grower Bulletin* (December).

Fein, David J. 1989. "The Social Sources of Census Omission: Racial and Ethnic Difference in Omission Rates in Recent U.S. Censuses." Ph.D. dissertation, Department of Sociology, Princeton University.

Figueroa, E. E., and P. Curry. 1993. "Survey of New York State Vegetable Producers." New York: Cornell University, Department of Agricultural Economics. Unpublished survey, March.

Fisher, Lloyd. 1952. *The Harvest Labor Market in California.* Cambridge, Mass.: Harvard University Press.

Fletcher, Peri, and J. Edward Taylor. 1990. "A Village Apart." *California Tomorrow* 5, No. 1: 8-17.

Florida Agricultural Statistics Service (FASS). 1993. *Citrus Summary, 1991-92.* Orlando, Fla.: U.S. Department of Agriculture, January.

The Flue-cured Tobacco Farmer. Various issues.

Fuller, Varden. 1940. "The Supply of Agricultural Labor as a Factor in the Evolution of Farm Organization in California." Exhibit 8762-A in *Violations of Free Speech and Rights of Labor, Part 54: Agricultural Labor in California.* Hearings Before the U.S. Senate Committee on Education and Labor (LaFollette Committee), 76th Cong., 3d Sess. (Jan. 13, 1940). Washington, D.C.: U.S. Senate.

Fuller, Varden. 1991. *Hired Hands in California's Farm Fields*. Giannini Foundation Special Report. Davis, Calif.: University of California, Division of Agriculture and Natural Resources.

Fuller, Varden, and Bert Mason. 1977. "Farm Labor." *Annals of the American Academy of Political and Social Science* 429 (January): 63-80.

Gabbard, Susan M., Edward Kissam, and Philip L. Martin. 1993. "The Impact of Migrant Travel Patterns on the Undercount of Hispanic Farm Workers." In U.S. Department of Commerce, Economics and Statistics Administration, *Proceedings of the Research Conference on Undercounted Ethnic Population*. Washington, D.C.: Bureau of the Census.

Gabbard, Susan M., Richard Mines, and Beatriz Boccalandro. Forthcoming. *Migrant Farm Workers in the United States: Findings from the National Agricultural Workers Survey*. Washington, D.C.: U.S. Department of Labor, Office of Program Economics, Office of the Assistant Secretary for Policy, Research Report No. 5.

Garcia, Victor Quiroz. 1992. "Surviving Farm Work: Economic Strategies of Mexican and Mexican American Households in a Rural California Community." Unpublished Ph.D. dissertation. Santa Barbara, Calif.: University of California, Department of Anthropology.

Garkey, Janet and Wen S. Chern. 1986. *Handbook of Agricultural Statistical Data*. Ames, Iowa: American Agricultural Economics Association, Economics Statistics Committee.

Gemoets, Lee. 1990. "Behind the Scene with Chile Picker." *Las Cruces Sun News*, August 31.

Griffith, David, and Jeronimo Camposeco. 1990. *Labor, Immigration Reform, and the Production of Winter Vegetables in South Florida*. Washington, D.C.: U.S. Commission on Agricultural Workers.

Griffith, David, and Jeronimo Camposeco. 1993. "The Winter Vegetable Industry in South Florida." In *Appendix I: Case Studies and Research Reports Prepared for the Commission on Agricultural Workers, 1989-1993, to Accompany the Report of the Commission*. Washington, D.C.: U.S. Government Printing Office, pp. 573-632.

Griffith, David, Ed Kissam, David Runsten, Anna Garcia, and Jeronimo Camposeco. 1990. *Farm Labor Supply Study: Second Interim Report*. Washington, D.C.: Department of Labor.

Gunter, Lewell F., Joseph C. Jarrett, and James A. Duffield. 1992. "Effect of U.S. Immigration Reform on Labor-Intensive Agricultural Commodities." *American Journal of Agricultural Economics* 74, No. 4 (November): 897-906.

Gwynn, Douglas B., Yoshio Kawamura, Edward Dolber-Smith, and Refugio I. Rochin. 1988. "California's Rural Poor: Trends, Correlates and Policies." Davis, Calif.: California Institute for Rural Studies, Working Paper No. 7.

Hanneman, Pat. 1990. "Citrus Outlook." Paper presented at California Agricultural Issues and Outlook Conference, Fresno, November 7.

Haydu, J. J. 1989. "What's Happening to Florida's Ornamental Industry?" *Florida Nurseryman* 36, No. 9: 25, 27, 29, 30.

Haydu, J. J. 1991. "Recent Financial Performance of Central and South Florida Foliage Nurseries." *Plant Market News* 1, No. 2 (Florida Cooperative Extension Service Bulletin). Gainesville, Fla.: University of Florida, Department of Food and Resource Economics.

Heppel, Monica, and Sandra L. Amendola, eds. 1991a. *Immigration Reform and Perishable Crop Agriculture.* Washington, D.C.: Center for Immigration Studies.

Heppel, Monica, and Sandra L. Amendola, eds. 1991b. *Immigration Reform and Perishable Crop Agriculture, Case Studies.* Washington, D.C.: Center for Immigration Studies.

Hodges, A. 1992. "Hurricane Andrew Crop Loss and Replanting Cost Estimate for the Ornamental Nursery Industry in Dade County." Unpublished report. Gainesville, Fla.: University of Florida, Department of Food and Resource Economics.

Hodges, A. W., and J. Haydu. 1989. "Challenges Facing the Florida Ornamental Industry." *Proc. Florida State Hort. Soc.* 102: 86-89.

Hodges, A. W., and J. Haydu. 1990. *An Economic Overview of Florida's Woody Ornamental Industry.* Gainesville, Fla.: University of Florida, Department of Food and Resource Economics, Economics Report No. 120.

Huffman, Wallace E. 1980. "Farm and Off-Farm Work Decisions: The Role of Human Capital." *Review of Economics and Statistics* 52: 14-23.

Huffman, Wallace E. 1986. *An Econometric Study of U.S.–Canadian Trade in Fresh Lettuce.* Ames, Iowa: Iowa State University, Department of Economics.

Huffman, Wallace E. 1993a. "An Assessment of the Process Underlying RAW Calculations." In *Appendix I: Case Studies and Research Reports Prepared for the Commission on Agricultural Workers, 1989-1993, to Accompany the Report of the Commission.* Washington, D.C.: U.S. Government Printing Office, pp. 739-782.

Huffman, Wallace E. 1993b. "Labor Markets, Human Capital, and the Human Agent's Share of Production." In *Essays in Agricultural*

Policy in Honor of D. Gale Johnson, edited by John Antle and Daniel Sumner. Chicago, Ill.: University of Chicago Press.

Huffman, Wallace E., and Robert E. Evenson. 1993. *Science for Agriculture: A Long-term Perspective.* Ames, Iowa: Iowa State University Press, p. 186.

Immigration Reform and Control Act of 1986. 1986. Ninety-Ninth Congress of the United States of America, Second Session, Public Law 99-603.

Jackson, J. 1992. "Du Pont Reports No More Threat from Fungicide." *Orlando Sentinel,* May 8.

Jackson, J., and M. Pankowski. 1992. "Storm Batters Foliage Industry for $30 Million." *Orlando Sentinel,* March 27.

Jesse, E. V., and M. J. Machado. 1975. *Trends in Production and Marketing of California Fresh Market Tomatoes.* Davis: University of California, Division of Agricultural Sciences.

Kawamura, Yoshio, Refugio I. Rochin, Douglas B. Gwynn, and Edward Dolber-Smith. 1989. "Rural and Urban Poverty in California: Correlations with 'Rurality' and Socioeconomic Structure." *Journal of Economic and Business Studies.* Ryukoku Island, Japan: Ryukoku University.

Kentucky Agricultural Statistics Service. 1992. *Kentucky Agricultural Statistics, 1991-1992.* Louisville, Ky.: USDA, NASS, and Kentucky Department of Agriculture.

Kissam, Edward. 1991. *Out in the Cold: The Causes and Consequences of Missing Farm Workers in the 1990 Census.* Sacramento, Calif.: La Cooperativa.

Kissam, Edward. 1993. "A Preliminary Assessment of the 1990 Census Undercount of America's Farmworkers. Paper presented to the Seventh Annual Conference Committee for Farmworker Programs, Arlington Va., March.

Kissam, Edward, Bernadette Dawson, and Jo Ann Intili. 1993. *State Plan for Workplace Learning: California's Plan for Adult Learning in the Workplace.* Adult Education Institute for Research and Planning.

Kissam, Edward, and Anna Garcia. 1993. "The Pickle Cucumber and Apple Industries in Southwest Michigan." In *Appendix I: Case Studies and Research Reports Prepared for the Commission on Agricultural Workers, 1989-1993, to Accompany the Report of the Commission.* Washington;, D.C.: U.S. Government Printing Office, pp. 297-368.

Kissam, Edward, Anna Garcia, and David Runsten. 1991. "Northward Out of Mexico: Migration Networks and Farm Labor Supply in Parlier, California." In *The Farm Labor Supply Study: 1989-1990, Vol. 2: Case Studies,* edited by Edward Kissam and David Griffith. Final Report to the U.S. Department of Labor. Berkeley, Calif.: Micro Methods.

Kissam, Edward, Anna Garcia, and David Runsten. 1993. "The Apple and Asparagus Industries in Washington." In *Appendix I: Case Studies and Research Reports Prepared for the Commission on Agricultural Workers, 1989-1993, to Accompany the Report of the Commission.* Washington;, D.C.: U.S. Government Printing Office, pp. 221-296.

Kissam, Edward, and David Griffith, eds. 1991. *The Farm Labor Supply Study: 1989-1990.* Final report to the U.S. Department of Labor. Berkeley, Calif.: Micro Methods.

Knight, Frank H. 1944. *Risk, Uncertainty, and Profit.* New York: Harper and Row, 1965.

Koch, Adrienne, and William Peden, eds. 1944. *The Life and Selected Writings of Thomas Jefferson.* New York: Modern Library.

Langan, Bud A., and Paul T. Melevin. 1987. *Yakima Valley Agricultural Survey.* Yakima, Ore.: Yakima Valley Community College, Summer/Fall.

Las Cruces Sun News. 1990. "Sweep Plows into Labor Violations: $48,500 in Fines," November 3, 1990.

Lewis, Oscar. 1961. *The Children of Sanchez.* New York: Random House.

Lomnitz, Larissa. 1977. *Networks and Marginality: Life in a Mexican Shantytown.* New York: Academic Press.

Love, John. 1991. "The Produce Industry's Challenge to Double Demand for Fresh Vegetables." *Vegetables and Specialties Situation and Outlook Yearbook.* USDA: Economic Research Service, No. TVS-255.

Mamer, John, and Alexa Wilkie. 1990. *Seasonal Labor in California Agriculture: Labor Inputs for California Crops.* Sacramento, Calif.: Employment Development Department, California Agricultural Studies No. 90-6.

Manta, Ben. 1976. "Toward Economic Development of the Chicano *Barrio:* Alternative Strategies and Their Implications," *Southwest Economy and Society* 1, No. 1 (Spring).

Marshall, David, et al. 1993. *Estimated Cost and Returns of Replanting an Apple Orchard to a Double Row V-Trellis High Density System in Central Washington.* Pullman, Wash.: Washington State University, Cooperative Extension Bulletin EB1735.

Martin, David A., and Philip L. Martin. 1992. "Coordination of Migrant and Seasonal Farm Worker Service Programs." Washington, D.C.: Report of the Administrative Conference of the United States.

Martin, Philip L. 1988. *Harvest of Confusion: Migrant Workers in U.S. Agriculture.* Boulder, Colo.: Westview Press.

Martin, Philip L. 1990. "Harvest of Confusion: Immigration Reform and California Agriculture." *International Migration Review* 24 (Spring): 69-95.

Martin, Philip L. 1991. "The Endless Debate: Immigration and U.S. Agriculture." Davis, Calif.: University of California.

Martin, Philip L. 1993. *Trade and Migration: NAFTA and Agriculture.* Washington, D.C., Institute for International Economics.

Martin, Philip L., and David A. Martin. 1994. *Helping America's Farmworkers: The Endless Quest.* Boulder, Colo.: Westview Press.

Martin, Philip L., and Gregory P. Miller. 1993. "Farmers Increase Hiring Through Labor Contractors." *California Agriculture* 47, No. 4 (July): 20-23.

Martin, Philip L., Richard Mines, and Angela Diaz. 1985. "A Profile of California Farmworkers," *California Agriculture* (May-June): 16-18.

Martin, Philip L., and J. Edward Taylor. 1989. "Has IRCA Reformed the Farm?" *California Farmer* (July 15): 14-31.

Martin, Philip L., and J. Edward Taylor. 1990. "Immigration Reform and California Agriculture a Year Later." *California Agriculture* 44, No. 1: 24-27.

Martin, Philip L., J. Edward Taylor, and Philip Hardiman. 1988. "California Farm Workers and the SAW Legalization Program." *California Agriculture* 42, No. 6: 4-6.

Martin, Philip L., and Suzanne Vaupel. 1987. "Evaluating Employer Sanctions: Farm Labor Contractor Experience," *Industrial Relations* 26, No. 3 (Fall): 304-313.

Mason, Herbert O., Andrew J. Alvarado, and Gary L. Riley. 1993. "The Citrus Industry in California and Arizona." *In Appendix I: Case Studies and Research Reports Prepared for the Commission on Agricultural Workers, 1989-1993, to Accompany the Report of the Commission.* Washington, D.C.: U.S. Government Printing Office, pp. 65-102.

Mason, Robert, and Tim Cross. 1992a. *Labor Demand, Productivity and Overhead Cost Estimates for Harvesting the 1990 Strawberry Crop.* Corvallis, Ore.: Oregon State University, Agricultural Experiment Station, Special Report No. 886.

Mason, Robert, and Tim Cross. 1992b. *Labor Demand and Productivity Estimates for the 1990 Cranberry Harvest Workforce.* Corvallis, Ore.: Oregon State University, Agricultural Experiment Station, Special Report No. 889.

Mason, Robert, Timothy Cross, and Carole Nuckton. 1993. *IRCA and Oregon Agricultural Industries.* Corvallis, Ore.: Oregon State University, Agricultural Experiment Station.

Massey, Douglas, Rafael Alarcon, Jorge Durand, and Humberto Gonzalez. 1987. *Return to Aztlan: The Social Process of International Migration from Western Mexico.* Berkeley, Calif.: University of California Press.

Meister, Dick, and Anne Loftis. 1977. *A Long Time Coming.* New York: Macmillan Co.

Migrant and Seasonal Agricultural Worker Protection Act. 1983. Ninety-Seventh Congress of the United States of America, Public Law 97-470, January 14.

Miles, Raymond E. 1989. "Adapting to Technology and Competition: A New Industrial Relations System for the 21st Century." *California Management Review* 31, No. 2 (Winter): 9-28.

Miles, Stanley D. 1992. "Oregon County and State Agricultural Estimates." Corvallis, Ore.: Oregon State University, Special Report No. 790, 1992 and prior years.

Mines, Richard. 1981. *Developing a Community Tradition of Migration to the United States: A Field Study in Rural Zacatecas, Mexico, and California Settlement Areas.* San Diego, Calif.: University of California, Monographs in U.S.–Mexican Studies.

Mines, Richard, and Ricardo Anzaldua. 1982. *New Migrants vs. Old Migrants: Alternative Labor Market Structures in the California Citrus Industry.* San Diego, Calif.: UC Center for U.S.–Mexican Studies, Monograph No. 9.

Mines, Richard, Susan Gabbard, and Beatriz Boccalandro. 1991. *Findings from the National Agricultural Workers Survey (NAWS), 1990: A Demographic and Employment Profile of Perishable Crop Farm Workers.* Washington, D.C.: Office of Program Economics, Office of the Assistant Secretary for Policy, U.S. Department of Labor, Research Report No. 1.

Mines, Richard, Susan Gabbard, and Ruth Samardick. 1993a. "U.S. Farmworkers in the Post-IRCA Period." In *Appendix I: Case Studies and Research Reports Prepared for the Commission on Agricultural Workers, 1989-1993, to Accompany the Report of the Commission.* Washington, D.C.: U.S. Government Printing Office, pp. 635-699.

Mines, Richard, Susan M. Gabbard, and Ruth Samardick. 1993b. *U.S. Farmworkers in the Post-IRCA Period: Based on Data from the National Agricultural Workers Survey (NAWS).* Washington, D.C.: U.S. Department of Labor, Office of Program Economics, Office of the Assistant Secretary for Policy, Research Report No. 4.

Mines, Richard, and Michael Kearney. 1982. *The Health of Tulare County Farmworkers.* Report submitted to Tulare County Health Department.

Mines, Richard, and Philip L. Martin. 1986a. *A Profile of California Farmworkers.* Giannini Foundation of Agricultural Economics. Monograph.

Mines, Richard, and Philip L. Martin. 1986b. "California Farmworkers: Survey Results of the UC-EDD Survey of 1983." Davis, Calif.: University of California, Department of Agricultural Economics. Mimeographed.

Mishel, Lawrence, and Jared Bernstein. 1993. *The State of Working America.* Armonk, N.Y.: M. E. Sharp, p. 166.

Morrow, Lance. 1993. "The Temping of America." *Time,* March 29, pp. 40-41.

National Advisory Commission on Food and Fiber (NACFF). 1967. *Final Report.* Washington, D.C.: U.S. Government Printing Office.

New York Agricultural Statistics Service. Various issues. *New York Agricultural Statistics.*

New York State Employment Service. 1991. *Agricultural Employment Bulletin,* October 15.

Oliveira, Victor J. 1992. *A Profile of Hired Farmworkers.* Washington, D.C.: U.S. Government Printing Office, AER No. 658.

Oliveira, Victor J., and Jane E. Cox. 1988. *The Agricultural Work Force of 1985.* Washington, D.C.: USDA-ERS, AAER No. 582.

Oliveira, Victor J., and Jane E. Cox. 1989. *The Agricultural Work Force of 1987: A Statistical Profile.* Washington, D.C.: USDA, Economics Research Service, Agricultural Economic Report No. 609.

Palerm, Juan Vicente. 1987/1988. "Transformations in Rural California" *UC MEXUS News,* Nos. 21/22 (Fall/Winter).

Palerm, Juan Vicente. 1991. *Farm Labor Needs and Farm Workers in California: 1970 to 1989.* Sacramento: Employment Development Department, California Agricultural Studies No. 91-2.

Polopolus, Leo C. 1989. "Agricultural Labor in the 1990s." Gainesville, Fla.: University of Florida, Food and Resource Economics Department, Staff Paper No. 367.

Polopolus, Leo C. 1991. "Florida Farm Labor and the Immigration Reform and Control Act." Gainesville, Fla.: University of Florida, Food and Resource Economics Department, Staff Paper No. 91-27.

Polopolus, Leo C. 1992. "North American Free Trade Agreement, Caribbean Basin Initiative, and Florida Agriculture." Paper presented before Florida Farm Managers and Rural Appraisers Conference, Tampa, Florida, October 21.

Polopolus, Leo C., Noy Chunkasut, and Robert D. Emerson. 1993. "Demographics of Florida Orange Harvest from Workers Prevailing

Wage Surveys, 1990-1993." Gainesville, Fla.: University of Florida, unpublished report.

Polopolus, Leo C., and Robert D. Emerson. 1975. "Florida Agricultural Labor and Unemployment Insurance." Gainesville, Fla.: University of Florida, Agricultural Experiment Station, Bulletin No. 767.

Polopolus, Leo C., and Robert D. Emerson. 1991. "Entrepreneurship, Sanctions, and Labor Contracting." *Southern Journal of Agricultural Economics* 23: 57-68.

Polopolus, Leo C., Susan Moon, and Noy Chunkasut. 1990. *Farm Labor in the Ornamental Industries of Florida.* Gainesville, Fla.: University of Florida, Food and Resource Economics Department, Economic Information Report No. 283.

Portes, Alejandro, and Robert D. Manning. 1986. "The Immigrant Enclave: Theory and Empirical Examples," In *Competitive Ethnic Relations,* edited by J. Nagel and S. Olzak. New York: Academic Press.

Ramos, George. 1989. "Fraud Charged as Disputed Amnesty Program Closes." *Los Angeles Times,* December 1, pp. 1, 34.

Rochin, Refugio I., and Monica D. Castillo. 1993. *Immigration, Colonia Formation and Latino Poor in Rural California: Evolving "Immiseration."* Claremont, Calif.: Tomas Rivera Center, Claremont University.

Rosen, Sherwin. 1983. "Specialization and Human Capital." *Journal of Labor Economics* 1: 43-49.

Rosen, Sherwin. 1986. "The Theory of Equalizing Differences." In *Handbook of Labor Economics,* by O. Ashenfelter and R. Layard. New York, N.Y.: North Holland, pp. 641-682.

Rosenberg, Gil. 1992. "They Are Just Like Family: Framing the Introduction of Hispanic Migrant Farmworkers into the Kentucky Tobacco Harvest." Masters thesis, University of Kentucky, Lexington.

Rosenberg, Howard R. 1988. *Emerging Outcomes in California Agriculture from the Immigration Reform and Control Act of 1986.* Davis, Calif.: University of California, Agricultural Issues Center, Issues Paper No. 88-3.

Rosenberg, Howard R. 1993. "Contractor Crackdown." *California Farmer,* January, pp. 19-22.

Rosenberg, Howard R., and Gregory E. Billikopf. 1989. "Verifying the Right to Work: The Paper Chase Comes to the Farm." *Choices* 4, No. 2: 34-35.

Rosenberg, Howard R., Daniel L. Egan, and Valerie J. Horwitz. 1995. *Labor Management Laws in California Agriculture,* 2d ed. Oakland, Calif.: University of California, ANR Publications.

Rosenberg, Howard R., and John W. Mamer. 1987. "The Impact of the New Immigration Reform Act." *California Agriculture* 41, Nos. 3-4 (March-April): 30-32.

Rosenberg, Howard R., and Jeffrey M. Perloff. 1988. "Initial Effects of the New Immigration Reform Law on California Agriculture." *California Agriculture* 42, No. 3 (May-June): 28-32.

Rosenberg, Howard R., Suzanne Vaupel, Don Villarejo, Jeffrey M. Perloff, and David Runsten. 1992. *Farm Labor Contractors in California*. Report to the Labor Market Information Division, California Employment Development Department.

Runsten, David, Roberta Cook, Anna Garcia, and Don Villarejo. 1993. "The Tomato Industry in California and Baja California: Regional Labor Markets and IRCA." In *Appendix I: Case Studies and Research Reports Prepared for the Commission on Agricultural Workers, 1989-1993, to Accompany the Report of the Commission*. Washington;, D.C.: U.S. Government Printing Office, pp. 3-64.

Runsten, David, and Michael Kearney. Forthcoming. *A Census of Mixtec Immigrants in Selected Regions of California, 1991*. Davis: California Institute for Rural Studies.

Runsten, David, and Phillip LaVeen. 1981. *The Mechanization of the California Farm Labor Market*. San Diego: University of California, Center for Mexico-U.S. Studies.

Runyan, Jack. 1994. *A Profile of Hired Farmworkers, 1992 Annual Averages* Washington, D.C.: U.S. Department of Agriculture, Economic Research Service.

Runyan, Jack, and Victor J. Oliveira. 1992. *A Geographic Analysis of Seasonal Agricultural Services Farms* Washington, D.C.: U.S. Department of Agriculture, Economic Research Service.

SAS/ETS Users Guide. 1988. Cary, N.C.: SAS Institute, pp. 112-113.

Schultz, Theodore W. 1975. "The Value of the Ability to Deal with Disequilibria." *Journal of Economic Literature* 13: 827-846.

Schultz, Theodore W. 1980. "Investment in Entrepreneurial Ability." *Scandanavian Journal of Economics* 82: 437-448.

Schumpeter, Joseph A. 1961. *The Theory of Economic Development*. New York: Oxford University Press.

SCR 43 Task Force. 1989. *The Challenge: Latinos in a Changing California*. Riverside, Calif.: University of California, UC MEXUS Program.

Smith, Leslie W., and Robert Coltrane. 1981. *Hired Farmworkers: Background and Trends for the Eighties*. Washington, D.C.: USDA, ERS), RDR Report No. 32.

Snow, David A., E. Burke Rochford, Jr., Steven K. Worden, and Robert D. Benford. 1986. "Frame Alignment Processes, Micromobilization and Movement Participation." *American Sociological Review* 5, No. 9: 414-448.

"Special Report: International Trade." 1991. *California Agriculture,* September-October.

Steinbeck, John. 1958 (originally published 1939). *The Grapes of Wrath.* New York, N.Y.: Viking Press.

Stigler, George J. 1951. "The Division of Labor Is Limited by the Extent of the Market." *Journal of Political Economy* 59: 185-193.

Stoddard, Ellwyn R. 1986. "Identifying Legal Mexican Workers in the U.S. Borderlands." *Southwest Journal of Business and Economics* 3: 11-26.

Strain, J. R., and A. Hodges. 1989. *Business Analysis of Container Nurseries in Florida.* Gainesville, Fla.: University of Florida, Department of Food and Resource Economics, Economic Information Report No. 266.

Street, Richard Steven. 1990. "Agriculture's Wild West Town." *California Farmer* 272, No. 6: 14-26.

Taylor, J. Edward. 1992. "Earnings and Mobility of Legal and Illegal Immigrant Workers in Agriculture." *American Journal of Agricultural Economics* 74, No. 4 (November): 889-896.

Taylor, J. Edward, and Thomas Espenshade. 1987. "Foreign and Undocumented Workers in California Agriculture." *Population and Development Review* 6: 223-229.

Taylor, J. Edward, and Dawn Thilmany. 1992. "California Farmers Still Rely on New Immigrants for Field Labor." *California Agriculture* 46, No. 5 (September-October): 4-6.

Texas Department of Human Services. 1988. *The Colonias Factbook.* Austin, Tex.

Thigpen, Jack, and C. Milton Coughenour. 1988. *A Study of Fruit, Vegetable, and Other Farmers in Twenty Kentucky Counties: Survey Methods.* Lexington, Ky.: University of Kentucky, Department of Sociology, No. RS-74.

Thompson, James F. 1992. "Harvesting Systems." In *Postharvest Technology of Horticultural Crops,* edited by Adel A. Kader. Davis, Calif.: University of California, Division of Agriculture and Natural Resources, No. 3311.

Tienda, Marta. 1989. "Looking to the 1990's: Mexican Immigration in Sociological Perspective." In *Mexican Migration to the United States: Origins, Consequences and Policy Options,* edited by W. A. Cornelius and J. A. Bustamente. San Diego, Calif.: University of California, Center for U.S.—Mexican Studies. pp. 109-147.

Tokle, Joanne, and Wallace E. Huffman. 1991. "Local Economic Conditions and Wage Labor Decisions of Farm and Rural Nonfarm

Couples." *American Journal of Agricultural Economics* 71 (August): 652-670.

Torok, S. J., and W. E. Huffman. 1986. "U.S.–Mexican Trade in Winter Vegetables and Illegal Immigration." *American Journal of Agricultural Economics* 68 (May): 246-260.

United States Bureau of the Census (USCensus). 1983. 1982 Census of Agriculture, Vol. 1, Part 17, Kentucky State and County Data. Washington, D.C: U.S. Government Printing Office.

United States Bureau of the Census (USCensus). 1989a. *1987 Census of Agriculture.* North Carolina State and County Data. Washington D.C.: U.S. Government Printing Office.

United States Bureau of the Census (USCensus). 1989b. *1987 Census of Agriculture, Vol. 1, Part 17, Kentucky State and County Data.* Washington, D.C.: U.S. Government Printing Office.

United States Bureau of the Census (USCensus). 1989c. *1987 Census of Agriculture. Geographic Area Series,* Vol. 1, Parts 3 and 5, California State and County Data. Washington, D.C.: U.S. Government Printing Office.

United States Bureau of the Census (USCensus). 1991. *Poverty in the United States.* Current Population Reports, Series P-60, No. 181.

United States Commission on Agricultural Workers (USCAW). 1993. *Report of the Commission on Agricultural Workers.* Washington, D.C.: U.S. Government Printing Office, No. 0-332-456: QL3.

United States Commission on Agricultural Workers (USCAW). 1993a. *Appendix I: Case Studies and Research Reports Prepared for the Commission on Agricultural Workers, 1989-1993, to Accompany the Report of the Commission.* Washington, D.C.: U.S. Government Printing Office.

United States Commission on Agricultural Workers (USCAW). 1993b. "IRCA and the Relationship Between Farm and Nonfarm Wages." In *Appendix I: Case Studies and Research Reports Prepared for the Commission on Agricultural Workers, 1989-1993, to Accompany the Report of the Commission.* Washington, D.C.: U.S. Government Printing Office, pp. 783-798.

United States Department of Agriculture (USDA), Agricultural Marketing Service (Ag. Mkt. Serv.). *Fruit and Vegetable Shipments by Commodities, States, and Months.* Various issues.

United States Department of Agriculture and New Mexico Agricultural Statistics Service. 1990. *Domestic Agricultural In-Season Wage Survey Reports (Chile).*

United States Department of Agriculture and New Mexico Agricultural Statistics Service. 1991a. *Domestic Agricultural In-Season Wage Survey Reports (Chile).*

United States Department of Agriculture and New Mexico Agricultural Statistics Service. 1991b. *New Mexico Agricultural Statistics.*

United States Department of Agriculture (USDA), Economic Research Service (ERS). *Economic Indicators of the Farm Sector.* Washington, D.C.: U.S. Government Printing Office. Annual.

United States Department of Agriculture (USDA), Economic Research Service (ERS). *Foreign Agricultural Trade of the United States.* Various issues.

United States Department of Agriculture (USDA), Economic Research Service (ERS). 1990. *Economic Indicators of the Farm Sector: State Financial Summary, 1989.* Washington, D.C.: U.S. Government Printing Office), No. ECIFS 9-3.

United States Department of Agriculture (USDA), Economic Research Service (ERS). 1994. *Economic Indicators of the Farm Sector: State Financial Summary, 1992.* Washington, D.C.: U.S. Government Printing Office), No. ECIFS 12-2.

United States Department of Agriculture (USDA), Economic Research Service (ERS). 1991. *Economic Indicators of the Farm Sector: National Financial Summary, 1990.* Washington, D.C.: U.S. Government Printing Office, No. ECIFS 10-2.

United States Department of Agriculture (USDA), Economic Research Service (ERS). 1994. *Economic Indicators of the Farm Sector: National Financial Summary, 1992.* Washington, D.C.: U.S. Government Printing Office, No. ECIFS 12-1.

United States Department of Agriculture (USDA), Economic Research Service (ERS). 1992a. *Fruit and Tree Nuts Situation and Outlook Yearbook.* TFS-263.

United States Department of Agriculture (USDA), Economic Research Service (ERS). 1992b. *Vegetables and Specialties Situation and Outlook Yearbook.* TVS-257.

United States Department of Agriculture (USDA), Economic Research Service (ERS). 1993. *Economic Indicators of the Farm Sector: National Financial Summary, 1991.* Washington, D.C.: U.S. Government Printing Office, No. ECIFS 11-1.

United States Department of Agriculture (USDA), Economic Research Service (ERS). 1993. "New Estimation Procedures for Farm Income Accounts," *Agricultural Income and Finance: Situation and Outlook Report.* Washington, D.C.: U.S. Government Printing Office, No. AFO-50.

United States Department of Agriculture (USDA), National Agricultural Statistical Service (NASS). 1975ff. The Quarterly Agricultural Labor Survey, published as *Farm Labor,* 1975ff. Various issues.

United States Department of Agriculture (USDA), National Agricultural Statistical Service (NASS). 1991. *Farm Employment and Wage Rates 1910-1990.* Washington, D.C.: U.S. Government Printing Office, Statistical Bulletin No. 822.

United States Department of Agriculture (USDA), National Agricultural Statistical Service (NASS). 1992. *Farm Production Expenditures: 1991 Summary.* Washington, D.C.: U.S. Government Printing Office, No. Sp Sy 9(92).

United States Department of Agriculture (USDA), National Agricultural Statistical Service (NASS). 1993. *Farm Labor.* Washington, D.C.: U.S. Government Printing Office.

United States Department of Commerce (USDC). 1982. *1980 Census of Population and Housing.* Washington, D.C.: U.S. Department of Commerce, Geographic Area Series.

United States Department of Commerce (USDC). 1989. *1987 Census of Agriculture.* Washington, D.C.: U.S. Government Printing Office.

United States Department of Commerce (USDC). 1992. *Census of Population and Housing, 1990.* Washington, D.C.: U.S. Government Printing Office.

United States Department of Commerce (USDC). 1992. *Detailed Occupation and Other Characteristics from the EEO File for the United States: 1990 Census of Population Supplementary Reports.* Washington, D.C.: U.S. Government Printing Office.

United States Department of Commerce (USDC). 1993. *Statistical Abstract of the United States, 1992.* Washington, D.C.: U.S. Department of Commerce.

United States Department of Commerce (USDC). 1994. *1992 Census of Agriculture.* Washington, D.C.: U.S. Government Printing Office.

United States Department of Health and Human Services. 1992. "Cost Reimbursement for IRCA Population Expenses." Unpublished report.

United States Department of Labor (USDOL). 1991. *Findings from the National Agricultural Workers Survey: A Demographic and Employment Profile of Perishable Crop Farm Workers.* Washington, D.C.: U.S. Government Printing Office, Research Report No. 1.

United States Department of Labor (USDOL), Bureau of Labor Statistics (BLS). 1992. *BLS Handbook of Methods.* Washington, D.C.: U.S. Government Printing Office, Belltin No. 2414.

United States Department of Labor (USDOL), Bureau of Labor Statistics (BLS). 1993. *Employment and Earnings.* Washington, D.C.: U.S.

Government Printing Office.

United States Department of Labor (USDOL), Bureau of Labor Statistics (BLS). 1994. *Employment and Earnings.* Washington, D.C.: U.S. Government Printing Office.

United States Department of Labor (USDOL), Secretary's Commission on Achieving Necessary Skills. 1992. *Learning a Living: A Blueprint for High Performance.* Washington, D.C.: U.S. Government Printing Office.

United States Immigration and Naturalization Service (USINS). 1992. *Statistical Yearbook, 1991.* Washington, D.C.: U.S. Government Printing Office.

United States Industrial Commission. 1901. Reports, Volume XV, Part III. Washington, D.C.: U.S. Government Printing Office.

United States President (USPres.). 1909. *Report of the Country Life Commission.* Washington, D.C.: U.S. Government Printing Office.

United States President (USPres.). 1992. *Economic Report of the President, 1992.* Washington, D.C.: U.S. Government Printing Office.

United States President (USPres.). 1993. *Economic Report of the President 1993.* Washington, D.C.: U.S. Government Printing Office.

United States President's Commission on Migratory Labor. 1951. *Migratory Labor in American Agriculture.* Washington, D.C. U.S. Government Printing Office.

United States Senate. 1981. *The Knowing Employment of Illegal Immigrants.* Hearing before the Subcommittee on Immigration and Refugee Policy of the Committee on the Judiciary, Serial J-97-61.

United States Senate, Committee on Education and Labor (LaFollette Committee). 1942. *Violations of Free Speech and the Rights of Labor.* Senate Report No. 1150. Washington, D.C.: U.S. Government Printing Office.

University of Arizona Cooperative Extension. 1990. "1989 Yuma County Agricultural Statistics." *Yuma County Farm Notes.*

University of Florida, Bureau of Economic and Business Research. 1992. *1992 Florida Statistical Abstract.* Gainesville, Fla.: University Press of Florida.

Vandeman, Ann Marie. 1988. "Labor Contracting in California Agriculture." Ph.D. dissertation, Agricultural and Resource Economics, University of California, Berkeley, p. 62.

Vaupel, Suzanne. 1991. "Farm Labor Contractors in the Fresh Market Tomato Harvest, Fresno County." In *Immigration Reform and Perishable Crop Agriculture, Vol. 2, Case Studies,* edited by Monica L. Heppel and Sandra L. Amendola. Washington, D.C.: Center for Immigration Studies.

"The Virtual Corporation." 1993. *Business Week*, February 8.

Vroomen, H., and James Duffield. 1992. "Testing for Impacts of Immigration Reform on Farm Employment and Wages." In Agricultural Resources, *Inputs Situation and Outlook Report. Economic Research Service*, USDA, No. AR-28, pp. 47-53.

Washington State Monitor Advocate. 1992. *Annual Summary Report of Employment Services to Migrant Seasonal Workers (MSFWS), Program Year 1991-1992*. Olympia, Wash.

Whitener, Leslie. 1984. "A Statistical Portrait of Hired Farm-workers," *Monthly Labor Review*, 107(June):49-53.

Wilson, William Julius. 1987. *The Truly Disadvantaged: The Inner City, the Underclass and Public Policy*. Chicago: University of Chicago Press.

Wrigley, H. S., and G. J. A. Guth. 1992. *Bringing Literacy to Life: Issues and Options in Adult ESL Literacy*. San Mateo, Calif.: Aguirre International.

Zabin, Carol, Michael Kearney, Anna Garcia, David Runsten, and Carole Nagengast. 1993. *Mixtec Migrants in California Agriculture: A New Cycle of Poverty*. Davis: California Institute for Rural Studies.

Index

A

acreage, in fruits and vegetables, 40–44
African-American farm workers
  in Florida, 87–88, 89
  history of, 445
  in Kentucky, 141
  in New York, 104, 301, 304, 315
  supplanting by Mexican farm workers, 458
  See also Black farm workers; slaves
age of farm workers
  in citrus industry, 342
  FLC vs. non-FLC workers, 198
  in Florida, 88, 89, 90
  in New York, 306
  in raisin industry, 388
  SAWs, 438
  in United States, 67
Agricultural Affiliates (New York), 99
agricultural exceptionalism, 137, 417, 448, 453–454
Agricultural Extension Service
  bias toward farmers, 53
  labor management programs and, 136
agriculture
  Congressional structure and priorities and, 454
  dependence on immigrant farm workers, 1–4, 11, 242, 445–446, 454–455, 458–459, 460
  farm labor policy issues, 133–137, 447–448
  GNP from, 37
  immigration policy and. *See* immigration policy
  infrastructure of, 137
  IRCA effects on, 10–17, 28–36, 142–143,

269–270, 295–297, 416, 425–434, 439–442, 444, 445, 446–450, 455
  "Latinization" of, 75, 112–113, 178–179, 447, 529
  number employed in, 37
  primacy of, 21
  subsectors of, 1
  United States as world leader in, 22
  work-place dynamics in, 117–118
alien farm workers. *See* immigrant farm workers
almonds, 49, 186
amnesty. *See* legalization programs
anchor households, defined, 527
apples
  consumption of, 50
  export of, 49, 74, 261, 299
  import of apple concentrate, 299
  international competition and, 439
  international trade agreements and, 74
  mechanization and, 53
  Michigan, 116
  national industry in, 299
  New York and Pennsylvania, 98–100, 299–318
    crew leaders and, 100, 110, 301, 302–303, 317–318
    demographics of workers, 100, 104–105, 110, 301, 303–308, 315
    earnings of workers, 309–312
    enforcement of regulations and, 316
    export of, 74

    farm labor contractors and, 97, 99, 100, 110
    farm labor market for, 99–100, 299–301
    H-2/H-2A programs and, 99, 109, 301, 303–304, 305, 309, 314, 315–316
    housing and, 309, 312–315, 316
    illegal workers and, 100, 305
    IRCA and, 98–100, 109, 304, 315–316
    migration patterns and, 105, 306–307
    overview of, 98–100, 299–300
    production of, 98–99, 100
    recruitment patterns for, 300–301
    SAW program and, 304–305
    social services and, 316–317
    stability of employment and, 112
    unemployment and, 307–308
    wages rates and, 100, 309–312, 318
    working conditions for, 309–312
  Washington
    acreage in, 259
    earnings of workers, 128–130
    expansion of, 118
    export of, 74, 261
    farm labor requirements for, 256–257
    Mexican settlements and, 126–127
    production of, 259
    unemployment and, 128
    value of, 255
apricots, 188, 259
Arizona
  benefits for workers in, 346–347, 350

551

E

Earned Income Credit
payments, 318
earnings
of *colonia* residents, in
California,
249–250
of crew leaders, in
Arizona, 340
of farm labor contrac-
tors, 232–233,
328–332, 383
of farm workers
in apple industry,
309–312
in Arizona, 311, 345
in California, 182,
183, 187, 196,
200–202, 204, 270,
311, 345, 390–392
in citrus industry,
345
farm labor contrac-
tors and, 35, 198,
200–202
in Florida, 85, 86
IRCA and, 38n.10,
428–434
in Michigan, 311
in New Mexico, 230,
239n.2
in New York,
309–312
in raisin industry,
183, 197, 390–392
SAW program and,
6, 25, 31
in tomato industry,
200–201,
205–206n.2,
366–367
in Washington,
128–130, 266
in Western states,
178
See also wage rates
Eastern states, overview of,
67, 73–76. *See also*
Florida; Kentucky;
Michigan; New
York; North
Carolina
educational level
in *colonias* in California,
248, 249

of farm labor contrac-
tors, 199, 280
of farm workers
in citrus industry, 342
FLC vs. non-FLC
workers, 198
in Florida, 91
IRCA and, 427, 434
in New Mexico,
228–229
occupational mobili-
ty and, 434
in Oregon, 218
in raisin industry, 388
in tomato industry,
362
in United States, 63,
67
in Washington, 262
educational services, 134,
218–219
employer sanctions. *See*
Immigration
Reform and
Control Act of
1986 (IRCA),
employer sanc-
tions
employment. *See*
Employment
Services; farm
labor contractors
(FLCs); recruit-
ment
employment eligibility
verification, in
Washington, 260,
262–264. *See also*
I-9 form
Employment Services,
recruitment
through, 142, 146
in California, 181, 196,
276, 282
by farm labor contrac-
tors, 282, 369
in Kentucky, 142, 146,
154
in New York, 301
in North Carolina, 169
in Oregon, 214
U.S. Employment
Service, 530
in Washington, 260–261,
262–264
employment training, 134,

191, 448
employment trends, 57–59
enforcement. *See*
Immigration
Reform and
Control Act of
1986 (IRCA),
employer sanc-
tions
English, ability to speak.
See educational
level
entrepreneurship, in
Florida citrus
labor market,
323–324, 326, 328
environmental controls, in
Florida, 84
ESA-92 form, 339, 528
ethnic economic enclaves,
colonias in
California as, 245,
247, 250, 251–252
European Community
(EC), 457, 458–459
export of fruits and veg-
etables
growth in, 47–49, 55,
438, 456–458
international competi-
tion and, 456, 457,
458
international trade
agreements and,
55, 74, 439
labor costs and, 38
recession and, 456
SAW program and, 31
Extension Service. *See*
Agricultural
Extension Service

F

Fair Labor Standards Act,
448
families of farm workers,
36, 66–67
cost of legalizing all
members of, 231
illegality of, 36, 231,
452n.6, 473
labor management and,
119, 128